Problem Solving Handbook in Computational Biology and Bioinformatics

Lenwood S. Heath • Naren Ramakrishnan
Editors

Problem Solving Handbook in Computational Biology and Bioinformatics

Editors

Lenwood S. Heath
Department of Computer Science
Virginia Tech
114 McBryde Hall
24061-0106 Blacksburg Virginia
USA
heath@vt.edu

Naren Ramakrishnan
Department of Computer Science
Virginia Tech
114 McBryde Hall
24061-0106 Blacksburg Virginia
USA
naren@cs.vt.edu

ISBN 978-1-4899-7348-1 ISBN 978-0-387-09760-2 (eBook)
DOI 10.1007/978-0-387-09760-2
Springer New York Dordrecht Heidelberg London

Printed on acid-free paper

Springer is part of Springer Science+Business Media (www.springer.com)

LSH acknowledges the loving support of Deanie, James, and Kaitlin, who are always his inspiration. NR dedicates this book to Kayar and Anant for their unwavering love.

Preface

Bioinformatics is today a breadth-wise subject spanning practically every aspect of the life sciences, from studying DNA sequences, to modeling the structure and function of proteins, to unraveling the interactions between proteins, and finally to capturing the relationship with phenotypes of organisms. While there are several excellent textbooks and monographs covering every aspect of the field, there is a need to collect together, in one place, the algorithms and methods that form the modern bioinformatician's toolkit. The *Problem Solving Handbook in Computational Biology and Bioinformatics* was conceived to fill this need.

In organizing this book, we have consciously chosen those topics that have strong algorithmic or methodological underpinnings and that are now widely used in a range of bioinformatics investigations, such as functional genomics, haplotype studies, and simulation of disease pathways. The intended audience for this book are practitioners of bioinformatics algorithms. The coverage is not (intended to be) exhaustive.

The chapters are written by experts in their respective disciplines and are closely organized with an introduction to the underlying problem/task/domain, detailed algorithmic descriptions, available software implementations, applications, and advanced topics. For the benefit of the reader, exercises and references to the literature for further reading are also provided.

The five sections of the handbook focus on algorithms for sequences, phylogenetics, proteins, networks, and biological data management/mining, respectively. The sequences section begins with an introduction to BLAST—arguably the most ubiquitous bioinformatics algorithm—by Jian Ma and Louxin Zhang, including a discussion of its recent incarnations. This is followed by a chapter on practical multiple sequence alignment algorithms by Tobias Rausch and Knut Reinert. In addition to the algorithmic details, this chapter presents the historical context surrounding this domain and how it has influenced the development of methods. The final chapter in this section, by John Spouge, focuses more generally on sequence alignment statistics, in particular how to assign p-values to alignment scores.

The second section of the handbook addresses the modeling of evolutionary relationships. The first chapter here, by Paul Marjoram and Paul Joyce, focuses on an introduction to coalescent theory and its applications, as a framework for evolutionary analysis. In the second chapter, Laxmi Parida dwelves further into coalescence and presents a graph-theoretic framework to understand 'recombinomics.' Paul Ryvkin and Li-San Wang then present a chapter on the more specific problem of phylogenetic tree reconstruction from sequence information. They cover the broad classes of algorithms in this domain and their modeling assumptions. These ideas are generalized in the next chapter on phylogenetic networks by Luay Nakhleh, which focuses on the modeling of non-treelike evolutionary behaviors. The final chapter on genome-wide association studies by Paola Sebastiani and Nadia Solovieff helps make the connection from sequences and phylogeny to phenotypes.

The third section focuses on the multifaceted world of proteins. Proteins are rich in multiple aspects of biology: structure, function, biochemistry, to name a few aspects. Our section focuses on two of these aspects. Bonnie Berger and Jérôme Waldispühl introduce the protein structure prediction problem and the latest developments on that front. Yang Cao then describes how networks of proteins can be simulated, specifically using stochastic methods.

The fourth section describes algorithms for networks, as broadly modeled in bioinformatics. Christopher Lasher, Christopher Poirel, and T. M. Murali introduce cellular response networks as a mechanism to both integrate diverse sources of data and to understand how distinct stresses manifest in the cellular state of a cell. Sinan Erten and Mehmet Koyutürk focus on protein-protein interaction networks and algorithms for identifying modules in them.

The final section is broadly about biological data management and mining. The first chapter here, by Paola Sebastiani, Jacqui Milton, and Ling Wang, is about the design and organization of microarray experiments. The second chapter, by Karthik Devarajan, focuses on matrix decompositions as a specific algorithmic technique for analyzing many kinds of bioinformatics data sets. The third chapter, by Rachael Huntley, Emily Dimmer, and Rolf Apweiler, describes the Gene Ontology Resource and the myriad uses it has come to serve in modern bioinformatics research.

We hope this book will serve as a useful companion to the bioinformatics practitioner. We would like to thank Springer publishing for encouraging us to undertake this project and providing constant guidance throughout the process. In particular, Susan Lagerstrom-Fife has been very supportive of this project every step of the way. Jennifer Maurer helped us organize the various chapters and indexes in a coherent manner. Many thanks to the authors for their enthusiastic participation and responding to our queries on time.

Blacksburg, VA,
July, 2010

Lenwood S. Heath
Naren Ramakrishnan

Contents

Part IV Networks

Part V Biological Data Management and Mining

List of Contributors

Rolf Apweiler
European Bioinformatics Institute, e-mail: apweiler@ebi.ac.uk

Bonnie Berger
Department of Mathematics & Computer Science and AI Lab, MIT, Cambridge, MA, USA, e-mail: bab@mit.edu

Yang Cao
Department of Computer Science, Virginia Tech, Blacksburg, VA, USA, e-mail: ycao@cs.vt.edu

Karthik Devarajan
Department of Biostatistics & Bioinformatics, Fox Chase Cancer Center, Philadelphia, PA USA, e-mail: karthik.devarajan@fccc.edu

Emily C. Dimmer
European Bioinformatics Institute, e-mail: edimmer@ebi.ac.uk

Sinan Erten
Department of Electrical Engineering & Computer Science, Case Western Reserve University, Cleveland, OH, USA, e-mail: sinan.erten@case.edu

Rachael P. Huntley
European Bioinformatics Institute, e-mail: huntley@ebi.ac.uk

Paul Joyce
University of Idaho, e-mail: joyce@uidaho.edu

Mehmet Koyutürk
Department of Electrical Engineering & Computer Science, Case Western Reserve University, Cleveland, OH, USA, e-mail: koyuturk@eecs.case.edu

Christopher D. Lasher
Genetics, Bioinformatics, and Computational Biology Program, Virginia Polytechnic Institute and State University, Blacksburg, VA, USA, e-mail: lasher@vt.edu

Jian Ma
University of California at Santa Cruz, e-mail: jianma@soe.ucsc.edu

Paul Marjoram
University of Southern California, e-mail: pmarjora@usc.edu

Jacqui Milton
Boston University, Boston MA, e-mail: jnmilton@gmail.com

T. M. Murali
Department of Computer Science, Virginia Polytechnic Institute and State University, Blacksburg, VA, USA, e-mail: murali@cs.vt.edu

Luay Nakhleh
Rice University, e-mail: nakhleh@cs.rice.edu

Laxmi Parida
IBM Thomas J. Watson Research Center, Yorktown Heights, NY, USA, e-mail: parida@us.ibm.com

Christopher L. Poirel
Department of Computer Science, Virginia Polytechnic Institute and State University, Blacksburg, VA, USA, e-mail: poirel@vt.edu

Tobias Rausch
Freie Universität, e-mail: rausch@inf.fu-berlin.de

Knut Reinert
Freie Universität, e-mail: reinert@inf.fu-berlin.de

Paul Ryvkin
Genomics and Computational Biology Graduate Program, University of Pennsylvania, e-mail: pry@mail.med.upenn.edu

Paola Sebastiani
Boston University, Boston MA, e-mail: sebas@bu.edu

Nadia Solovieff
Boston University, Boston MA, e-mail: ntimofee@bu.edu

John L. Spouge
National Center for Biotechnology Information, National Library of Medicine, National Institutes of Health, Department of Health and Human Services, Bethesda, MD 20894, e-mail: spouge@ncbi.nlm.nih.gov

Jérôme Waldispühl
School of Computer Science, McGill University, Montreal, QC, Canada, e-mail: jeromew@cs.mcgill.ca

Ling Wang
Novartis Vaccines and Diagnostics, Emeryville CA, e-mail: ling-1.wang@novartis.com

Li-San Wang
Penn Center for Bioinformatics, e-mail: lswang@mail.med.upenn.edu

Louxin Zhang
National University of Singapore, e-mail: matzlx@nus.edu.sg

Part I
Sequences

This first part of the book covers basic algorithms that work with biological sequences, from sequence search to multiple sequence alignment.

Modern BLAST Programs

Jian Ma and Louxin Zhang

Abstract The Basic Local Alignment Search Tool (BLAST) is arguably the most widely used program in bioinformatics. By sacrificing sensitivity for speed, it makes sequence comparison practical on huge sequence databases currently available. The original version of BLAST was developed in 1990. Since then it has spawned a variant of specialized programs. This chapter surveys the development of BLAST and BLAST-like programs for homology search, discusses alignment statistics that are used in assessment of reported matches in BLAST, and provides the reader with guidance to select appropriate programs and set proper parameters to match research requirements.

1 Introduction

The sequence structures of genes and proteins are conserved in nature. It is common to observe strong sequence similarity between a protein and its counterpart in another species that diverged hundreds of millions of years ago. Accordingly, the best method to identify the function of a new gene or protein is to find its sequence-related genes or proteins whose functions are already known.

The Basic Local Alignment Search Tool (BLAST) is a computer program for finding regions of local similarity between two DNA or protein sequences. It is designed for comparing a query sequence against a target database. It is a heuristic that finds short matches between query and database sequences and then attempts to start alignments from these 'seed hits'. By sacrificing sensitivity for speed, it makes sequence comparison practical on huge sequence databases currently available, such as GenBank, which has over 80 million sequence records as of August 2008. In

Jian Ma
University of California at Santa Cruz, e-mail: jianma@soe.ucsc.edu

Louxin Zhang
National University of Singapore, e-mail: matzlx@nus.edu.sg

L.S. Heath and N. Ramakrishnan (eds.), *Problem Solving Handbook in Computational Biology and Bioinformatics*, DOI 10.1007/978-0-387-09760-2_1,

addition to generating local alignments, BLAST also provides statistical assessment of reported alignments. Because of these powerful features, BLAST is one of the most widely used bioinformatics tools.

BLAST analysis is often used to identify conserved sequence patterns and to establish functional or evolutionary relationships among proteins. It finds numerous applications in molecular biology, evolutionary biology, and drug discovery.

The original version of BLAST [6] was developed by Altschul, Gish, Lipman, Miller and Myers in 1990. The improved version PSI-BLAST [7] was made available in 1997. Over the intervening years, the original version has been customized into a set of specialized programs. These new variants of BLAST handle homology searches on different types of databases. They were designed to find gapped local alignments and to detect weak signals in sequence alignment. Table 3 lists the popular BLAST programs together with their functions.

The rest of this chapter is divided into seven sections. Section 2 describes the available BLAST programs and other BLAST-like programs. Sections 3 and 4 present the algorithmic and statistical aspects of BLAST, respectively. Section 5 describes two practical examples of using BLAST. Through these examples, we examine the biological and statistical information output from BLAST. Section 6 addresses three advanced issues of using BLAST homology search. Section 7 lists some exercises for the reader to master BLAST programs. Finally, we summarize the most relevant and useful references on BLAST for further reading in Section 8.

2 Available Implementations

From a user point of view, based on different purposes, a BLAST search generally involves three important parts: input, database searched against, and a particular BLAST program.

On NCBI BLAST web site, the available databases can be categorized into protein databases and nucleotide databases. Frequently used databases are summarized in Table 1 and Table 2. In addition, NCBI also provides specialized BLAST databases, e.g. genome databases for different species, trace databases, as well as various databases for model organisms.

Table 1 Main protein sequence databases for BLAST

Database	Description
nr	Non-redundant collections from GenBank CDS translations, PDB, SwissProt, PIR, and PRF
month	The nr updates in the last 30 days
refseq	Protein sequences from the RefSeq project
swissprot	SWISS-PROT protein sequence database
pdb	Sequences from the 3-dimensional structure records in PDB

Table 2 Main nucleotide sequence databases for BLAST

Database	Description
nr	All sequences in GenBank, EMBL, DDBJ, PDB except EST, STS, GSS, etc.
month	The nr updates in the last 30 days
refseq_mrna	mRNA sequences from the RefSeq Project
refseq_genomic	Genomic sequences from the RefSeq Project
est	EST sequences in GenBank, EMBL, and DDBJ
gss	Genome Survey Sequence

A family of BLAST programs have been developed since its original version was launched in 1990. The difference mainly comes from the input type and the databases that the input is searched against. For example, BLASTN is useful to identify an unknown nucleotide sequence or to search homologous genomic sequences in different organisms. The major BLAST programs on the NCBI web site are summarized in Table 3. These programs can be used via a web interface (http://www.ncbi.nlm.nih.gov/blast) or as stand-alone tools.

Apart from the set of programs on the NCBI BLAST server, there are other BLAST-like homology search programs and web servers. Table 4 lists a few widely-used tools. On WU-BLAST server, BLAST programs available on NCBI are also available, but all the programs were implemented differently. WU-BLAST also includes other tools developed in Warren Gish's lab.

FASTA is a sequence similarity search program first developed by Lipman and Pearson in NCBI. Its sequence format, called the FASTA format, has been widely adopted for sequence comparison. It uses a multiple-step approach to aligning the query and target sequences. It first finds runs of ktup or more identities, which are called word matches. Here ktup is a program parameter used for controlling the sensitivity and speed of the program. From these identified word matches, it determines a band in which good alignments likely locate and then calculates the optimal alignment in the band using the dynamic programming method.

Sequence Search and Alignment by Hashing Algorithm (SSAHA) is developed to search a large DNA database efficiently. The essential idea is to preprocess the sequences in a database by breaking them into consecutive k-tuples of k contiguous bases and then using a hash table to store them. Therefore, searching for a query sequence in the database is done by obtaining from the hash table the 'hits' for each k-tuple in the query sequence and then performing a sort on the results.

Sim4 employs a BLAST-based approach. It first determines the maximal scoring gap-free segments and then extends these segments into the adjacent regions greedily. It can be downloaded from Webb Miller's lab and installed in a standalone workstation. It can also be run through the web server http://pbil.univ-lyon1.fr/sim4.php.

BLAT is an alignment tool like BLAST, and it is extremely efficient, developed by Jim Kent. On DNA sequences, BLAT works by keeping an index of an entire genome, consisting of all non-overlapping 11-mers, which makes BLAT quickly find sequences of 95% and greater similarity of length 40 bases or more. However,

Table 3 Major BLAST programs on the NCBI web site

Program	Description
BLASTN	Search a nucleotide sequence against a nucleotide sequence database
BLASTP	Search an amino acid sequence against a protein sequence database
BLASTX	Search a nucleotide sequence translated in all reading frames against a protein sequence database
TBLASTN	Search a protein sequence against a nucleotide sequence database dynamically translated in all reading frames
TBLASTX	Search the six-frame translations of a nucleotide sequence against the six-frame translations of a nucleotide sequence database
MEGABLAST [29]	Find long alignments between very similar sequences more efficiently
PSI-BLAST [7]	Find members of a protein family or build a custom position-specific score matrix
PHI-BLAST [28]	Find proteins similar to the query around a given pattern

it is less sensitive to more divergent or short sequence alignments. On protein sequences, BLAT uses 4-mers, rapidly finding protein sequences of 80% and greater similarity to the query of length longer than 20 amino acids. However, it is far less sensitive than BLAST and PSI-BLAST at NCBI.

3 Algorithm Description

Alignment is a way of arranging two DNA or protein sequences to identify regions of similarity that are conserved among species. Each aligned sequence appears as a row within a matrix. Gaps are inserted between the residues of each sequence so that identical or similar bases in different sequences are aligned in successive positions. Each gap spans one or more columns within the alignment matrix. The score of an alignment is calculated by summing the rewarding scores for match columns that contain the same bases and the penalty scores for gaps and mismatch columns

Table 4 Other BLAST-like programs

Program	Description	URL	Refs
WU-BLAST	Washington University BLAST	http://blast.wustl.edu/	[17]
FASTA	Homology search against Protein or DNA databases	http://fasta.bioch.virginia.edu/	[26]
SSAHA	Fast matching and alignment of DNA sequences	http://www.sanger.ac.uk/	[24]
Sim4	Homology search of an expressed DNA sequence (EST, cDNA, mRNA) with a genomic sequence	http://www.bx.psu.edu/miller_lab/	[15]
BLAT	BLAST-Like Alignment Tool	http://genome.ucsc.edu/	[21]

that contain different bases. A scoring scheme specifies the scores for matches and mismatches, which form the scoring matrix, and the scores for gaps, called the gap cost. There are two types of alignments for sequence comparison. Given a scoring scheme, calculating a global alignment is a kind of global optimization that 'forces' the alignment to span the entire length of two query sequences, whereas local alignments just identify regions of high similarity within two sequences.

The original version of BLAST finds good ungapped local alignments between the query and database sequences [6]. Accordingly, it is also called ungapped BLAST. Database sequences are usually called target sequences. To speed up the homology search process, BLAST employs a filtration strategy: It first scans the database for length-w word matches of alignment score at least T between the query and target sequences and then extends each match in both ends to generate local alignment (in the sequences) whose alignment score is larger than a threshold S. The matches are called *high-scoring segment pairs (HSPs)*. BLAST outputs a list of HSPs together with E-values that measure how frequent such HSPs would occur by chance.

3.1 Phase 1: Scan the Database for Match Hits

Consider a set of parameters w, T and S. A sequence of length w is called a w-mer. For a query sequence, a w-mer is called a *neighborhood sequence* if it forms a match of alignment score at least T with some w-mer in the query sequence. We illustrate this concept using a DNA query sequence.

Consider query sequence Q: GCATTGACCC and parameters $w = 8, T = 6$. Under a simple scoring scheme by which matches and mismatches score 1 and -1 respectively, the neighborhood sequences that match 8-mer GCATTGAC in the query sequence are all 1-mismatch 8-mers:

```
.CATTGAC,  G.ATTGAC, GC.TTGAC, GCA.TGAC,
GCAT.GAC,  GCATT.AC, GCATTG.C, GCATTGA.,
```

where '.' stands for any letter of A, G, C, and T. Similarly, the set of neighborhood sequences also include the following sequences:

```
.ATTGACC, C.TTGACC, CA.TGACC, CAT.GACC,
CATT.ACC, CATTG.CC, CATTGA.C, CATTGAC.,
.TTGACCC, A.TGACCC, AT.GACCC, ATT.ACCC,
ATTG.CCC, ATTGA.CC, ATTGAC.C, ATTGACC.,
```

which match 8-mers CATTGACC or ATTGACCC.

The set of neighborhood sequences is efficiently constructed from the query sequence since there are at most 4^w neighborhood sequences. Having the set of neighborhood sequences, the next task is to check whether each neighborhood sequence occurs in the target sequence or not. Such an occurrence of a neighborhood sequence is called a seed hit. For example, for target sequence

T: ATAGCATGGACTTGACCCCGGCATTGTCATCG,

the 8-mer GCATTGAC hits T at positions 4 and 21, whereas the 8-mer ATTGACCC hits T at position 11. Here seed hits are not perfect. In fact, BLAST programs use perfect hits for DNA sequence search and imperfect hits whose score is higher than a threshold. All the hits can be identified using an efficient data structure such as hash table, suffix tree, or suffix array. The reader is referred to the book [12] of Chao and Zhang for implementation details.

The sensitivity and speed of BLAST search are closely related to the match size w. When w is large, the BLAST search is fast but has low sensitivity in the sense that it may miss short homologous sequences. In contrast, when w is small, it is slower, but has high sensitivity. The w is set by default to 11 and 3 for BLASTN and BLASTP, respectively. To achieve the optimal balance between sensitivity and speed, the discontiguous MEGABLAST finds l-mer pairs that match in w discontiguous positions specified by a fixed pattern. Such a pattern is called a spaced seed. For example, one default spaced seed used for searching non-coding sequences is $111*1*11**1*11*111$. When such a spaced seed is used, two 18-mers match if they have identical nucleotides in the positions indicated by the 1s: 1, 2, 3, 5, 7, 8, 11, 13, 14, 16, 17, 18. It is first observed by Ma, Tromp, and Li that an optimally spaced seed significantly improves homology search sensitivity [23].

3.2 Phase 2: Hit extension

In the second phase, ungapped BLAST extends each 'seed' hit in both directions to generate a HSP and outputs this HSP if its alignment score is S or greater. At each end, the extension includes aligned pairs in successive positions, with corresponding increments to the alignment score. It continues until the alignment score drops more than X below the maximum score that has attained up to that position.

It was observed that ungapped BLAST consumes more than 90% of the running time in hit extension. It was also observed that an HSP usually contains multiple hits that are close to one another. Accordingly, Gapped BLAST uses double hits to trigger hit extension to generate high-scoring gapped local alignments. It starts the extension process only if there are two non-overlapping hits within D_g positions, where the subscript g indicates that it is a parameter for Gapped BLAST. These adjacent non-overlapping hits can be detected if all hit positions are maintained.

In Gapped BLAST, gap extension is done by using the dynamic programming approach. Since the approach takes quadratic time, the extension process is much slower than ungapped one. Here two more ideas are employed in order to handle gap extension more efficiently. One idea is only to extend those HSPs that have alignment score S_g or greater. The threshold S_g is determined in such a way that only one gap extension is invoked on average in per 50 database sequences. Another idea for handling the extension is to restrict gapped extension to those positions in which the optimal local alignment score drops no more than X_g below the maximum local alignment score attained up to the position.

4 BLAST Statistics

An important feature of BLAST is that it rank-orders the reported HSPs by E-values. For a local alignment of score s, an E-value of 10^{-5} is often used as a cutoff for BLAST homology search. It means that with a collection of random query sequences, only once in a hundred thousand instances would an alignment with that score or greater occur by chance. The smaller the E-value, the greater the belief that the aligned sequences are homologous.

The E-values for HSPs in BLAST printout are calculated based on the seminal work of Karlin and Altschul on the distribution of optimal ungapped local alignment scores [19]. Both theoretical and empirical studies suggest that the distributions of optimal local alignment scores with or without gaps are accurately described by an extreme value distribution.

Assume that we search a query sequence Q against a database. Let l_Q be the length of Q. For each database sequence T, the mean number E_T of HSPs with score s or greater occurring in the comparison of Q and T is

$$E_T = K(l_Q - \bar{l}(s))(l_T - \bar{l}(s))e^{-\lambda s}, \tag{1}$$

where K and λ are constants independent of T and $\bar{l}(s)$ is the length adjustment. K and λ are the two parameters of the extreme value distribution of optimal local alignment scores. Their values are efficiently calculated from the letter composition of the database sequences and the scoring scheme used for the search. The values of K and λ are listed in the BLAST search printout.

The length adjustment $\bar{l}(s)$ is equal to the mean length of HSPs with score s or greater. It is used to eliminate the 'edge' effect of the fact that optimal local alignments are unlikely to occur at the end of both query and target sequences. Let N and M be the numbers of sequences and letters in the database. The current BLASTP (version 2.2.18) calculates the length adjustment $\bar{l}(s)$ for score s as an integer-valued approximation to the unique root of the following functional equation

$$x = \alpha \frac{\ln(K(l_Q - x)(M - Nx))}{\lambda} + \beta. \tag{2}$$

For ungapped alignment, $\alpha = \lambda/H$, and $\beta = 0$, where H is the relative entropy of the scoring matrix used for the database search. For gapped alignment, the values of α and β depend on scoring matrix and affine gap cost. Take BLOSUM62 as an example. We have that $\alpha = 1.90$ and $\beta = -29.70$ for the affine gap cost in which the gap opening and extension costs are 11 and 1 respectively.

We define the effective size of the search space as

$$\text{eff-searchSP} = \sum_{T \in \mathscr{D}} l_T - N\bar{l}(s). \tag{3}$$

By the linearity property of means, the expected number of high-scoring alignments with score s or greater found in the entire database is

$$\text{E-value} = \sum_{T \in \mathscr{D}} E_T = K \times (l_Q - \bar{l}(s)) \times \text{eff-searchSP} \times e^{-\lambda s}. \quad (4)$$

When two sequences are aligned, insertions and deletions can break a long alignment into several parts. If this is the case, focusing on the single highest-scoring segment could lose useful information. As an option, one may consider the scores of the multiple highest-scoring segments.

Assessing multiple highest-scoring segments is more involved than it might first appear. Suppose, for example, comparison X reports two highest scores 88 and 68, whereas comparison Y reports 79 and 75. One can say that Y is not better than X, because its high score is lower than that of X. But neither is X considered better, because the second high score of X is lower than that of Y. The natural way to rank all the possible results is to consider the sum of the alignment scores of the HSPs as suggested by Karlin and Altschul [20]. This sum is now called the Karlin-Altschul sum statistic.

In earlier versions of BLAST, the Karlin-Altschul sum statistic was only used for ungapped alignments as an alternative to performing gapped alignment. Now, it is applied to any HSP. The Karlin-Altschul sum statistics is too involved to be described here.

Finally, we must warn that formulas for P-value and E-value in BLAST are evolving. The above calculations are used in the current version of BLAST (version 2.2). They are different from the calculations used in earlier versions. The length adjustment was calculated as the product of λ and the raw score divided by H in earlier versions. Accordingly, they might be modified again in the future.

5 Examples

5.1 A BLASTP Search Example

As an example of using BLASTP, we will consider the capsid protein of the West Nile Virus (WNV)[1]. This virus mainly infects birds, but occasionally infects humans through the bite of an infected mosquito. The WNV is a positive-sense, single strand of RNA, having about 11,000 nucleotides. There are 7 non-structural proteins and 3 structural proteins in the RNA. The capsid protein of the WNV has sequence [10]:

```
MSKKPGGPGK SRAVNMLKRG MPRVLSLIGL KRAMLSLIDG KGPIRFVLAL LAFFRFTAIA
PTRAVLDRWR GVNKQTAMKH LLSFKKELGT LTSAINRRSS KQKKR
```

whose GenBank accession id is YP_001527877. We compare this sequence against the non-redundant GenBank database by using BLASTP available at the NCBI server with default settings. A BLAST printout contains (a) the information on the program, (b) a set of local alignments together with the statistical scores, and (c) a set of parameters used for the statistical analysis. A partial printout from our search follows:

[1] This example first appeared in the article of Casey [11].

BLASTP 2.2.18+

```
   ...

Database: All non-redundant GenBank CDS
translations+PDB+SwissProt+PIR+PRF excluding environmental samples from WGS
  projects
       7,036,788 sequences; 2,431,208,758 total letters

   ...

Query= gi|158516889|ref|YP_001527878.1| capsid protein [West Nile virus]
Length=105

   ...

Alignments

>gb|ABD67759.1| polyprotein precursor [West Nile virus]
Length=3433

   Score = 203 bits (517), Expect = 3e-51, Method: Composition-based stats.
   Identities = 100/100 (100%), Positives = 100/100 (100%), Gaps = 0/100 (0%)

Query 1    MSKKPGGPGKSRAVNMLKRGMPRVLSLIGLKRAMLSLIDGKGPIRFVLALLAFFRFTAIA 60
           MSKKPGGPGKSRAVNMLKRGMPRVLSLIGLKRAMLSLIDGKGPIRFVLALLAFFRFTAIA
Sbjct 1    MSKKPGGPGKSRAVNMLKRGMPRVLSLIGLKRAMLSLIDGKGPIRFVLALLAFFRFTAIA 60

Query 61   PTRAVLDRWRGVNKQTAMKHLLSFKKELGTLTSAINRRSS 100
           PTRAVLDRWRGVNKQTAMKHLLSFKKELGTLTSAINRRSS
Sbjct 61   PTRAVLDRWRGVNKQTAMKHLLSFKKELGTLTSAINRRSS 100

   ...

>gb|ACA28703.1| polyprotein [Japanese encephalitis virus]
Length=3432

   Score = 164 bits (414), Expect = 2e-39, Method: Composition-based stats.
   Identities = 71/105 (67%), Positives = 90/105 (85%), Gaps = 0/105 (0%)

Query 1    MSKKPGGPGKSRAVNMLKRGMPRVLSLIGLKRAMLSLIDGKGPIRFVLALLAFFRFTAIA 60
           M+KKPGGPGK+RA+NMLKRG+PRV  L+G+KR ++SL+DG+GP+RFVLAL+ FF+FTA+A
Sbjct 1    MTKKPGGPGKNRAINMLKRGLPRVFPLVGVKRVVMSLLDGRGPVRFVLALITFFKFTALA 60

Query 61   PTRAVLDRWRGVNKQTAMKHLLSFKKELGTLTSAINRRSSKQKKR 105
           PT+A+L RWR V K  AMKHL SFK+ELGTL  A+N+R  KQ KR
Sbjct 61   PTKALLGRWRAVEKSVAMKHLTSFKRELGTLIDAVNKRGKKQNKR 105

   ...

  Database: All non-redundant GenBank CDS translations+PDB+SwissProt+PIR+PRF
    excluding environmental samples from WGS projects

    Posted date: Sep 9, 2008 5:57 PM

  Number of letters in database: -1,863,758,534

  Number of sequences in database: 7,036,788


Lambda     K       H
    0.324    0.137    0.389
Gapped
Lambda     K       H
    0.267    0.0410   0.140
Matrix: BLOSUM62
Gap Penalties: Existence: 11, Extension: 1
Number of Sequences: 7036788
...
```

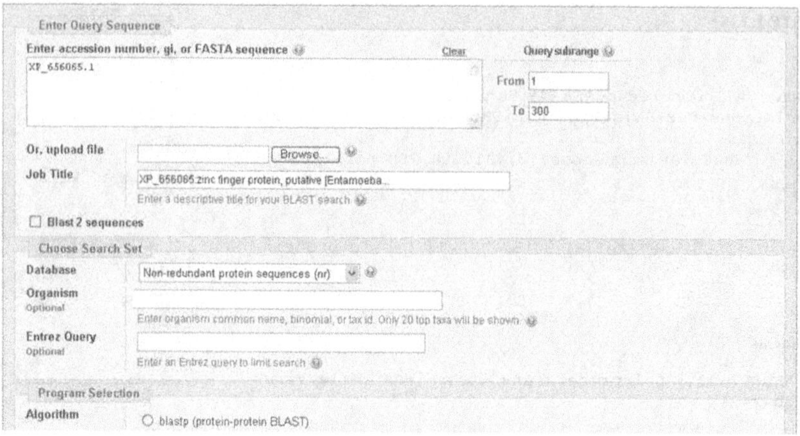

Fig. 1 The web page for launching a PSI-BLAST search.

```
Length of query: 105
Length of database: 2431208758
Length adjustment: 111

  ...
```

The second local alignment in the printout shows that the capsid protein of the WNV has a significant similarity with a domain region of the Japanese encephalitis virus. Our BLAST search reveals correctly the fact that the Japanese encephalitis virus and the WNV share similar proteins in their protein coats.

The statistical analysis associated with each alignment in the printout is done as follows. As shown in the printout, the query sequence has 105 letters; the target database contains 7,036,788 sequences and 2,431,208,758 letters; and the length adjustment displayed in the printout is 73. Since the local alignment involving the Japanese encephalitis virus is gapped, the following values are used in the calculation of the E-value:

$$\lambda = 0.267, \quad K = 0.041.$$

The raw score of the alignment is 414 and hence its bit score is

$$\frac{\lambda \times S_{raw} - \ln(K)}{\ln(2)} = \frac{0.267 \times 414 - \ln(0.041)}{\ln(2)} = 164.080851,$$

which agrees with 164 in the printout [16]. By Equation (4), the E-value is

$$0.041 \times (105 - 73) \times (2,431,208,758 - 7,036,788 \times 73) \times e^{-0.267 \times 414},$$

which is 2.481035e-39, in agreement with the printout value 2e-39.

5.2 A PSI-BLAST Search Example

The Position-Specific Iterated (PSI)-BLAST was designed to identify subtle homologous protein relationships that might be missed by other BLAST programs [7]. It searches a protein database iteratively. At each iteration step, PSI-BLAST generates a profile, or a position specific scoring matrix (PSSM), based on a multiple alignment of the identified high scoring hits to a given query sequence. The PSSM is calculated by considering position-specific scores for each position in the alignment. Highly conserved positions receive high scores, whereas weakly conserved positions receive low scores. The profile is then used to perform subsequent rounds of BLAST search. The strategy is to use the results of each iteration to refine the profile progressively. When such a profile is used to search a database, it can often detect distantly homologous, in structure or function, relationships between proteins.

We illustrate how to operate PSI-BLAST by searching part (the first 300 bp) of a putative zinc finger protein (XP_656065.1) in Entamoeba histolytica against the non-redundant protein sequence database as an example. We run the online version of PSI-BLAST available at http://ncbi.nlm.nih.gov/BLAST. The search is done in the following steps:

1. Paste the query sequence into the query box of the PSI-BLAST Web page and choose the searched database. Here, we used the accession id XP_656065.1 of the putative zinc finger protein (see Figure 5.1).
2. Set the algorithm parameters. If one chooses to use the default parameters, this step is skipped. In our example, we changed the maximum number of target sequences from 500 to 1000, the expected threshold from 10 to 4, and the PSI-BLAST threshold from 0.005 to 0.05.
3. Format to get the results. We ticked the box next to the BLAST button for the results to be retrieved in a new web page. The hits will be displayed into two sections. The hits with E-value smaller than the threshold s, 0.05 in our search, are listed first; those with E-value larger than s but smaller than the expected threshold, 4 in our case, are listed further down the page. The hits listed in the first section will be used in forming the profile that will be used in the next iteration step.
4. Click repeatedly the 'run PSI-BLAST iteration' button until the user decides to stop the search process or the search result cannot be improved. By clicking the 'Taxonomy reports' link on the top of the result window, one can view the distribution of the hits and decide to stop the search or not.

We obtained three significant hits (shown in Figure 2) in the initial search and surprisingly nine more significant hits (Figure 3) after the first iteration. The following several iterations generated even more significant hits.

PSI-BLAST is a powerful tool. Many important but subtle relationships that previously were detectable only by structural comparison can now be uncovered by a simple PSI-BLAST search. However, the user must use it with caution. A false relationship can easily be amplified by iteration. As a result, different queries that belong to the same family of proteins can perform differently in searches against

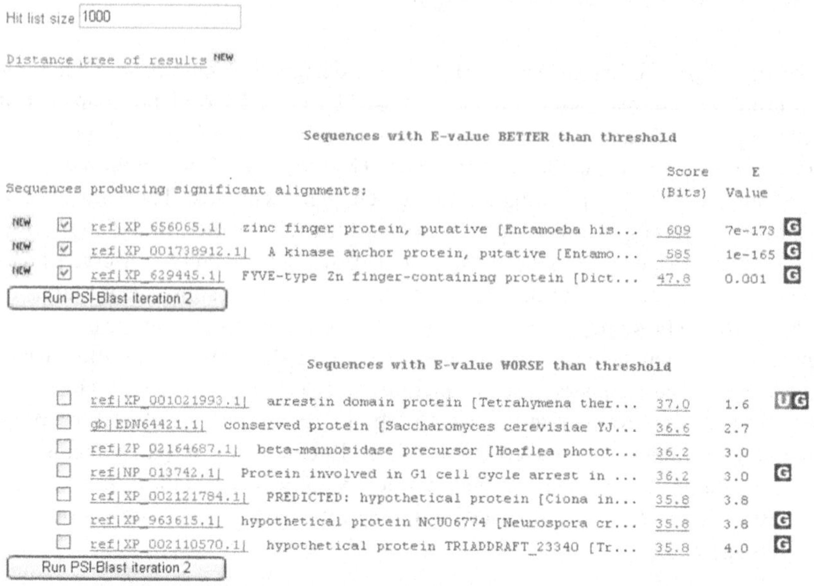

```
Hit list size 1000

Distance tree of results NEW

                     Sequences with E-value BETTER than threshold

                                                                        Score     E
Sequences producing significant alignments:                             (Bits)  Value

NEW  ref|XP_656065.1|  zinc finger protein, putative [Entamoeba his...    609   7e-173  G
NEW  ref|XP_001738912.1|  A kinase anchor protein, putative [Entamo...    585   1e-165  G
NEW  ref|XP_629445.1|  FYVE-type Zn finger-containing protein [Dict...   47.8   0.001   G
Run PSI-Blast iteration 2

                     Sequences with E-value WORSE than threshold

ref|XP_001021993.1|  arrestin domain protein [Tetrahymena ther...   37.0   1.6   UG
gb|EDN64421.1|  conserved protein [Saccharomyces cerevisiae YJ...   36.6   2.7
ref|ZP_02164687.1|  beta-mannosidase precursor [Hoeflea photot...   36.2   3.0
ref|NP_013742.1|  Protein involved in G1 cell cycle arrest in ...   36.2   3.0   G
ref|XP_002121784.1|  PREDICTED: hypothetical protein [Ciona in...   35.8   3.8
ref|XP_963615.1|  hypothetical protein NCU06774 [Neurospora cr...   35.8   3.8   G
ref|XP_002110570.1|  hypothetical protein TRIADDRAFT_23340 [Tr...   35.8   4.0   G
Run PSI-Blast iteration 2
```

Fig. 2 The initial PSI-BLAST search results.

the same database. It is recommenced that the user run PSI-BLAST search with different query sequences to obtain reliable homology relationships.

6 Advanced Topics

The NCBI web interface provides biologists easy access to BLAST homology search against different databases. It has a simple search form on which a dozen default values can be overwritten and displays aligned sequences together with significance analysis. But, using BLAST effectively requires knowledge of alignment statistics and insights on the algorithmic details of the program.

6.1 Scoring Matrices

The statistical significance of HSPs listed in BLAST printout is calculated mainly based on the internal scoring matrix. For protein sequence comparison, BLOSUM or PAM matrices are usually used; for DNA sequence comparison, similar but simpler substitution matrices are used. Although these scoring matrices are derived in

Sequences with E-value BETTER than threshold

	Sequences producing significant alignments:	Score (Bits)	E Value	
	ref|XP_656065.1| zinc finger protein, putative [Entamoeba his...	547	4e-154	G
	ref|XP_001738912.1| A kinase anchor protein, putative [Entamo...	546	6e-154	G
	ref|XP_629445.1| FYVE-type Zn finger-containing protein [Dict...	265	3e-69	G
NEW	ref|XP_001021993.1| arrestin domain protein [Tetrahymena ther...	51.2	9e-05	UG
NEW	ref|XP_001021994.1| arrestin domain protein [Tetrahymena ther...	48.1	9e-04	G
NEW	ref|XP_001436511.1| hypothetical protein GSPATT00037550001 [P...	45.4	0.005	G
NEW	ref|XP_001453022.1| hypothetical protein GSPATT00019342001 [P...	45.0	0.008	G
NEW	ref|XP_001450786.1| hypothetical protein GSPATT00039538001 [P...	44.3	0.012	G
NEW	ref|XP_001952503.1| PREDICTED: similar to arrestin homolog [A...	43.1	0.023	UG
NEW	ref|NP_001098299.1| arrestin [Oryzias latipes] >dbj|BAA21719....	43.1	0.024	UG
NEW	ref|XP_001447043.1| hypothetical protein GSPATT00014576001 [P...	42.4	0.045	G
NEW	gb|AAK84368.1|AF393635_1 visual arrestin [Loligo pealei]	42.4	0.047	

Run PSI-Blast iteration 3

Sequences with E-value WORSE than threshold

ref|XP_001431708.1| hypothetical protein GSPATT00006163001 [P...	42.0	0.059	G
ref|XP_001423611.1| hypothetical protein GSPATT00004307001 [P...	42.0	0.061	G
ref|XP_002116059.1| hypothetical protein TRIADDRAFT_60106 [Tr...	42.0	0.066	G

Fig. 3 The result after the first iteration of PSI-BLAST search.

different ways, they take essentially the same log-odds form: the score for aligning bases *a* and *b* is basically the logarithm of the ratio of the probability that *a* and *b* are aligned in homologous sequences to the probability that we expect to observe *a* and *b* aligned in random sequences with the same background letter composition.

The choice of scoring matrix for homology search can have a profound effect on the search output. Choosing PAM120 will generate one HSP, whereas choosing BLOSUM62 will generate another. In other words, scoring matrices detect different classes of alignments. For example, in searching a protein database containing 10,000,000 letters, the length range of the local alignments that PAM120 can detect is roughly from 15 to 50 (see [1]). Accordingly, choosing PAM120 may miss short but strong or long but weak alignments.

To obtain better results, one should choose the proper scoring matrix rather than just using the default matrix listed on the web interface. In PAM matrix naming system, higher numbers denote larger evolutionary distance. Hence, PAM120 is generally more appropriate than PAM30 for finding remotely-related homologous sequences. In contrast, larger numbers in BLOSUM matrices denote higher sequence similarity and so BLOSUM45 is generally better than BLOSUM62 for studying sequence relationship among divergent species.

6.2 Gap Penalties

Another important issue of BLAST search is to choose the gap cost. It is rare for two gene sequences to align perfectly with one another. Gapped BLAST introduces gaps between residues to bring up matches in the following positions. The way Gapped BLAST treats gaps can significantly affect its output. The user is allowed to choose different costs for gap opening and gap extension in a BLAST web page.

It is observed that optimal local alignments do not usually contain gaps of more than 1 residues [5] if the gap extension cost is relatively large. Hence, it is not rewarding to use any gap extension cost that is too close to the gap opening one.

On the other hand, if the gap extension cost is too small, the optimal local alignment scores might not follow the extreme value distribution. If this happens, the statistical analysis done by BLAST will no longer be meaningful.

6.3 Should DNA or Protein Sequence Be Used?

One natural question often asked by a BLAST user is: Should I compare gene sequences or the corresponding protein sequences? This can be answered by the following analysis.

Synonymous mutations are nucleotide substitutions that do not result in a change to the amino acids sequence of a protein. Evolutionary study suggests that there tend to be approximately 1.5 synonymous point mutations for every nonsynonymous point mutation. Because each codon has 3 nucleotides, each protein PAM translates into roughly $\frac{1+1.5}{3} \approx 0.8$ PAMs at the DNA level.

The substitution scores in the scoring matrix are implicitly log-odds scores. By multiplying by a constant factor, a scoring matrix is normalized to the logarithms of odds-ratios to base 2. The alignment score obtained with such a normalized scoring matrix is called the bit score and considered as bit information. In the alignment of two proteins that have diverged by 120 PAMs, each residue carries on average 0.98-bit information, whereas in the alignment of two DNA sequences that are diverged at 96 (or 120×0.8) PAMs, every three residues (a codon) carry only about 0.62-bit information [27]. Hence, at this evolutionary distance, 37% of the information available in protein comparison will be lost in DNA sequence comparison.

In a nutshell, protein sequence comparison is generally more sensitive than that of DNA sequences.

7 Exercises

Here we list 7 exercises for the reader to gain mastery of BLAST and BLAST-like programs.

1. Which of the following in the BLAST output provides an estimate of the false positive rate of the BLAST search: E-value, Score, or Identity?

2. What are the major advantages of Gapped-BLAST over BLASTP?

3. Use one of the BLAST programs to determine the frame shift of the following sequence.

```
ATGAGAGTGAAGGAGAAATATCAGCACTTGTGGAGATGGGGCACCATGCTCCTTGGGTTGT
TGATGATCCGTAGTGCTGCAGACCAATTGTGGGTCACAGTCTATTATGGGGTACCTGTGTG
GAAAGAAGCAACCACCACTCCATTTTGTGCATCAGATGCTAAAGCATATGATACAGAGGTA
CATAATGTTTGGGCCACACACGCCTGTGTACCCACAGACCCCAACCCACAAGAAGTAGTAT
TGGCAAATGTGGCAGAAAATTTTAACATGTG
```

4. Use BLASTP to align protein sequence P23749 and P16235 with different matrices, PAM30, BLOSUM45, and BLOSUM80. Which matrix gives the best alignment score?

5. Higher eukaryotic genomes contain large amounts of repetitive DNA. The most abundant interspersed repeat in the human genome is the Alu element. Alus tend to occur near genes, within the introns of genes, or in the regions between genes. In some cases, their presence and absence can fairly accurately show the intron-exon structure of a gene. Demonstrate this by performing a nucleotide-nucleotide BLAST search against the Alu database (alu_repeats) with the genomic sequence of the human Von Hippel Lindau syndrome gene (Accession AF010238). Note that the exons appear in the BLAST graphic as places where the Alu elements do not align.

6. A PSI-BLAST search is most useful when you want to

1. Extend a database search to find additional proteins,
2. Extend a database search to find additional DNA sequences,
3. Find the mouse ortholog of a human protein, or
4. Use a pattern to extend a protein search.

The human fragile histidine triad protein (FHIT, Accession P49789) is structurally related to galactose-1-phosphate uridylyltransferases. However, this relationship is not apparent in an ordinary BLAST search. Perform a protein-protein BLAST search against the SWISS-PROT database with P49789 and search your results for galactose-1-phosphate uridylyltransferases. Now use PSI-BLAST to verify the relationship between these two protein families.

7. On UCSC Genome Browser (http://genome.ucsc.edu/), find the protein sequence for rat leptin. BLAT this sequence against the human genome to find the human homology. Look for SNPs in the coding region of this gene. Are there any?

8 Further Reading

Here we point out some useful references on the topics covered in this chapter for the reader to consult. Sequence alignment has been extensively studied by biologists, computer scientists, and mathematicians in the past four decades. There is a large body of literature on this subject matter. For general treatment of sequence alignment, we refer the reader to the survey papers of Batzoglou [9] and Altschul et al. [3] and the book of Chao and Zhang [12].

For further consultation on how to use BLAST, we refer the reader to the online tutorial on the BLAST server at NCBI or the book of Korf, Yandell, and Bedell [22].

In 1990, Karlin and Altschul published their seminal work [19] on the distribution of optimal ungapped local alignment scores. Later, Altschul and Gish [5] and Pearson [25] investigated empirically the distribution of optimal gapped local alignment scores. Karlin-Altschul statistics of local alignment scores are surveyed by Pearson and Wood in [8] and Karlin in [18] and covered in the books of Chao and Zhang [12] and Ewens and Grant [14].

For the general theory of scoring matrices, the reader is referred to the papers of Altschul [1, 2] and Eddy [13]. The information on parameters and formulas used for statistical analysis in BLAST can be found in the paper [4] of Altschul et al. and the note (ftp://ftp.ncbi.nlm.nih.gov/blast/documents/developer/scoring.pdf) of Gertz [16].

Acknowledgements LX Zhang was partially supported by Singapore ARF grant R146-000-109-112. He would also like to thank Stephen Altschul for useful communication (through e-mail) on scoring matrices and alignment statistics and Kun-Mao Chao, Wayne Matten and Scott D. McGinnis for discussion of the implementation issues of BLAST programs.

References

1. Altschul, S.F.: Amino acid substitution matrices from an information theoretic perspective. Journal of Molecular Biology **219**(3), 555–565 (1991)
2. Altschul, S.F.: A protein alignment scoring system sensitive at all evolutionary distances. Journal of Molecular Evolution **36**, 290–300 (1993)
3. Altschul, S.F., Boguski, M.S., Gish, W., Wootton, J.C.: Issues in searching molecular sequence databases. Nature Genetics **6**, 119–129 (1994)

4. Altschul, S.F., Bundschuh, R., Olsen, R., Hwa, T.: The estimation of statistical parameters for local alignment score distributions. Nucleic Acids Research **29**, 351–361 (2001)
5. Altschul, S.F., Gish, W.: Local alignment statistics. Methods in Enzymology **266**(2), 460–480 (1996)
6. Altschul, S.F., Gish, W., Miller, W., Myers, E.W., Lipman, D.J.: Basic local alignment search tool. Journal of Molecular Biology **215**(3), 403–410 (1990)
7. Altschul, S.F., Madden, T.L., Schaffer, A.A., Zhang, J.H., Zhang, Z., Miller, W., Lipman, D.J.: Gapped BLAST and PSI-BLAST: a new generation of protein database search programs. Nucleic Acids Research **25**(17), 3389–3402 (1997)
8. Balding, D., Bishop, M., Cannings, C. (eds.): Handbook of Statistical Genetics, chap. 2, pp. 39–65. John Wiley & Sons (2003)
9. Batzoglou, S.: The many faces of sequence alignment. Briefings in Bioinformatics **6**(1), 6–22 (2005)
10. Borisevich, V., Seregin, A., Nistler, R., Mutabazi, D., Yamshchikov, V.: Biological properties of chimeric West Nile viruses. Virology **349**(2), 371–381 (2006)
11. Casey, R.M.: Blast sequences aid in genomics and proteomics. http://www.b-eye-network.com/print/1730 (2005)
12. Chao, K.M., Zhang, L.: Sequence Comparison: Theory and Methods. Springer (2008)
13. Eddy, S.R.: Where did the BLOSUM62 alignment score matrix come from? Nature Biotechnology **22**(8), 1035–1036 (2004)
14. Ewens, W.J., Grant, G.R.: Statistical Methods in Bioinformatics: An Introduction. Springer-Verlag (2001)
15. Florea, L., Hartzell, G., Zhang, Z., Rubin, G., Miller, W.: A computer program for aligning a cDNA sequence with a genomic DNA sequence. Genome Research **8**(9), 967–974 (1998)
16. Gertz, E.M.: Blast scoring parameters (2005)
17. Gish, W.: personal communication
18. Karlin, S.: Statistical signals in bioinformatics. Proc Nat'l Acad Sci USA **102**, 13,355–13,362 (2005)
19. Karlin, S., Altschul, S.F.: Methods for assessing the statistical significance of molecular sequence features by using general scoring schemes. Proc Nat'l Acad Sci USA **87**, 2264–2268 (1990)
20. Karlin, S., Altschul, S.F.: Applications and statistics for multiple high-scoring segments in molecular sequences. Proc Nat'l Acad Sci USA **90**, 5783–5877 (1993)
21. Kent, W.J.: BLAT — The BLAST-like alignment tool. Genome Research **12**(4), 656–664 (2002)
22. Korf, I., Yandell, M., Bedell, J.: BLAST. O'Reilly Media, Inc. (2003)
23. Ma, B., Tromp, J., Li, M.: PatternHunter — Faster and more sensitive homology search. Bioinformatics **18**, 440–445 (2005)
24. Ning, Z., Cox, A.J., Mullikin, J.C.: SSAHA: A fast search method for large DNA databases. Genome Research **11**(10), 1725–1729 (2001)
25. Pearson, W.R.: Empirical statistical estimates for sequence similarity searches. Journal of Molecular Biology **276**(1), 71–84 (1998)
26. Pearson, W.R., Lipman, D.J.: Improved tools for biological sequence comparison. Proceedings of the National Academy of Sciences **85**(8), 2444–2448 (1988)
27. States, D.J., Gish, W., Altschul, S.F.: Improved sensitivity of nucleic acid databases searches using application-specific scoring matrices. Methods **3**(1), 61–71 (1991)
28. Zhang, Z., Schaffer, A.A., Miller, W., Madden, T.L., Lipman, D.J., Koonin, E.V., Altschul, S.F.: Protein sequence similarity searches using patterns as seeds. Nucleic Acids Research **26**(17), 3986–3990 (1998)
29. Zhang, Z., Schwartz, S., Wagner, L., Miller, W.: A greedy algorithm for aligning DNA sequences. Journal of Computational Biology **7**(1-2), 203–214 (2000)

Practical Multiple Sequence Alignment

Tobias Rausch and Knut Reinert

Abstract Multiple sequence alignment as a means of comparing DNA, RNA, or amino acid sequences is an essential precondition for various analyses, including structure prediction, modeling binding sites, phylogeny, or function prediction. This range of applications implies a demand for versatile, flexible, and specialized methods to compute accurate alignments. This chapter summarizes the key algorithmic insights gained in the past years to facilitate an easy understanding of the current multiple sequence alignment literature and to enable the readers to use and apply current tools in their own research.

1 History of the Problem

The problem of comparing multiple sequences is a long-standing brainteaser of molecular biology. The research was sparked by a simple insight: Weak and faint biologically important sequence similarities vanish in a pairwise alignment but stand out in a multiple sequence alignment (MSA). For the last 20 years, the driving force behind MSA is the assumption that sequence similarity or sequence conservation implies structural, evolutionary, or functional correspondence. In other words, biologically important residues or nucleotides are assumed to be less likely to mutate than unimportant ones. In a MSA, we thus rewrite the sequences in such a way that conserved residues or nucleotides appear in the same column (see Table 1). MSA problems are characterized by (1) the number of sequences, (2) the length of the sequences, (3) the alphabet of the sequences (usually DNA, RNA, or amino acids), and (4) the relatedness of the sequences. Here, relatedness refers to both

Tobias Rausch
Freie Universität, e-mail: rausch@inf.fu-berlin.de

Knut Reinert
Freie Universität, e-mail: reinert@inf.fu-berlin.de

L.S. Heath and N. Ramakrishnan (eds.), *Problem Solving Handbook in Computational Biology and Bioinformatics*, DOI 10.1007/978-0-387-09760-2_2,

the divergence of the sequences and whether the sequences are globally or locally related. Many applications rely on accurate MSAs. The most prominent ones are phylogeny, functional predictions, domain identification, modeling binding sites, sequence consensus, and structure prediction (see Table 1 and Figure 1). With the benefit of hindsight, it is obvious that MSA algorithms and applications mutually fueled each other. Applications and progress in sequencing technologies created a continuous demand for new and more efficient alignment algorithms. In return, the progress in algorithms opened up unforeseen possibilities in terms of applications. For years, for instance, research has focused on sequence comparisons where the order of characters in the sequences is preserved. This colinearity condition was in fact the defining property of an alignment. In recent years, however, with an increasing number of genomic sequences at hand, sequence comparison involves the identification of the classical alignment operations, namely substitutions, deletions and insertions *and* more complex operations such as transpositions, translocations, duplications and inversions. This chapter is an attempt to capture the essential developments in the past years, starting from the first programs developed in the late 1980s to the first genome aligners of recent years.

HBA_HUMAN	.MVLSPADKTNVKAAWGKVGAHAGEYGAEALERMFLSFPTTKTYFPHF.DLSH
HBB_HUMAN	MVHLTPEEKSAVTALWGKV..NVDEVGGEALGRLLVVYPWTQRFFESFGDLST
HBA_HORSE	.MVLSAADKTNVKAAWSKVGGHAGEYGAEALERMFLGFPTTKTYFPHF.DLSH
HBB_HORSE	.VQLSGEEKAAVLALWDKV..NEEEVGGEALGRLLVVYPWTQRFFDSFGDLSN
MYG_PHYCA	.MVLSEGEWQLVLHVWAKVEADVAGHGQDILIRLFKSHPETLEKFDRFKHLKT
LGB2_LUPLU	MGALTESQAALVKSSWEEFNANIPKHTHRFFILVLEIAPAAKDLFSFLKGTSE
HBA_HUMAN	.....GSAQVKGHGKKVADALTNAVAHVDD...M..PNALSALSDLHAHKLRVD
HBB_HUMAN	PDAVMGNPKVKAHGKKVLGAFSDGLAHLDN...L..KGTFATLSELHCDKLHVD
HBA_HORSE	.....GSAQVKAHGKKVGDALTLAVGHLDD..L..PGALSNLSDLHAHKLRVD
HBB_HORSE	PGAVMGNPKVKAHGKKVLHSFGEGVHHLDN..L..KGTFAALSELHCDKLHVD
MYG_PHYCA	EAEMKASEDLKKHGVTVLTALGAILKKKGH...H..EAELKPLAQSHATKHKIP
LGB2_LUPLU	VPQ..NNPELQAHAGKVFKLVYEAAIQLQVTGVVVTDATLKNLGSVHVSK.GVA
HBA_HUMAN	PVNFKLLSHCLLVTLAAHLPAEFTPAVHASLDKFLASVSTVLTSKYR......
HBB_HUMAN	PENFRLLGNVLVCVLAHHFGKEFTPPVQAAYQKVVAGVANALAHKYH......
HBA_HORSE	PVNFKLLSHCLLSTLAVHLPNDFTPAVHASLDKFLSSVSTVLTSKYR......
HBB_HORSE	PENFRLLGNVLVVVLARHFGKDFTPELQASYQKVVAGVANALAHKYH......
MYG_PHYCA	IKYLEFISEAIIHVLHSRPGDFGADAQGAMNKALELFRKDIAAKYKELGYQG
LGB2_LUPLU	DAHFPVVKEAILKTIKEVVGAKSEELNSAWTIAYDELAIVIKKEMNDA...

Table 1 MSA of 6 globin sequences: Human hemoglobin subunit alpha (UniProt accession: P69905), human hemoglobin subunit beta (P68871), horse hemoglobin subunit alpha (P01958), horse hemoglobin subunit beta (P02062), sperm whale myoglobin (P02185) and European yellow lupin leghemoglobin-2 (P02240). The helix secondary structure annotation from UniProt is shown in bold font.

Throughout the history of MSA, one can distinguish two types of algorithms, optimal ones and heuristics. The former algorithms compute an optimal alignment with respect to some scoring function such as the sum of pairs score. The latter algorithms compute an alignment based on some kind of biologically sound procedure

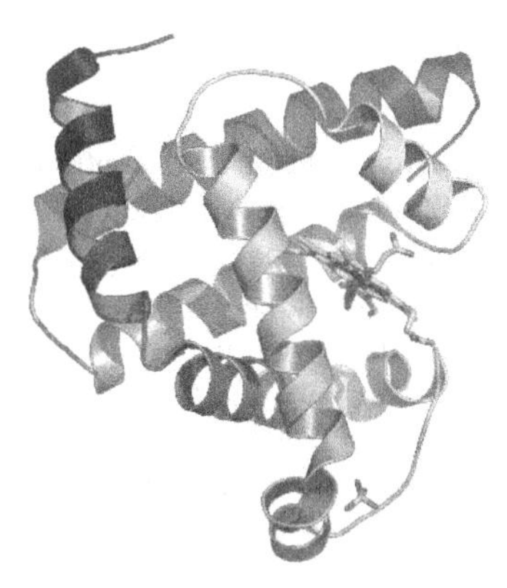

Fig. 1 A 3D model showing the helical domains of myoglobin.

such as progressive alignment. Both classes of algorithms are reviewed in Section 2. The first optimal methods could align three sequences simultaneously using standard dynamic programming [32, 56]. A few years later, the program MSA [35, 52] could align up to eight sequences of average protein length by using a clever bounding technique for the dynamic programming lattice. Time and space was further reduced using the A* algorithm [51, 72] and (partly heuristic) divide and conquer techniques [73]. Besides bounding techniques for the dynamic programming formulation, other algorithms used an alignment graph or trace graph [43, 77]. This alignment graph was used in an integer linear programming (ILP) formulation [71] extended by various branch-and-cut techniques [2, 3, 4].

Computing an optimal alignment for an arbitrary number of sequences is, however, NP-complete using the sum of pairs score [93]. That is why a vast number of heuristics has been developed enabling the alignment of more sequences of greater length. Heuristic methods were difficult to compare in the beginning but gained enormous leverage with the advent of protein benchmark data sets of sometimes manually refined MSAs such as BAliBASE [87, 89], PREFAB [24], OXBENCH [67], SABmark [92] and IRMBASE [85]. For protein alignments, these benchmarks are the de facto standard for judging the performance of individual methods. The first heuristic *progressive* aligner was published in 1987 [27] followed by a great variety of other heuristics, most prominently the Clustal series of programs [38, 47, 88]. More recently, the progressive alignment paradigm has been extended using approaches outlined in the next section, such as consistency [20, 59], refinement [24, 41], and segmentation [70, 85]. The increasing number of genomic sequences stimulated the development of genome aligners or genome comparison tools in the past 10 years. The MUMmer series of programs [18, 19, 46] remarkably pioneered this research area, but lately a number of other interesting anchor-based alignment tools appeared [11, 16].

2 Algorithm Description

Over the years, numerous research projects have contributed to steady progress in the area of MSA. Despite such a long history, methods are still far away from being optimal in a biological sense. The main obstacles are (1) that we still lack a precise mathematical formulation of such a biologically optimal alignment and (2) that the problem is already NP-hard if we use a very simplified formulation such as alignment score maximization. This very question of finding an alignment of maximum score has driven the field significantly in the past years and many sequence based methods, both heuristics and optimal ones, have been developed to solve this problem. The nuts and bolts of these methods are described in depth in Section 2.1. Recently, the sequence based methods have been complemented by methods that go beyond the raw sequence data. These structure based methods use a great variety of structure prediction methods and databases with structural information. The goal is either to substantiate a possibly weak signal of sequence similarity or to identify novel domains where conservation only manifests itself on a structural level. The basics of these methods are investigated in Section 2.2. The predominant representation of an alignment is the well-known alignment matrix. An example is shown on the right in Figure 2. Based upon that matrix, we can formally define the properties of a multiple alignment of a set S $= \{S^0, S^1, \dots, S^{n-1}\}$ of n sequences.

- $S^i \in$ S is a string over the finite ordered alphabet Σ that is $S^i \in \Sigma^*$. Σ is, for instance, the DNA or amino acid alphabet. Each string S^i is a sequence of letters $s_0^i s_1^i \dots s_{|S^i-1|}^i$ of length $|S^i|$ where $s_u^i \in \Sigma$.
- The alphabet $\tilde{\Sigma} = \Sigma \cup \{-\}$ is the extended alphabet including a gap character '$-$'.

A multiple alignment A of the strings in S is an $n \times l$ matrix consisting of n strings $\tilde{S}^0, \tilde{S}^1, \dots, \tilde{S}^{n-1} \in \tilde{\Sigma}^*$ such that

- The strings $\tilde{S}^0, \tilde{S}^1, \dots, \tilde{S}^{n-1}$ are of length l.
- The string $\tilde{S}^i$ with gaps removed is equal to S^i.
- The matrix entry a_u^i in row i and column u is either from the alphabet Σ or equal to the gap character '$-$': $\quad a_u^i \in \tilde{\Sigma} \quad \forall\, 0 \le i < n,\ 0 \le u < l$
- No column consists entirely of gap characters. This implies:

$$\max_{i=0,\dots,n-1} |S^i| \le l \le \sum_{i=0,\dots,n-1} |S^i|$$

Alternatively, one can think of an alignment as a path through an n-dimensional hypercube as shown in Figure 2.

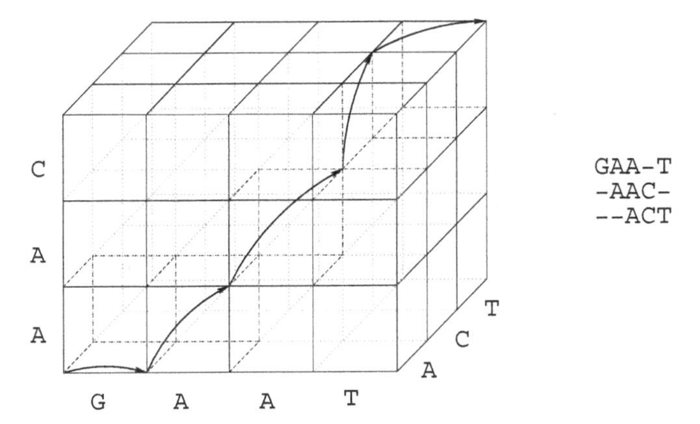

Fig. 2 The MSA path in a 3-dimensional lattice corresponding to the alignment shown on the right.

2.1 *Sequence based methods*

Given a number of different multiple alignments for a set of sequences, we somehow need a quantitative measure to decide which one is the best. For sequence based methods the ubiquitously used measure is the sum of pairs multiple alignment score, which is an extension of the pairwise alignment score to more than 2 sequences. The score of a multiple alignment is simply the sum over all alignment scores of each and every possible induced pairwise alignment. This notion can be formalized using (1) the projection A^I of a multiple alignment onto a set of sequences and (2) the definition of pairwise alignment scores. A pairwise alignment projection is a mere selection of 2 distinct rows in a given alignment and a subsequent removal of all columns containing only gaps. For example, the projection $A^{\{0,1\}}$ of the alignment in Figure 2 results in the pairwise alignment:

$$\begin{array}{ccccc} G & A & A & - & T \\ - & A & A & C & - \end{array}$$

A projection A^I for an index set $I \subset \{0,\ldots,n-1\}$ is obtained from A by

1. Selecting row i in A if and only if $i \in I$.
2. Deleting column u in A^I if and only if column u contains only gap characters.

This formulation respects the requirement that $A^{\{i\}} = S^i$. Similarly, one can project the path through the n-dimensional space onto a subspace as shown in Figure 3. The most common pairwise scoring function uses linear gap costs. Linear gap costs penalize a gap of length γ with a cost of $g + e \cdot (\gamma - 1)$ where g is the constant gap opening penalty, e is the constant gap extension penalty and $g \leq e$ with $g, e \leq 0$. If $g = e$ the number of gap openings is irrelevant and such gap costs are called constant hereafter. Using linear gap costs, the score of a pairwise alignment is

$$Score(\mathrm{A}^{\{i,j\}}) = (\sum_{\substack{u=0,\dots,\tilde{l}-1 \\ a_u^i \neq -;\ a_u^j \neq -}} \delta(a_u^i, a_u^j)) + g \cdot \#\text{GapOpen} + e \cdot \#\text{GapExtension}$$

where $\tilde{l}$ is the length of the projected alignment and δ a scoring function or substitution matrix for all pairs of characters $a_u^i, a_u^j \in \Sigma$. The BLOSUM [37] and PAM [17] matrices are commonly used substitution matrices for protein alignments. For DNA alignments, most tools use a simple match / mismatch scoring function. The alignment

$$\begin{array}{cccccccc} G & A & T & A & T & A & - & - & T \\ - & A & T & G & T & A & C & C & - \end{array}$$

evaluated with linear gap costs (gap opening penalty $g = -4$, gap extension penalty $e = -1$) and a scoring function defined by a match score of $\delta(x,x) = 4$ and a mismatch score of $\delta(x,y) = -2$ results in a total score of $14 + (-4) \cdot 3 + (-1) \cdot 1 = 1$. Finally, the sum of pairs multiple alignment score SP_{Score} can be simply defined as

$$SP_{Score}(\mathrm{A}) = \sum_{0 \leq i < j < n} Score(\mathrm{A}^{\{i,j\}})$$

For the sake of completeness, one should note that, besides the sum of pairs score, other quantitative alignment quality measures are available, most notably the weighted sum of pairs score, the tree alignment score, and the consensus score [34]. In this chapter, we focus on the well-established sum of pairs score. For the heuristic algorithms, all of these measures became less important with the publication of reference MSA benchmarks such as BAliBASE [87] but research devoted to exact algorithms relies on such quantitative measures.

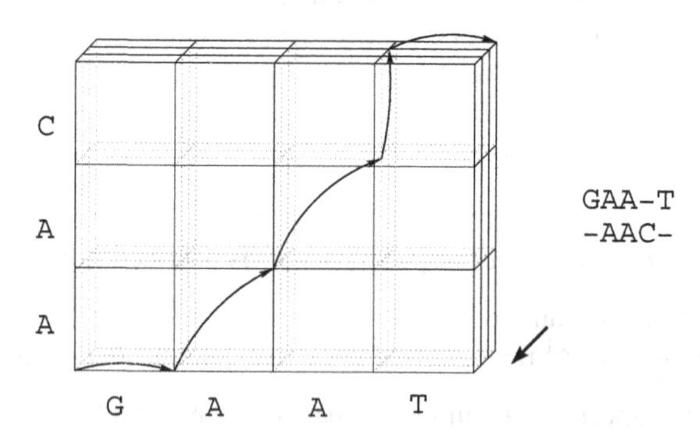

Fig. 3 A projection of a 3-dimensional lattice to a 2-dimensional matrix corresponding to the projection of an alignment of 3 sequences onto a subset of 2 sequences.

2.1.1 Exact algorithms

The multiple alignment score maximization problem can be solved optimally using either natural extensions of the dynamic programming algorithm [35, 51, 52, 72, 73] or exact algorithms based on graph-theoretic models [2, 3, 4, 71].

Dynamic programming

The dynamic programming recursion to compute the optimal pairwise alignment between sequence $S^0 = s_0^0 s_1^0 \dots s_{|S^0-1|}^0$ and sequence $S^1 = s_0^1 s_1^1 \dots s_{|S^1-1|}^1$ is

$$M_{u,v} = \max \begin{cases} M_{u-1,v-1} & + \delta(s_u^0, s_v^1) \\ M_{u-1,v} & + \delta(s_u^0, -) \\ M_{u,v-1} & + \delta(-, s_v^1) \end{cases}$$

where $M_{u,v}$ is the 2-dimensional dynamic programming matrix and δ is the scoring function. For a constant gap penalty $g = e$ and the Blosum62 substitution matrix one could define δ as $\delta(s_u^0, s_v^1) = \text{Blosum}_{62}(s_u^0, s_v^1)$ and $\delta(s_u^0, -) = \delta(-, s_v^1) = e$. The extension to 3 sequences involves two changes. First, a 3-dimensional dynamic

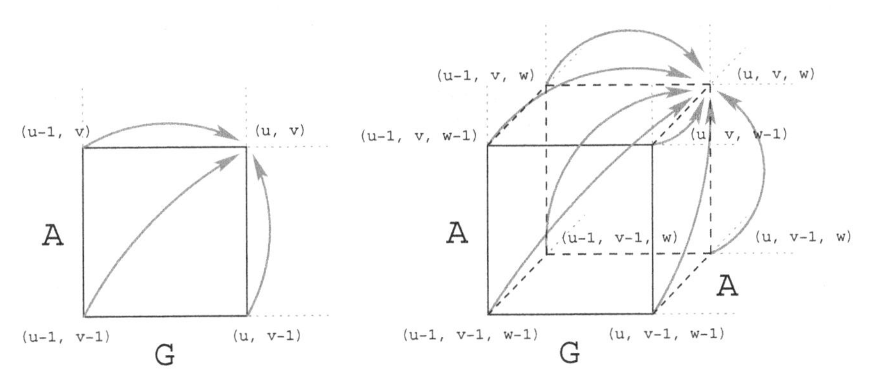

Fig. 4 In each cell of the dynamic programming matrix / cube $(2^n - 1)$ predecessor have to be evaluated where n is the number of sequences.

programming hypercube has to be computed and second, for each entry we have to evaluate $(2^n - 1) = (2^3 - 1) = 7$ predecessors as shown in Figure 4. The recursion is

$$M_{u,v,w} = \max \begin{cases} M_{u-1,v-1,w-1} & + \tilde{\delta}(s_u^0, s_v^1, s_w^2) \\ M_{u,v-1,w-1} & + \tilde{\delta}(-, s_v^1, s_w^2) \\ M_{u-1,v,w-1} & + \tilde{\delta}(s_u^0, -, s_w^2) \\ M_{u-1,v-1,w} & + \tilde{\delta}(s_u^0, s_v^1, -) \\ M_{u,v,w-1} & + \tilde{\delta}(-, -, s_w^2) \\ M_{u-1,v,w} & + \tilde{\delta}(s_u^0, -, -) \\ M_{u,v-1,w} & + \tilde{\delta}(-, s_v^1, -) \end{cases}$$

For the sum of pairs score with constant gap costs, $\tilde{\delta}$ can be defined in terms of δ as $\tilde{\delta}(a,b,c) = \delta(a,b) + \delta(b,c) + \delta(a,c)$ with $a,b,c \in \hat{\Sigma}$ and $\delta(-,-) = 0$. This can be extended to higher dimensions d. As in the pairwise case, the key idea is that larger alignments are constructed from already computed subsolutions. Any $M_{u,v,\ldots,z}$ is the best score of aligning the prefixes $s_0^0 s_1^0 \ldots s_u^0$, $s_0^1 s_1^1 \ldots s_v^1$, $\ldots$, $s_0^{n-1} s_1^{n-1} \ldots s_z^{n-1}$. In addition, the optimal alignment can be retrieved through the standard traceback operations extended to the d-dimensional hypercube. Note that it is also possible to apply Gotoh's algorithm [31] for linear gap costs to more than two sequences. Similar to the pairwise case, we then require additional hypercubes for the best gapped alignment in each dimension. The size of the hypercube is exponential in the number of sequences $O(\prod_{i=0}^{n-1} |S^i|)$. For each cell of this hypercube, $(2^n - 1)$ predecessor cells have to be evaluated. Thus, the time complexity is $O((2^n - 1) \cdot \prod_{i=0}^{n-1} |S^i|)$ if and only if the computation of the δ function is constant $O(1)$. This is roughly $O((2\tilde{n})^n)$ where $\tilde{n}$ is the average sequence length. Bounding techniques try to minimize the actually computed hypercube alignment space by using lower and upper bounds [35, 51, 52, 72] or a combination of an exact algorithm with a heuristic divide and conquer approach [73].

Graph based models

An alignment can be visualized as an alignment graph of sequence segments as shown in Figure 5. The alignment edges of this graph represent matches or possible mismatches. Gaps are implicitly represented by the topology of the graph. For instance, a vertex without any outgoing edge is aligned to gaps in all other sequences. An alignment graph can be easily converted into an alignment matrix using standard graph algorithms, namely connected components and topological sort [14]. If we allow arbitrary alignment edges as shown in Figure 5 an actual alignment can only realize a subset of the given edges. This subset is called a trace [77]. The graph can be extended by edge weights that capture some kind of quantitative measure of alignment quality. A possible measure is, for instance, the pairwise BLOSUM score of two aligned segments. Given such a problem, the trace problem can be rephrased as a maximum weight trace problem and also naturally extended to multiple sequences [42, 43]. The graph formulation translates easily into an integer linear program, giving rise to the possibility of applying techniques from combinatorial optimization such as branch-and-cut [3, 4]. The alignment graph can be extended with gap arcs to incorporate positional gap penalties [71]. Recently, a Lagrangian approach was proposed to solve the integer linear programming formulation more efficiently [2].

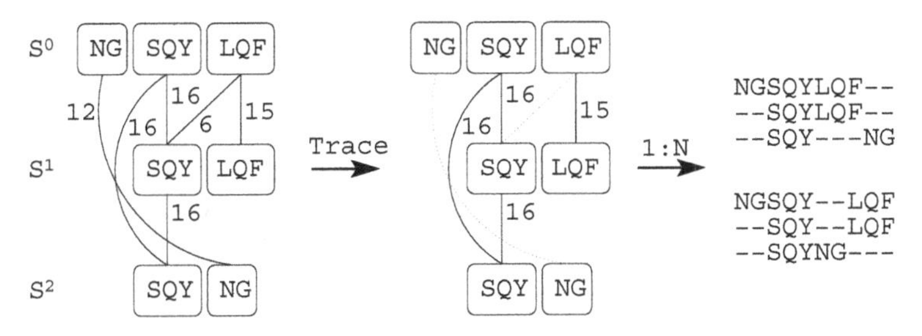

Fig. 5 An alignment graph with arbitrary alignment edges (left), its heaviest trace (middle) and its conversion to an alignment matrix (right). The graph does not impose an order on adjacent indels and hence, there is a 1:N relation between alignment graphs and alignment matrices.

2.1.2 Heuristic algorithms

In practice, optimal methods are only practical for a few, relatively short sequences. Hence, the development of fast and accurate heuristics for MSA problems is a very active research field. In this chapter, we can only review the most important heuristics, which is first and foremost the progressive alignment strategy [27].

Progressive alignment

A sound multiple alignment of n sequences should induce $(\frac{n\cdot(n-1)}{2})$ projected pairwise alignments that are as close as possible to optimal pairwise alignments. Unfortunately, pairwise alignments may be incompatible as shown in Figure 6. Progressive alignment resolves these inconsistencies in a greedy manner. The multi-

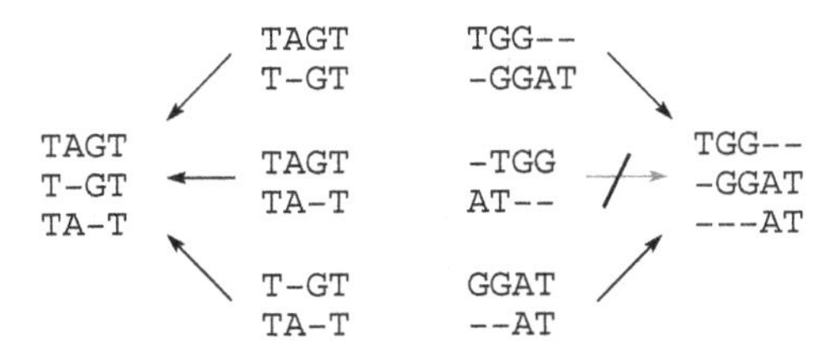

Fig. 6 A set of pairwise alignments that are compatible (left) or incompatible (right).

ple alignment is started from the most similar pair and then gradually, the other less similar sequences are added to the growing alignment. The intuitive assumption is

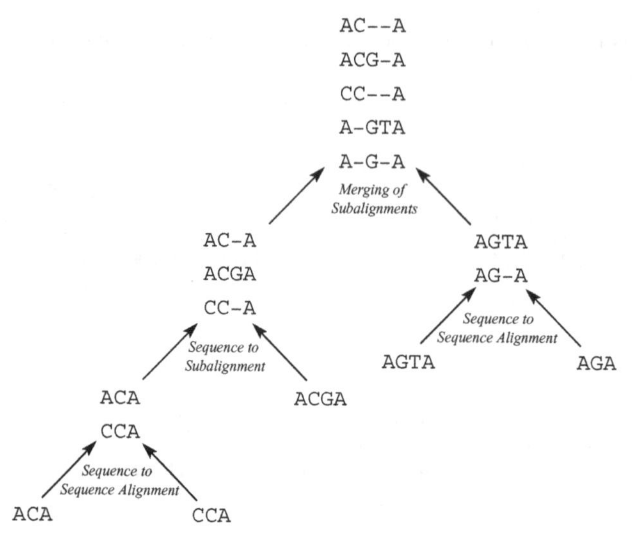

Fig. 7 The progressive alignment greedily builds a final alignment along the guide tree using a given method to merge subalignments.

that a pairwise alignment of closely related sequences is more to be trusted than an alignment of distantly related sequences [27]. The method thus requires 2 things. First, a binary tree, called a guide tree, that indicates when every sequence (a leaf of the tree) is merged into a growing multiple alignment and second, a means of aligning already finished subalignments with another sequence or another subalignment. The later situation arises if the progressive alignment is started from multiple seeding alignments as shown in Figure 7.

The guide tree can be obtained in 2 steps. First, a distance score between all pairs of sequences is computed, and, second, the phylogenetic tree is reconstructed [28] using clustering methods such as UPGMA [82] or neighbor-joining [76]. Several distance measures for two sequences are based upon simple similarity scores. Examples are the percent identity between two sequences or the fractional number of common k-mers where a k-mer is a contiguous substring of length k. For large alphabets, the percent identity and the number of common k-mers are less applicable, unless the sequences are closely related or both measures are applied over a compressed alphabet [23]. More precise measures are based upon pairwise global or local alignment scores [58, 81], which are usually normalized by alignment length. UPGMA is, besides neighbor-joining, a widely used distance based tree reconstruction method. The algorithm requires a set of n elements (e.g. sequences) and all pairwise distances $d_{i,j}$. Initially, each element is in its own group, and, thus, the sequences are the leaves of the tree. The algorithm proceeds in 4 steps:

1. Select the minimum distance $d_{i,j}$.
2. Create a new group u that joins i and j.
3. Compute the distances $d_{k,u}$ of any group k to the new group u.

4. Remove i, j from the set of elements and go to 1.

The UPGMA algorithm reconstructs the correct tree only for ultrametric distances. Such distances imply that all sequences have evolved from a common ancestor at a constant rate. This assumption is, in general, not true, and, thus, UPGMA is not used very often in phylogenetic studies. It is, however, widely used in progressive alignment tools because some authors argue [24] that a reliable evolutionary tree is not as important as a tree that guarantees that the subalignments with the fewest differences are merged first. In Step (3) the new distance $d_{k,u}$, from any group k to the new group u that joined i and j, can be computed using different methods:

1. Single linkage clustering: $d_{k,u} = \min(d_{k,i}, d_{k,j})$
2. Complete linkage clustering: $d_{k,u} = \max(d_{k,i}, d_{k,j})$
3. Average linkage clustering: $d_{k,u} = \frac{d_{k,i}+d_{k,j}}{2}$
4. Weighted average linkage clustering: $d_{k,u} = \frac{n_i \cdot d_{k,i}+n_j \cdot d_{k,j}}{n_i+n_j}$

In the last method n_i and n_j are the number of elements in group i and j, respectively. The neighbor joining method is a different method to reconstruct a tree. It has a higher time complexity of $O(n^3)$, where n is the number of sequences, compared to $O(n^2)$ for the UPGMA algorithm. The guide tree obtained with that method is, however, regarded as a better evolutionary tree because the neighbor joining method does not assume a molecular clock. The idea of the method is to start with a star tree and then to gradually group pairs of sequences so that the overall tree length is minimized.

Aligning the children of an internal node in the guide tree either involves an ordinary sequence alignment or an alignment of subalignments. In the latter case, one possible objective is to optimize the already mentioned sum of pairs multiple alignment score.

$$SP_{Score}(\mathrm{A}) = \sum_{0 \leq i < j < n} Score(\mathrm{A}^{\{i,j\}})$$

Using linear gap costs, an optimal merging of subalignments is NP-complete [44, 53]. However, in Kececioglu and Starrett [44], the authors propose an algorithm that is exact and quite fast in practice. Other methods favor speed over optimality and use approximations of gap opening counts [45]. More often, however, practical tools use their own way of merging subalignments with quite distinct objective functions [26]. These methods are usually subsumed under the generic term profile-profile alignments. A profile of a multiple alignment A of length l is a $|\tilde{\Sigma}| \times l$ matrix P, where $\mathrm{P}_{a,u}$ is the frequency of character $a \in \tilde{\Sigma}$ in column u of A.

```
A G C T
A G C C
A - C A
C G C A
```

P	1	2	3	4
A	0.75	0	0	0.5
C	0.25	0	1.0	0.25
G	0	0.75	0	0
T	0	0	0	0.25
–	0	0.25	0	0

Assuming constant gap costs, a string $S = s_0 s_1 \cdots s_{|S|-1}$ can be quickly aligned to a profile with a standard pairwise dynamic programming algorithm. Only the scoring function δ has to be adapted.

$$\delta_{\text{New}}(s_w, u) = \sum_{a \in \tilde{\Sigma}} \mathrm{P}_{a,u} \cdot \delta(s_w, a)$$

In this case, δ_{New} scores a column u against a character $s_w \in \Sigma$. The δ function has to be extended to handle the special case of scoring a gap character against another gap character.

$$\delta(a,b) = \begin{cases} \text{Blosum}_{62}(a,b) & \text{if and only if } a, b \in \Sigma \\ e & \text{if and only if } a = \text{"}-\text{"} \text{ or } b = \text{"}-\text{"} \\ 0 & \text{if and only if } a = b = \text{"}-\text{"} \end{cases}$$

Note that in a projected alignment gap columns are removed and hence, the score for two aligned gaps is set to 0. For instance, a final string to profile alignment of the string $S = ACCA$ can be scored as shown below, assuming $\delta(x,x) = 4, \delta(x,y) = -3, \delta(x,-) = \delta(-,x) = -2$ and $\delta(-,-) = 0$.

```
A G - C T
A G - C C
A - - C A
C G - C A
---------
A - C C A
```

P	1	2	–	3	4
A	0.75	0		0	0.5
C	0.25	0		1.0	0.25
G	0	0.75		0	0
T	0	0		0	0.25
–	0	0.25	1.0	0	0

S	A	–	C	C	A
δ_{New}	2.25	−1.5	−2	4	0.5

Hence, the score of the full string to the profile alignment is 3.25. Note that non-linear or constant gap penalties simplify the sum of pairs score of a multiple alignment A of length l to

$$SP_{Score}(\mathrm{A}) = \sum_{0 \leq i < j < n} Score(\mathrm{A}^{\{i,j\}}) = \sum_{i,j} \sum_{u=0}^{l-1} \delta(\tilde{s}_u^i, \tilde{s}_u^j) = \sum_{u=0}^{l-1} \sum_{i,j} \delta(\tilde{s}_u^i, \tilde{s}_u^j)$$

The last equality stems from the independence of the alignment columns using the δ scoring function with constant gap penalties. Using dynamic programming, the optimal string to profile alignment can be found in quadratic time $O(|\tilde{\Sigma}| \cdot l \cdot |S|)$ where l is the length of the profile, $|S|$ the length of the sequence and $|\tilde{\Sigma}|$ a small constant, e.g., 5 for the DNA alphabet or 21 for the amino acid alphabet. Similarly, a profile-profile alignment can be carried out. The only difference is an extra sum over the alphabet $\tilde{\Sigma}$.

$$\delta_{\text{New}}(u, w) = \sum_{a \in \tilde{\Sigma}} \sum_{b \in \tilde{\Sigma}} \mathrm{P}_{a,u} \cdot \mathrm{P}_{b,w} \cdot \delta(a,b)$$

Numerous other profile-profile column scoring functions have been published [24, 26, 41].

In summary, an optimal merging of subalignments with linear gap costs $g + e \cdot (\gamma - 1)$ is NP-complete. A merging with $g = e$ remains polynomial because gap opening counts are irrelevant. In this case, each column can be treated as a meta-character in an extended alphabet. Given a scoring function for such meta-characters, the problem is to find an alignment of two strings of meta-characters, which is clearly solvable with a pairwise dynamic programming algorithm.

Consistency and refinement

The choice of the binary tree and the method to merge subalignments has great influence on the final alignment. Once a new sequence is added to the growing alignment all the aligned characters and inserted gaps are fixed ("Once a gap, always a gap." [27]). But this is also true for alignment errors: once made they are preserved and they may even cause new alignment errors in the subsequent progressive steps. There are two strategies, "consistency" and "refinement", to handle alignment errors; one aims to prevent errors and the other one aims to correct errors [94]. The prevention approach tries to substantiate pairwise alignments by multiple sequence information. That is, it tries to make pairwise alignments consistent with all the other sequences, and, hence, the name consistency [33, 59]. The refinement approach takes a possibly erroneous alignment, iteratively splits this alignment into two subalignments and merges these alignments together again. These methods, thus, iteratively "refine" or "realign" a given alignment. In other publications authors sometimes use the term "Iterative Alignment" to describe such techniques [66].

Although current algorithms use slightly different means of consistency, the basic idea is always the same: the confidence of aligning substrings of a pair of sequences S^0 and S^1 is greater the more intermediate sequences S^i support this alignment. In other words, the alignments $S^0 \leftrightarrow S^i$ and $S^i \leftrightarrow S^1$ induce a putative transitive alignment $S^0 \leftrightarrow S^1$ that is either consistent or inconsistent with a precomputed alignment of S^0 and S^1. If it is consistent, greater confidence in the alignment of these substrings of S^0 and S^1 is established, and the scores are somehow increased. In an alignment graph, this consistency extension or triplet extension corresponds to a search for three-way cliques (see Figure 8).

The refinement approach [24, 41] splits a full alignment randomly or following a deterministic order into subalignments and then merges these subalignments using, for example, profile to profile alignment methods. Random cutting usually halts if no improvement in alignment score is observed during the last iterations.

Anchor-based alignment

Even the heuristic progressive alignment becomes prohibitively expensive when aligning *genomic* DNA sequences. In these cases, any approach involving a full pairwise dynamic programming is impossible. Nevertheless, genome alignments or genome comparisons are more important than ever before because of several

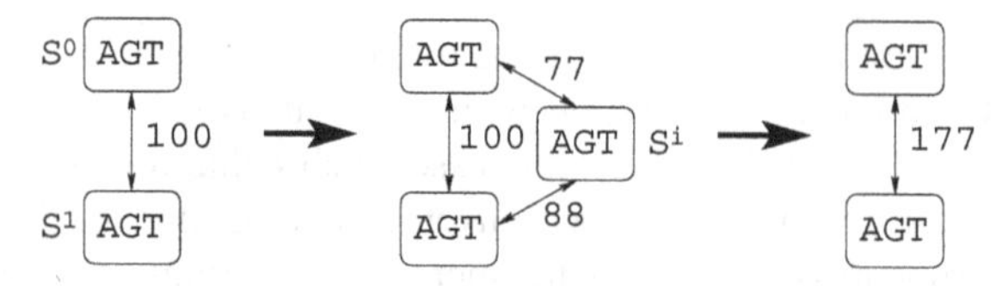

Fig. 8 A possible means of consistency extension: Every supported alignment is increased by the minimum of the two connecting edges.

vertebrate genomes at hand and thousands of on-going sequencing projects. The applications are numerous, ranging from the comparison of different assemblies, annotation tasks, regular elements identification, and phylogenetic studies to analyzing principal questions addressing mechanisms of genome evolution. Almost all genome aligners make use of the same strategy: anchor-based alignment or synonymously seeded alignment. Anchor-based alignment has three steps: (1) the computation of small segment matches of high similarity shared by multiple sequences, (2) the ordering of these segment matches into a collinear chain of non-overlapping segment matches (the fixed alignment anchors) and (3) closure of gaps between the anchors. The sole purpose of Step 1 and Step 2 is to abandon a large chunk of the possible alignment space as shown in Figure 9. Only small indels are allowed within the anchors and thus, full dynamic programming is only required between the anchors. Some programs also try to extend anchors first to the left and right to further reduce the search space. Note that Step 1 does not yet imply colinearity as shown in Figure 9. The initial segment matches can be, for example, maximal unique or exact matches [46], maximal multiple exact matches [39] or exact or hashed k-mers [12, 85]. Segment matches are optionally extended, and, finally, the quality of a segment match is assessed using some weight function. Chaining algorithms [1, 57] can be applied to compute the heaviest (best) collinear chain of these segment matches. The resulting list of anchors is refined by applying the above procedure iteratively (e.g. by using a smaller k-mer) or by filling the gaps between the anchors using more sensitive approaches such as pairwise dynamic programming. Since genomic rearrangements such as transposition, duplication, or inversion are rather likely, novel methods try to cover at least some of these operations, for example, by computing only local chains [16, 61].

Others

Another option for aligning large sequences is to consider gapless sequence segments instead of single characters. The segment matches can be derived from pairwise alignments, BLAST [5] matches or any other local comparison tool. The local alignments are usually scored and subdivided into gapless segment matches to simplify the subsequent alignment. The problem with these methods is, however, that segment matches might overlap and intersect each other. SeqAn::T-Coffee [70] refines the set of segment matches so that all parts of all segment matches can be

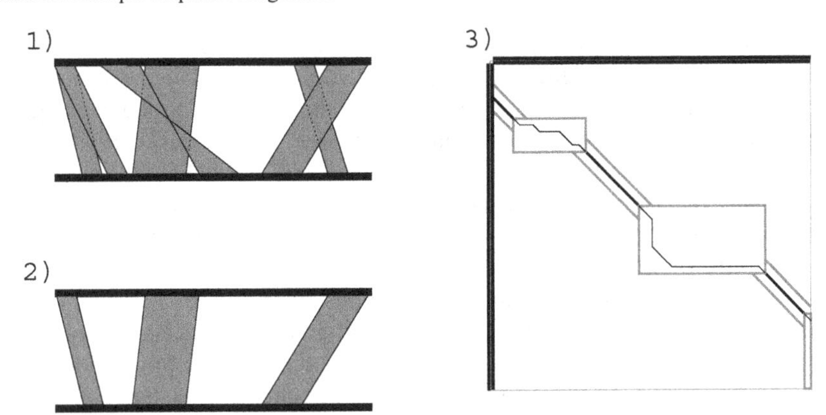

Fig. 9 Anchor-based alignment: (1) computation of initial segment matches, (2) collinear chaining of non-overlapping segment matches and (3) dynamic programming to close the alignment gaps.

used. In the subsequent progressive alignment, the program makes use of the alignment graph introduced in Section 2.1.1 and shown in Figure 5. The DIALIGN series of programs [55, 84, 85] takes a different approach by leaving the set of segment matches unchanged. This implies that overlapping segment matches involving the same pair of sequences must be greedily resolved. The objective function of DIALIGN is to find a consistent, maximum score subset of segment matches.

POA [50] uses partial order graphs to represent multiple sequence alignments. Each individual sequence is a trivial partial order graph where each character is a node connected to the subsequent node for the following character. The final partial order graph is obtained by successively aligning such a trivial graph to the growing partial order graph MSA. The key idea is that aligned nodes are merged to a new node whereas the graph bifurcates for unaligned regions. Thus in comparison to the segment based aligners, POA does not reduce the complexity depending on the length of the sequences but on the number of sequences.

2.2 Methods using structure and sequence homologs

The improvements in de-novo structure prediction methods and the growth of sequence and structural databases opened up new possibilities to extend sequence based alignment methods. These extended methods tend to deliver more accurate alignments on standard benchmarks, especially in the twilight zone of highly diverged sequences with less than 20% identity. Three combinable techniques are in common use: homology extension [40, 64, 80, 95], secondary structure prediction [63, 64, 79, 95], and the use of a known 3D structure [60, 65].

Homology extension augments the raw sequence information using database searches such as PSI-BLAST [6]. Given such a set of retrieved database homologs, a profile can be built for each input sequence. The profiles can then be readily used in progressive alignment, as outlined in Section 2.1.2. The use of profiles turned out to be beneficial because profiles differentiate between conserved and variable sites.

Predicted or known secondary structures can further improve the alignment quality because, in most cases, structure is more conserved than sequence information. Structural elements can be predicted, for instance, with PSIPRED [54] and many other tools [75]. The pairwise sequence alignment is then carried out under structural constraints. For instance, one could add a simple secondary structure weight function to the profile to profile alignment that indicates if the two corresponding structural elements at a given position match or mismatch.

Similarly, a known 3D structure eases the alignment of highly diverged sequences. Methods such as SAP [86] employ a double dynamic programming algorithm to compute a structural alignment. The time complexity is, however, $O(\tilde{n}^4)$ where $\tilde{n}$ is the average sequence length. Hence, structure based methods are usually significantly slower than sequence based heuristics. Results are, however, highly accurate because the structural constraints are of great value to build the final sequence alignment. The consistency-based methods usually employ these constraints during the consistency extension. That is, the weights of aligned substrings are adapted depending on intermediate sequences *and* structural information.

3 Available Implementations

In Table 2, we compile a list of current multiple sequence alignment tools. Given the plethora of available tools, this list is necessarily incomplete but should include most of the frequently used programs. Online web servers hosting the different alignment algorithms are frequently available, except for the genome aligners. Nevertheless, we restrained ourselves from providing web addresses of these servers because they tend to change frequently and can be easily found online by searching the name of the tool and the word "alignment". It is hard to recommend a specific tool because none of them is superior in all cases. For a classical protein alignment of less than a hundred sequences the most accurate aligners are probably MAFFT, ProbCons, MUSCLE, T-Coffee and SeqAn::T-Coffee. For huge numbers of sequences, MAFFT and MUSCLE seem to scale the best. For long sequences, the previously mentioned aligners tend to have memory problems due to the pairwise dynamic programming, except for MAFFT and if long segments are provided DIALIGN-TX and SeqAn::T-Coffee. ABA seems to be an interesting choice if there are repeated or shuffled elements. Genome aligners are still in its infancy but MUMmer is certainly the most widely used tool for all kinds of genomic analyses.

Category	Name	Method	Protein / DNA
Sequence-based exact	LASA [2]	Lagrangian ILP approach	Both
	MSA [52]	Bounded dynamic programming	Both
Sequence-based heuristic	ABA [68]	A-Bruijn alignment	Both
	AMAP [78]	Sequence annealing	Both
	CLUSTAL W [88]	Progressive alignment	Both
	DIALIGN-T [85]	Segment-based alignment	Both
	DIALIGN-TX [84]	Progressive, segment-based	Both
	Kalign [48]	Progressive alignment	Both
	POA [50]	Partial order alignment	Protein
	MAFFT [41]	Progressive with refinement	Both
	MUSCLE [24]	Progressive with refinement	Both
	ProbCons [20]	Progressive with consistency	Protein
	SeqAn::T-Coffee [70]	Progressive, segment-based	Both
	T-Coffee [59]	Progressive with consistency	Both
Sequence-based meta-alignment	M-Coffee [91]	Progressive with consistency	Both
	SeqAn::T-Coffee [70]	Progressive, segment-based	Both
Using secondary structure and database homologs	MUMMALS [63]	Progressive with consistency	Protein
	PRALINE [79]	Progressive alignment	Protein
	PROMALS [64]	Progressive with consistency	Protein
	SPEM [95]	Progressive with consistency	Protein
Using 3D structure	3D-Coffee [91]	Progressive with consistency	Protein
	Expresso [7]	Progressive with consistency	Protein
	PROMALS3D [65]	Progressive with consistency	Protein
Genome aligners	M-GCAT [90]	Anchor-based alignment	DNA
	Mauve [16]	Anchors, local collinear blocks	DNA
	MGA [39]	Anchor-based, chaining	DNA
	Mulan [61]	Anchor-based alignment	DNA
	Multi-LAGAN [11]	Anchor-based alignment	DNA
	MUMmer [46]	Anchor-based, suffix-tree	DNA
	TBA [10]	Anchor-based alignment	DNA

Table 2 Available MSA programs, categorized according to the used information sources (sequence / structure), the nature of the algorithm (exact / heuristic) and the ability to align genomic sequences. The method column highlights only the *predominant* technique. Thus, a progressive aligner using refinement might also use some kind of consistency extension.

4 Advanced Topics

This chapter focused on the classical alignment problems and described the predominant algorithms in detail. This demanded excluding some other topics. Most importantly, we could not touch some algorithms for finding conserved motifs in multiple sequences. Even if sequences look completely unrelated, they still might share a conserved pattern such as a regulatory binding site. This sort of local multiple sequence alignment problem is addressed in numerous programs such as the Gibbs Sampler [49] or MEME [8].

Related RNA sequences quite often have low sequence but high structural sequence similarity. An RNA sequence folds onto itself and the pairwise interactions induced by such a fold are characteristic for related RNA sequences, e.g. the cloverleaf shape of transfer-RNA. Hence, an accurate multiple sequence alignment of RNA sequences either requires (1) a sequence alignment followed by a folding step, (2) a de-novo folding followed by an alignment of fixed structures or (3) a simultaneous sequence-structure alignment. Similar to BAliBASE, RNA alignment algorithms can be evaluated using BRAliBase [30].

Another special problem occurs in genome assemblies. Genome assemblers usually follow a three phase methodology: an overlap phase, a layout phase and a consensus phase. In the consensus phase, the problem is to compute a multi-read alignment given a large set of reads and their approximate layout positions. That is, numerous short sequences overlap only a few bases and global relatedness cannot be assumed. Likewise a local approach is not enough, since all reads must be placed in a multi-read alignment to retrieve a consensus sequence. Tools from the AMOS and SeqAn library [21, 69, 83] address this kind of alignment problem.

Finally, we did not address the various alignment formats because Fasta seems to be the de facto standard. If you do need another format such as MSF, PIR, or Phylip, it is very likely that there exists already a converter in the EMBOSS [74] suite of programs. Similarly, we did not mention alignment viewers and editors. Interesting options are, for instance, Jalview [13] and SEAVIEW [29] or sequence logo generators [15].

5 Exercises

1. Calculate the sum of pairs score of the following MSA. (a) Assume a match score of 4, a mismatch score of -3, a gap extension score of -2 and a gap opening score of -4. (b) Recalculate the sum of pairs score for constant gap costs of -2.

$$\begin{array}{cccc} A & A & T & G \\ A & - & T & G \\ - & - & T & G \end{array}$$

2. The EMBOSS explorer is a graphical user interface to a number of bioinformatics tools from the EMBOSS project (http://emboss.sourceforge.net/). One of these programs, "fneighbor", computes a phylogeny from a distance matrix by neighbor-joining or UPGMA

   ```
   http://emboss.bioinformatics.nl/cgi-bin/emboss/fneighbor
   ```

   Use the Phylip distance matrix below to compute a phylogeny using both methods and compute the tree manually using UPGMA and weighted average linkage clustering.

```
6
A 0.0000 0.5000 0.4000 0.7000 0.6000 0.8000
B 0.5000 0.0000 0.7000 1.0000 0.9000 1.1000
C 0.4000 0.7000 0.0000 0.7000 0.6000 0.8000
D 0.7000 1.0000 0.7000 0.0000 0.5000 0.9000
E 0.6000 0.9000 0.6000 0.5000 0.0000 0.8000
F 0.8000 1.1000 0.8000 0.9000 0.8000 0.0000
```

3. Get the sequences shown in Table 1 from UniProt (http://www.uniprot.org/). Add other globin sequences and align all of them using different multiple sequence alignment programs. Compare the alignments using Jalview or any other alignment viewer of your choice.
4. How many distinct pairwise comparisons can be done for n sequences?
5. Given n sequences, what is the minimum height of a perfectly balanced binary guide tree? What is the height of a guide tree that never requires a merging of two subalignments but only sequence to sequence or subalignment to sequence alignments?
6. Build a multiple sequence alignment of the following sequences using MUSCLE and MAFFT. Do you get an alignment where all the characters are matched, that is, all columns contain only one specific DNA nucleotide and possibly gaps? If this is not the case build such an alignment yourself. Why do the programs fail to deliver such an alignment?

```
CTTCGCGTCATCATCACT
CTTGAGTCATCATCACC
TCATCATCACTTGA
TCATCATCACCTCGGA
```

7. Implement the exact multiple alignment algorithm using dynamic programming for n DNA sequences. Assume constant gap penalties and the sum of pairs score with a constant match and mismatch score.

6 Further Reading

A number of review articles cover certain aspects of multiple sequence alignment in more depth than the preceding chapter. For the following list of topics, we can point the reader to interesting articles: (1) computational methods for genomic alignments [9], (2) accurate protein sequence alignments for divergent protein sequences [62], (3) evaluation of parameter choices in progressive alignment methods [94], (4) algorithms for multiple string comparisons [36], and (5) two more program-centered multiple sequence alignment review articles [25, 66]. Another in-depth discussion of multiple sequence alignment problems can be found in the book "Biological sequence analysis" [22]. Besides introducing probabilistic alignment algorithms, the book provides extensive information on Profile HMMs to represent sequence families and search databases for new family members. It also covers stochastic context-free grammars (SCFGs) for RNA structure analysis.

References

1. Abouelhoda, M.I., Ohlebusch, E.: Multiple genome alignment: Chaining algorithms revisited. In: Proc. 14th Annual Symposium on Combinatorial Pattern Matching, Lect. Notes Comput. Sci., pp. 1–16 (2003)
2. Althaus, E., Canzar, S.: Bioinformatics research and development, chap. LASA: A tool for non-heuristic alignment of multiple sequences, pp. 489–498. Springer (2008)
3. Althaus, E., Caprara, A., Lenhof, H.P., Reinert, K.: Multiple sequence alignment with arbitrary gap costs: Computing an optimal solution using polyhedral combinatorics. Bioinformatics **18 Suppl 2**, S4–S16 (2002)
4. Althaus, E., Caprara, A., Lenhof, H.P., Reinert, K.: A branch-and-cut algorithm for multiple sequence alignment. Math. Programm. **105**, 387–425 (2006)
5. Altschul, S.F., Gish, W., Myers, E.W., Lipman, D.J.: Basic local alignment search tool. J. Mol. Biol. **215**(3), 403–410 (1990)
6. Altschul, S.F., Madden, T.L., Schaffer, A.A., Zhang, J., Zhang, Z., Miller, W., Lipman, D.J.: Gapped BLAST and PSI-BLAST: A new generation of protein database search programs. Nucleic Acids Res. **25**(17), 3389–3402 (1997)
7. Armougom, F., Moretti, S., Poirot, O., Audic, S., Dumas, P., Schaeli, B., Keduas, V., Notredame, C.: Expresso: Automatic incorporation of structural information in multiple sequence alignments using 3D-Coffee. Nucleic Acids Res. **34**, W604–608 (2006)
8. Bailey, T.L., Williams, N., Misleh, C., Li, W.W.: MEME: Discovering and analyzing DNA and protein sequence motifs. Nucleic Acids Res. **34**(suppl 2), W369–373 (2006)
9. Blanchette, M.: Computation and analysis of genomic multi-sequence alignments. Annu. Rev. Genomics Hum. Genet. **8**(1), 193–213 (2007)
10. Blanchette, M., Kent, W.J., Riemer, C., Elnitski, L., Smit, A.F., Roskin, K.M., Baertsch, R., Rosenbloom, K., Clawson, H., Green, E.D., Haussler, D., Miller, W.: Aligning multiple genomic sequences with the threaded blockset aligner. Genome Res. **14**(4), 708–715 (2004)
11. Brudno, M., Do, C.B., Cooper, G.M., Kim, M.F., Davydov, E., Green, E.D., Sidow, A., Batzoglou, S.: LAGAN and Multi-LAGAN: Efficient tools for large-scale multiple alignment of genomic DNA. Genome Res. **13**, 721–731 (2003)
12. Buhler, J.: Efficient large-scale sequence comparison by locality-sensitive hashing. Bioinformatics **17**(5), 419–428 (2001)
13. Clamp, M., Cuff, J., Searle, S.M., Barton, G.J.: The Jalview Java alignment editor. Bioinformatics **20**(3), 426–427 (2004)
14. Cormen, T.H., Leiserson, C.E., Rivest, R.L., Stein, C.: Introduction to algorithms. MIT Press, Cambridge, MA (2001)
15. Crooks, G.E., Hon, G., Chandonia, J.M., Brenner, S.E.: WebLogo: A sequence logo generator. Genome Res. **14**(6), 1188–1190 (2004)
16. Darling, A.C., Mau, B., Blattner, F.R., Perna, N.T.: Mauve: Multiple alignment of conserved genomic sequence with rearrangements. Genome Res. **14**(7), 1394–1403 (2004)
17. Dayhoff, M.O., Schwartz, R.M., Orcutt, B.C.: A model of evolutionary change in proteins. In: M.O. Dayhoff (ed.) Atlas of Protein Structure, vol. 5(Suppl. 3), pp. 345–352. National Biomedical Reasearch Foundataion, Silver Spring, Md. (1979)
18. Delcher, A.L., Kasif, S., Fleischmann, R.D., Peterson, J., White, O., Salzberg, S.L.: Alignment of whole genomes. Nucleic Acids Res. **27**(11), 2369–2376 (1999)
19. Delcher, A.L., Phillippy, A., Carlton, J., Salzberg, S.L.: Fast algorithms for large-scale genome alignment and comparison. Nucleic Acids Res. **30**(11), 2478–2483 (2002)
20. Do, C.B., Mahabhashyam, M.S., Brudno, M., Batzoglou, S.: ProbCons: Probabilistic consistency-based multiple sequence alignment. Genome Res. **15**, 330–340 (2005)

21. Döring, A., Weese, D., Rausch, T., Reinert, K.: SeqAn - An efficient, generic C++ library for sequence analysis. BMC Bioinformatics **9**, 11 (2008)
22. Durbin, R., Eddy, S., Krogh, A., Mitchison, G.: Biological sequence analysis: Probabilistic models of proteins and nucleic acids. Cambridge University Press (1998)
23. Edgar, R.C.: Local homology recognition and distance measures in linear time using compressed amino acid alphabets. Nucleic Acids Res. **32**(1), 380–385 (2004)
24. Edgar, R.C.: MUSCLE: Multiple sequence alignment with high accuracy and high throughput. Nucleic Acids Res. **32**(5), 1792–1797 (2004)
25. Edgar, R.C., Batzoglou, S.: Multiple sequence alignment. Curr. Opin. Struct. Biol. **16**(3), 368 – 373 (2006)
26. Edgar, R.C., Sjolander, K.: A comparison of scoring functions for protein sequence profile alignment. Bioinformatics **20**(8), 1301–1308 (2004)
27. Feng, D.F., Doolittle, R.F.: Progressive sequence alignment as a prerequisite to correct phylogenetic trees. J. Mol. Evol. **25**, 351–360 (1987)
28. Fitch, W.M., Margoliash, E.: Construction of phylogenetic trees. Science **155**(760), 279–84 (1967)
29. Galtier, N., Gouy, M., Gautier, C.: SEAVIEW and PHYLO WIN: Two graphic tools for sequence alignment and molecular phylogeny. Comput. Appl. Biosci. **12**(6), 543–548 (1996)
30. Gardner, P.P., Wilm, A., Washietl, S.: A benchmark of multiple sequence alignment programs upon structural RNAs. Nucleic Acids Res. **33**(8), 2433–2439 (2005)
31. Gotoh, O.: An improved algorithm for matching biological sequences. J. Mol. Biol. **162**(3), 705–708 (1982)
32. Gotoh, O.: Alignment of three biological sequences with an efficient traceback procedure. J. Theor. Biol. **121**(3), 327–37 (1986)
33. Gotoh, O.: Consistency of optimal sequence alignments. Bull. Math. Biol. **52**, 509–525 (1990)
34. Gotoh, O.: Multiple sequence alignment: Algorithms and applications. Adv. Biophys. **36**, 159–206 (1999)
35. Gupta, S.K., Kececioglu, J.D., Schffer, A.A.: Improving the practical space and time efficiency of the shortest-paths approach to sum-of-pairs multiple sequence alignment. J. Comput. Biol. **2**, 459–472 (1995)
36. Gusfield, D.: Algorithms on strings, trees, and sequences: Computer science and computational biology. Cambridge University Press, New York, NY, USA (1997)
37. Henikoff, S., Henikoff, J.G.: Amino acid substitution matrices from protein blocks. Proc. Natl. Acad. Sci. U.S.A. **89**(22), 10,915–10,919 (1992)
38. Higgins, D.G., Sharp, P.M.: CLUSTAL: A package for performing multiple sequence alignment on a microcomputer. Gene **73**(1), 237–244 (1988)
39. Hohl, M., Kurtz, S., Ohlebusch, E.: Efficient multiple genome alignment. Bioinformatics **18**(suppl 1), S312–320 (2002)
40. Katoh, K., Kuma, K., Toh, H., Miyata, T.: MAFFT version 5: Improvement in accuracy of multiple sequence alignment. Nucleic Acids Res. **33**(2), 511–518 (2005)
41. Katoh, K., Misawa, K., Kuma, K., Miyata, T.: MAFFT: A novel method for rapid multiple sequence alignment based on fast Fourier transform. Nucleic Acids Res. **30**, 3059–3066 (2002)
42. Kececioglu, J.D.: Exact and approximation algorithms for DNA sequence reconstruction. Ph.D. thesis, University of Arizona, Tucson, AZ, USA (1992)
43. Kececioglu, J.D.: The maximum weight trace problem in multiple sequence alignment. In: Proc. 4th Annual Symposium on Combinatorial Pattern Matching, Lect. Notes Comput. Sci., pp. 106–119. Springer-Verlag, London, UK (1993)
44. Kececioglu, J.D., Starrett, D.: Aligning alignments exactly. In: Proc. 8th Annual International Conference on Research in Computational Molecular Biology, RECOMB, pp. 85–96. ACM, New York, NY, USA (2004)
45. Kececioglu, J.D., Zhang, W.: Aligning alignments. In: Proc. 9th Annual Symposium on Combinatorial Pattern Matching, Lect. Notes Comput. Sci., pp. 189–208. Springer Verlag (1998)
46. Kurtz, S., Phillippy, A., Delcher, A., Smoot, M., Shumway, M., Antonescu, C., Salzberg, S.: Versatile and open software for comparing large genomes. Genome Biol. **5**(2), R12 (2004)

47. Larkin, M.A., Blackshields, G., Brown, N.P., Chenna, R., McGettigan, P.A., McWilliam, H., Valentin, F., Wallace, I.M., Wilm, A., Lopez, R., Thompson, J.D., Gibson, T.J., Higgins, D.G.: Clustal W and Clustal X version 2.0. Bioinformatics **23**(21), 2947–2948 (2007)
48. Lassmann, T., Sonnhammer, E.: Kalign - An accurate and fast multiple sequence alignment algorithm. BMC Bioinformatics **6**(1), 298 (2005)
49. Lawrence, C.E., Altschul, S.F., Boguski, M.S., Liu, J.S., Neuwald, A.F., Wootton, J.C.: Detecting subtle sequence signals: A Gibbs sampling strategy for multiple alignment. Science **262**(5131), 208–214 (1993)
50. Lee, C., Grasso, C., Sharlow, M.F.: Multiple sequence alignment using partial order graphs. Bioinformatics **18**(3), 452–464 (2002)
51. Lermen, M., Reinert, K.: The practical use of the A* algorithm for exact multiple sequence alignment. J. Comput. Biol. **7**, 655–671 (2000)
52. Lipman, D.J., Altschul, S.F., Kececioglu, J.D.: A tool for multiple sequence alignment. Proc. Natl. Acad. Sci. U.S.A. **86**, 4412–4415 (1989)
53. Ma, B., Wang, Z., Zhang, K.: Alignment between two multiple alignments. In: Proc. 14th Annual Symposium on Combinatorial Pattern Matching, Lect. Notes Comput. Sci., *Lect. Notes Comput. Sci.*, vol. 2676, pp. 254–265. Springer (2003)
54. McGuffin, L.J., Bryson, K., Jones, D.T.: The PSIPRED protein structure prediction server. Bioinformatics **16**(4), 404–405 (2000)
55. Morgenstern, B., Frech, K., Dress, A., Werner, T.: DIALIGN: Finding local similarities by multiple sequence alignment. Bioinformatics **14**(3), 290–294 (1998)
56. Murata, M., Richardson, J.S., Sussman, J.L.: Simultaneous comparison of three protein sequences. Proc. Natl. Acad. Sci. U.S.A. **82**(10), 3073–3077 (1985)
57. Myers, G., Miller, W.: Chaining multiple-alignment fragments in sub-quadratic time. In: Proc. 6th Annual ACM-SIAM Symposium, pp. 38–47. Soc. Ind. Appl. Math., Philadelphia, PA, USA (1995)
58. Needleman, S.B., Wunsch, C.D.: A general method applicable to the search for similarities in the amino acid sequence of two proteins. J. Mol. Biol. **48**, 443–453 (1970)
59. Notredame, C., Higgins, D., Heringa, J.: T-Coffee: A novel method for fast and accurate multiple sequence alignment. J. Mol. Biol. **302**, 205–217 (2000)
60. O'Sullivan, O., Suhre, K., Abergel, C., Higgins, D.G., Notredame, C.: 3DCoffee: Combining protein sequences and structures within multiple sequence alignments. J. Mol. Biol. **340**(2), 385 – 395 (2004)
61. Ovcharenko, I., Loots, G.G., Giardine, B.M., Hou, M., Ma, J., Hardison, R.C., Stubbs, L., Miller, W.: Mulan: Multiple-sequence local alignment and visualization for studying function and evolution. Genome Res. **15**(1), 184–194 (2005)
62. Pei, J.: Multiple protein sequence alignment. Curr. Opin. Struct. Biol. **18**(3), 382 – 386 (2008)
63. Pei, J., Grishin, N.V.: MUMMALS: Multiple sequence alignment improved by using hidden Markov models with local structural information. Nucleic Acids Res. **34**, 4364–4374 (2006)
64. Pei, J., Grishin, N.V.: PROMALS: Towards accurate multiple sequence alignments of distantly related proteins. Bioinformatics **23**, 802–808 (2007)
65. Pei, J., Kim, B.H., Grishin, N.V.: PROMALS3D: A tool for multiple protein sequence and structure alignments. Nucleic Acids Res. **36**(7), 2295–2300 (2008)
66. Pirovano, W., Heringa, J.: Multiple sequence alignment. Methods Mol. Biol. **452**, 143–61 (2008)
67. Raghava, G.P., Searle, S., Audley, P., Barber, J., Barton, G.: OXBench: A benchmark for evaluation of protein multiple sequence alignment accuracy. BMC Bioinformatics **4**(1), 47 (2003)
68. Raphael, B., Zhi, D., Tang, H., Pevzner, P.: A novel method for multiple alignment of sequences with repeated and shuffled elements. Genome Res. **14**(11), 2336–2346 (2004)
69. Rausch, T., Emde, A.K., Reinert, K.: Robust consensus computation. BMC Bioinformatics **9**(Suppl 10), P4 (2008)
70. Rausch, T., Emde, A.K., Weese, D., Döring, A., Notredame, C., Reinert, K.: Segment-based multiple sequence alignment. Bioinformatics **24**(16), i187–192 (2008)
71. Reinert, K.: A polyhedral approach to sequence alignment problems. Ph.D. thesis, Universität Saarbrücken (1999)

72. Reinert, K., Lenhof, H.P., Mutzel, P., Mehlhorn, K., Kececioglu, J.: A branch-and-cut algorithm for multiple sequence alignment. In: Proc. 1st Annual International Conference on Research in Computational Molecular Biology, RECOMB, pp. 241–249 (1997)
73. Reinert, K., Stoye, J., Will, T.: An iterative method for faster sum-of-pairs multiple sequence alignment. Bioinformatics **16**(9), 808–814 (2000)
74. Rice, P., Longden, I., Bleasby, A.: EMBOSS: The european molecular biology open software suite. Trends Genet. **16**(6), 276 – 277 (2000)
75. Rost, B.: Review: Protein secondary structure prediction continues to rise. J. Struct. Biol. **134**(2-3), 204 – 218 (2001)
76. Saitou, N., Nei, M.: The neighbor-joining method: A new method for reconstructing phylogenetic trees. Mol. Biol. Evol. **4**, 406–425 (1987)
77. Sankoff, D., Kruskal, J.B.: Time warps, string edits, and macromolecules: The theory and practice of sequence comparison. Addison-Wesley, Reading, MA (1983)
78. Schwartz, A.S., Pachter, L.: Multiple alignment by sequence annealing. Bioinformatics **23**, e24–29 (2007)
79. Simossis, V.A., Heringa, J.: PRALINE: A multiple sequence alignment toolbox that integrates homology-extended and secondary structure information. Nucleic Acids Res. **33**, W289 (2005)
80. Simossis, V.A., Kleinjung, J., Heringa, J.: Homology-extended sequence alignment. Nucleic Acids Res. **33**(3), 816–824 (2005)
81. Smith, T.F., Waterman, M.S.: Identification of common molecular subsequences. J. Mol. Biol. **147**(1), 195–197 (1981)
82. Sokal, R.R., Michener, C.D.: A statistical method for evaluating systematic relationships. Univ. Kansas Sci. Bull. **38**, 1409–1438 (1958)
83. Sommer, D., Delcher, A., Salzberg, S., Pop, M.: Minimus: A fast, lightweight genome assembler. BMC Bioinformatics **8**(1), 64 (2007)
84. Subramanian, A., Kaufmann, M., Morgenstern, B.: DIALIGN-TX: Greedy and progressive approaches for segment-based multiple sequence alignment. Algorithms Mol. Biol. **3**(1), 6 (2008)
85. Subramanian, A., Weyer-Menkhoff, J., Kaufmann, M., Morgenstern, B.: DIALIGN-T: An improved algorithm for segment-based multiple sequence alignment. BMC Bioinformatics **6**(1), 66 (2005)
86. Taylor, W.: Protein structure comparison using iterated double dynamic programming. Protein Sci. **8**(3), 654–665 (1999)
87. Thompson, J., Plewniak, F., Poch, O.: BAliBASE: A benchmark alignment database for the evaluation of multiple alignment programs. Bioinformatics **15**, 87–88 (1999)
88. Thompson, J.D., Higgins, D.G., Gibson, T.J.: CLUSTAL W: Improving the sensitivity of progressive multiple sequence alignment through sequence weighting, position-specific gap penalties and weight matrix choice. Nucleic Acids Res. **22**, 4673–4680 (1994)
89. Thompson, J.D., Koehl, P., Ripp, R., Poch, O.: BAliBASE 3.0: Latest developments of the multiple sequence alignment benchmark. Proteins **61**, 127–136 (2005)
90. Treangen, T., Messeguer, X.: M-GCAT: Interactively and efficiently constructing large-scale multiple genome comparison frameworks in closely related species. BMC Bioinformatics **7**(1), 433 (2006)
91. Wallace, I.M., O'Sullivan, O., Higgins, D.G., Notredame, C.: M-Coffee: Combining multiple sequence alignment methods with T-Coffee. Nucleic Acids Res. **34**, 1692–1699 (2006)
92. Walle, I.V., Lasters, I., Wyns, L.: SABmark - A benchmark for sequence alignment that covers the entire known fold space. Bioinformatics **21**(7), 1267–1268 (2005)
93. Wang, L., Jiang, T.: On the complexity of multiple sequence alignment. J. Comput. Biol. **1**, 337–348 (1994)
94. Wheeler, T.J., Kececioglu, J.D.: Multiple alignment by aligning alignments. Bioinformatics **23**, 559–568 (2007)
95. Zhou, H., Zhou, Y.: SPEM: Improving multiple sequence alignment with sequence profiles and predicted secondary structures. Bioinformatics **21**(18), 3615–3621 (2005)

Sequence Alignment Statistics

John L. Spouge

Abstract This chapter gives some simple, useful techniques for approximating the p-values of various types of optimal alignment scores. It starts with general techniques: if, e.g., a dynamic programming computation has probabilistically independent inputs, its successive states form a Markov chain. Thus, if the states are not too numerous, a "Markov computation" yields their distribution. The chapter reviews the three extreme-value distributions, which are relevant to approximating the distribution of random maxima, in the same way the normal distribution is relevant to approximating the distribution of random sums. In general, convergence to an extreme-value distribution is often painfully slow, so the Poisson approximation for counting rare and weakly dependent events can be a more flexible tool for approximating the distribution of maxima. In particular, the extreme-value and Poisson distributions yield an approximate distribution for the optimal local alignment score of two random sequences, and a finite-size correction can increase the accuracy of statistical approximations if the sequences are relatively short. Moreover, the concept of "islands" permits many statistical approximation problems in local alignment to be transformed to combinatorial problems. Finally, the "Independent Diagonals Approximation" broadens the application of many of the previous methods, and an "Independent Alignments Approximation" converts many alignment variants into the combinatorial problem of determining an "effective length".

John L. Spouge
National Center for Biotechnology Information, National Library of Medicine, National Institutes of Health, Department of Health and Human Services, Bethesda, MD 20894, e-mail: spouge@ncbi.nlm.nih.gov

L.S. Heath and N. Ramakrishnan (eds.), *Problem Solving Handbook in Computational Biology and Bioinformatics*, DOI 10.1007/978-0-387-09760-2_3,

1 Introduction

This chapter gives techniques for approximating p-values of various types of optimal alignment scores. The range of techniques is incredibly varied, so the chapter mentions most of them only in passing.

Simulations are adequately reviewed elsewhere [22]. They deserve a brief mention here, because extremely small p-values are critical to statistical applications in bioinformatics. Generally, small p-values can be calculated with numerical transforms [31], combinatorial methods [8, 9], and by specialized simulation techniques [23, 45]. More specifically, in sequence alignment, small p-values are usually calculated with crude sampling, but occasionally (and usually more effectively) with importance sampling. Importance sampling uses a biased "trial distribution" to generate random samples. It then corrects the bias to generate results relevant to the actual distribution of interest, the "target distribution" [29]. In alignment problems, hybrid alignment can provide a useful trial distribution [33, 46, 47]. More generally, if a log-odds statistic is the basis of a statistical test, extremely small p-values can be estimated by importance sampling from the alternative (non-null) hypothesis, an area of active statistical research [36, 37, 38].

The present chapter also deliberately avoids tangential statistical topics such as small sample approximations [13, 14] or more general saddle-point approximations and asymptotic expansions, which are useful for approximating the distribution of sums of independent random variates [11]. Bioinformatics in general, and sequence alignment in particular, typically deal with maxima, not sums.

Hidden Markov Models (HMMs) are very important in sequence alignment, but the author has little to add to their literature and the many reviews available [18, 34]. The estimation of unknown parameters also falls outside present purview. Thus, the expectation-maximization (EM) method is useful for fitting HMM parameters by maximum likelihood [17], but it is not mentioned again.

In bioinformatics, computational speed and storage are often important, so results are rarely expressed as full probability models. To give an example of the concept of a full probability model, an alignment algorithm finds optimal alignments by maximizing a score. The resulting alignment is therefore an "unknown parameter" estimated by the optimization; a full probability model would explicitly assign a probability to each alignment. The method of "centroids" is the cutting edge of a backlash against the neglect of full probability models in bioinformatics [12, 44]. Having acknowledged the backlash, however, this chapter gives techniques for determining the distribution of optimal alignment statistics and does not consider full probability models.

The present chapter is broken into sections, as follows. It assumes familiarity with the basics of pairwise sequence alignment [32, 39] and the manipulation of position-specific scoring matrices (PSSMs) [42]. Section 2 shows how to compute the distribution of a dynamic programming state when the computation has probabilistically independent inputs. (More generally, the inputs could be conditional on the dynamic programming state.) Section 3 describes the basics of extreme-value distributions, which approximate the distribution of random maxima, just as the normal distribution approximates the distribution of random sums. The techniques of simulating a distribution and curve-fitting its parameters are commonplace, so Section 3 omits mention of them. Section 4 discusses the Poisson approximation for counting rare (weakly dependent) events. Section 5 starts to focus specifically on alignment by giving the relevance of the extreme-value and Poisson distributions to pairwise local alignment. Section 6 discusses the concept of "islands" in a local alignment, which permits many statistical approximation problems to be transformed to combinatorial problems. Section 7 refines some notions about the finite-size correction, which can be necessary if sequences are not "infinitely long". (The finite-size correction is in fact a subdominant term in an asymptotic expansion of local alignment p-values.) Section 8 considers the independent diagonals approximation, which broadens the application of the methods of Section 2. Finally, Section 9 gives a technique that converts many alignment variants into the combinatorial problem of determining an "effective length".

Regrettably, this chapter is too short to mention every topic relevant to sequence alignment, so some appear now as an afterthought. "Algorithmic significance" [30] is an interesting concept related to Lund's inequality in renewal theory [6]. Renewal theory itself is essential to the formal mathematical treatment of sequence alignment, but is far too technical to receive appropriate attention in this chapter. Finally, the Neyman-Pearson Lemma establishes an optimality property of log-odds statistics. It reduces the construction of statistics to the construction of statistical null and alternative probability distributions [28]. If the lemma were widely appreciated, it would spare much useless ingenuity in constructing statistics, both in sequence alignment and elsewhere.

2 Dynamic Programming with Independent Random Inputs

The following defines a dynamic programming computation [10]. Without loss of generality, it starts in a deterministic state D_0, and its progress is determined inductively as follows. At the n-th step, where $n = 1, 2, \ldots$, the computation receives an external input I_n. A pre-determined update function f_n is applied to the current state D_{n-1} and the input I_n to produce the next state $D_n = f_n(D_{n-1}, I_n)$.

If the input I_n has a known distribution when conditioned on D_{n-1}, then

$$\Pr[D_n = d_n] = \sum_{d_n = f_n(d_{n-1}, i_n)} \Pr[D_{n-1} = d_{n-1}] \Pr[I_n = i_n \mid D_{n-1} = d_{n-1}], \quad (1)$$

where the sum is over all states d_{n-1} and inputs i_n satisfying $d_n = f_n(d_{n-1}, i_n)$. In bioinformatics, the probability models commonly assume that the inputs are independent of the current dynamic programming state, so $\Pr[I_n = i_n \mid D_{n-1} = d_{n-1}] = \Pr[I_n = i_n]$, simplifying Equation (1) accordingly.

The sequence $\{D_n\}$ of random states constitutes a non-homogeneous Markov chain [6], as a change in notation clarifies. Denote the set of all possible (dynamic programming) states by $\mathscr{D}$ and all possible inputs by $\mathscr{I}$. Let

$$p_{d,d'}^{(n)} = \sum_{\{i_n \in \mathscr{I} | d' = f_n(d, i_n)\}} \Pr[I_N = i_n \mid D_{n-1} = d]$$

be the sum of the probabilities over all inputs that cause the update to change state d to state d'. Note that the probability $p_d^{(n)} = \Pr[D_n = d]$ of the current state being d after step n satisfies $p_{d'}^{(n)} = \sum_{d \in \mathscr{D}} p_{d,d'}^{(n)} p_d^{(n-1)}$, the familiar matrix multiplication equation for computing probabilities in a non-homogeneous Markov chain. Essentially, because the matrix multiplication equation can itself be described as a dynamic programming computation, Equation (1) shows that the distribution of dynamic programming states can itself be computed with dynamic programming. In practice of course, the number of states in the original dynamic programming computation must not be too great.

Sequence alignment statistics frequently involve dynamic programming algorithms and independent letters, so to connect dynamic programming algorithms with random inputs to the corresponding Markov chains can be quite fruitful. Consider, e.g., the following problem of computing the distribution of a sum of position-specific scores for a word consisting of independent letters.

Exercise 0.1. Let $A_1 \cdots A_m \in \Sigma^m$ be a sequence of independent random letters chosen from an alphabet Σ with probability $\Pr[A_1 \cdots A_m] = p_1(A_1) \cdots p_m(A_m)$. Assign the score $X_n(A_n)$ to the letter A_n. Determine the distribution of $S_n = \sum_{i=1}^n X_i(A_i)$.

In many examples, the position-specific scores $X_n(A_n)$ for a sequence $A_1 \cdots A_m$ of single letters are replaced by scores for letter-pairs $X_i(A_i, B_{i+k})$ offset by some fixed k in two sequences, $A_1 \cdots A_m$ and $B_1 \cdots B_n$. (The numbers i and k are subject to appropriate restrictions, omitted here, and similarly below). The alphabet Σ then consists of letter-pairs (A, B), not letters A. The case of greatest interest is $X_i(A_i, B_{i+k}) = X(A_i, B_{i+k})$, corresponding to a pairwise letter-score independent of i. Several exercises below determine the distribution of various quantities associated with diagonals in a pairwise alignment matrix, e.g., Exercise 0.1 then examines $S_j = \sum_{i=1}^j X(A_i, B_{i+k})$, the running total of scores on a diagonal of an alignment matrix.

Solution 0.1. Let $S_0 = 0$ be the initial state, the random input at the n-th step be A_n with distribution $\{p_n(A_n)\}$, and the update function be $S_n = f_n(S_{n-1}, A_n) = S_{n-1} + X_n(A_n)$. In bioinformatics, when the letter frequencies $p_1(A) = \cdots = p_m(A)$ are independent of the index m, the solution is known as "Staden's method" [42].

Exercise 0.2. Let $A_1 \cdots A_m \in \Sigma^m$, $X_n(A_n)$, and S_n be as in Exercise 0.1. Determine the distribution of $\hat{S}_n = \max\{S_n - S_i \mid 0 \le i \le n\}$. (Note: $\hat{S}_n = \max\{\hat{S}_{n-1} + X_n(A_n), 0\}$, another exercise for the reader.)

Solution 0.2. Let $\hat{S}_0 = 0$ be the initial state, the random input at the n-th step be A_n with distribution $\{p_n(A_n)\}$, and the update function be $\hat{S}_n = f_n(\hat{S}_{n-1}, A_n) = \max\{\hat{S}_{n-1} + X_n(A_n), 0\}$.

Exercise 0.3. Let $A_1 \cdots A_m \in \Sigma^m$, $X_n(A_n)$, and $\hat{S}_n$ be as in Exercise 0.2. Determine the distribution of $\hat{M}_n = \max\{\hat{S}_i \mid 0 \le i \le n\}$ [26]. (Hint: let the dynamic programming states be $(\hat{S}_n, \hat{M}_n)$. With the distribution of $(\hat{S}_n, \hat{M}_n)$ in hand, sum over values of $\hat{S}_n$.)

Solution 0.3. Let $(\hat{S}_0, \hat{M}_0) = (0,0)$ be the initial state and the random input at the n-th step be A_n with distribution $\{p_n(A_n)\}$, and let the update function

$$(\hat{S}_n, \hat{M}_n) = f_n((\hat{S}_{n-1}, \hat{M}_{n-1}), A_n)$$

satisfy $\hat{S}_n = \max\{\hat{S}_{n-1} + X_n(A_n), 0\}$ and $\hat{M}_n = \max\{\hat{M}_{n-1}, \hat{S}_n\}$.

Exercise 0.3 is a reminder that dynamic programming states can be more than just numbers. They can be complicated data structures, e.g, vectors. Modern computers have enough memory and are fast enough to calculate useful distributions for some two-dimensional random vectors.

3 The Extreme-Value Distributions for Maxima

Most database searches produce many statistical variates $\{Y_1, \ldots, Y_n\}$ (e.g., alignment scores), with only a few variates worthy of closer examination. Consider the following idealization of a database search. Let $Y_1^* \le \cdots \le Y_n^*$ represent the order statistics corresponding to $\{Y_1, \ldots, Y_n\}$, i.e., they are the same values $\{Y_1, \ldots, Y_n\}$ but in increasing order. Thus, $Y_n^* = \max\{Y_1, \ldots, Y_n\}$is the maximum of $\{Y_1, \ldots, Y_n\}$. Assume $\{Y_1, \ldots, Y_n\}$ have a known joint distribution. What can one say about the distribution of the maximum Y_n^*?

If $\{Y_1, \ldots, Y_n\}$ are mutually independent, the distribution satisfies

$$\Pr[Y_n^* \le y] = \Pr[Y_1 \le y] \cdots \Pr[Y_n \le y].$$

But what if the variates $\{Y_1, \ldots, Y_n\}$ are dependent? Consider the analogy with the sum $S_n = \sum_{i=1}^n Y_i$, when the variates $\{Y_1, \ldots, Y_n\}$ are independent, identically distributed random variates. Normalize S_n to the standardized variate $Z_n = (S_n - \mathbf{E}[S_n])/\sigma(S_n)$. (A "standardized variate" has mean 0 and standard deviation 1.) The Lindeberg-Feller central limit theorem gives hypotheses implying that a standard Gaussian distribution still approximates the distribution of Z_n, even if $\{Y_1, \ldots, Y_n\}$ have different distributions and small dependencies [20].

In bioinformatics, maxima are much more important than sums, so Gaussian approximations are rarely appropriate (and greatly over-used). By analogy to the central limit theorem, one can ask, however: Are there two real parameters, a location parameter m_n and a scaling parameter s_n (analogous to $\mathbf{E}[S_n]$ and $\sigma(S_n)$ in the central limit theorem), so the scaled variates $(Y_n^* - m_n)/s_n$ have a limiting distribution? Based on a lengthy but logically uncomplicated algebraic proof, if $\{Y_1, \ldots, Y_n\}$ are independent and identically distributed variates, (loosely) the classical extreme-value theorem states: For large n, if $\eta = (Y_n^* - m_n)/s_n$ is not almost constant, it approximates one of only three limiting extreme-value distributions (EVDs) [21]. With $\alpha > 0$ arbitrary, the EVDs are: (Type 1) the Gumbel distribution

$$\Pr[\eta_1 \leq y] = \exp(-e^{-y});$$

(Type 2) the Frechét distribution

$$\Pr[\eta_2 \leq y] = \exp(-y^{-\alpha})$$

for $y > 0$, and 0 otherwise; or (Type 3) the Reversed Weibull distribution

$$\Pr[\eta_3 \leq y] = \exp[-(-y)^{\alpha}]$$

for $y < 0$, and 1 otherwise.

Let $\{Y_1, \ldots, Y_n\}$ be mutually independent, all with the same distribution as the variate Y. The right-hand tail of the distribution of Y determines how fast the maximum Y_n^* increases with n, determining which of the three EVDs pertains. In bioinformatics, the order of importance of the EVD distributions is the Gumbel, the Reversed Weibull, and the Frechét. The Gumbel distribution pertains if Y has a normal or exponential distribution; the Reversed Weibull, if the value of Y is bounded above.

Like the central limit theorem, the EVD approximation often holds, even if the variates $\{Y_1, \ldots, Y_n\}$ have differing distributions and are weakly dependent. Empirical fits to the parameters m_n and s_n can then yield appropriate thresholds for statistical significance in the extreme-value distribution. Classical EVD theory therefore provides practical statistical methods for some database searches.

Consider a database search producing variates $\{Y_1, \ldots, Y_n\}$, where most of the set outside a few true positives represents unwanted background, and a p-value is required to flag true positives. Robust regression, which ignores contaminating outliers, can fit any background distribution, because it can ignore a few true positives [25]. Thus, an EVD fit with robust regression can sometimes give serviceable E-values (see Section 4) from empirical results [21]. (It is wise to recall that when fitting distributional parameters, maximum likelihood generally produces better approximations than fitting moments.)

Unfortunately, the classical theory of EVDs is not as useful as suggested, because n is often impractically large before an EVD approximates the distribution of Y_n^* closely. Modern EVD theory improves the approximations by emphasizing the rich combinatoric structure of extreme values, relating them to the Poisson distribution.

4 The Poisson Approximation for Counting Rare Events

If $\Pr[N = i] = e^{-\lambda}(\lambda^i/i!)$, where $i = 0, 1, \ldots$, then N has a Poisson distribution with mean λ. The Poisson distribution derives its importance as an approximation to the distribution of the count of rare events with at most weak dependences. (The events need not necessarily have equal probability.) As a memorable example, the

Poisson distribution made its first appearance in a study by Ladislaus Josephovich Bortkiewicz (1868–1931), counting Prussian troops kicked by their horses.

In the database search above, consider the rare events $[Y_i \geq y]$, where $i = 1, \ldots, n$ and y is large. An extreme value $Y_n^* \geq y$ means a rare event $[Y_i \geq y]$ occurred. If $\{Y_1, \ldots, Y_n\}$ are only weakly dependent (and not necessarily identically distributed), the number N counting the events $[Y_i \geq y]$ that occur is approximately Poisson distributed, where the mean $\lambda = \sum_{i=1}^{n} \Pr[Y_i \geq y]$. The events $[Y_n^* < y]$ and $[N = 0]$ are identical, so the Poisson distribution yields $\Pr[Y_n^* < y] = \Pr[N = 0] \approx e^{-\lambda}$. Likewise, $Y_{n-1}^* < y, Y_{n-i+1}^* \geq y, \ldots, Y_n^* \geq y$ if and only if $N = i$, since the i largest Y's exceed y if and only if i of the Y's exceed y. Thus,

$$\Pr\left[Y_{n-1}^* < y, Y_{n-i+1}^* \geq y, \ldots, Y_n^* \geq y\right] = \Pr[N = i] \approx e^{-\lambda} \lambda^i / i!,$$

an approximation that can appear somewhat mysterious when the Poisson context is not emphasized. Through the Chen-Stein method, modern probability theory can often bound the error in a Poisson approximation, even when $\{Y_1, \ldots, Y_n\}$ are dependent variates [7].

The usage of "E-value" ("E" for "expectation") in bioinformatics derives from the Poisson distribution; the expectation $\lambda = \mathbf{E}[N]$ of a Poisson distribution determines all its probabilities, and hence all relevant p-values. In particular, for small λ, $\Pr[Y_n^* \geq y] = \Pr[N > 0] \approx 1 - e^{-\lambda} \approx \lambda = \mathbf{E}[N]$. Thus, for the Poisson distributions most useful to bioinformatics, where λ is small, p- and E-values are practically the same. Even if a non-negative integer variate N is not Poisson distributed, however,

$$\Pr[N > 0] = \sum_{i=1}^{n} \Pr[N = i] \leq \sum_{i=1}^{n} i \Pr[N = i] = \mathbf{E}[N], \tag{2}$$

so a small E-value still implies a small p-value. The term "E-value" should be reserved for Poisson approximations (because "expectation" or "mean" is perfectly expressive otherwise), but current usage of "E-value" is regrettably ambiguous. Even without the Poisson distribution, however, the mean $\mathbf{E}[N]$ is often available when the p-value is not. Equation (2) can provide a practical bound for the p-value.

5 Pairwise Sequence Alignment

A genetic algorithm compares different combinations of subsolutions to find the optimal solution of a problem. Similarly, biological sequences (e.g., a protein or nucleic acid sequence) combine subsequences to solve evolutionary "problems". Thus, a "local alignment", which compares all subsequences of two sequences [39], is often more sensitive in determining evolutionary relationships than a "global alignment", which compares entire sequences [32].

Local alignment is therefore more important in database applications than global alignment [4, 5, 35]. The mathematics and statistics of sequence alignment are related to path optimization, as follows [32].

Let $\mathbf{A} = A_0A_1 \cdots A_{m-1}$ and $\mathbf{B} = B_0B_1 \cdots B_{n-1}$ be two sequences drawn from a finite alphabet Σ, e.g., the amino acid alphabet

$$\{\mathrm{A,C,D,E,F,G,H,I,K,L,M,N,P,Q,R,S,T,V,W,Y}\}$$

or the nucleotide alphabet $\{\mathrm{A,C,G,T}\}$. Let $X : \Sigma \times \Sigma \to \mathbb{R}$ denote a "scoring matrix", where $X(a,b)$ quantifies some type of similarity between a and b, e.g., the PAM and BLOSUM scoring matrices quantify the evolutionary similarity between two amino acids [15, 24]. In practice, X is symmetric, i.e., $X(a,b) = X(b,a)$, and although unnecessary below, symmetry is assumed for ease of exposition.

The alignment graph Γ of the sequence-pair $(\mathbf{A},\mathbf{B})$ is a directed, weighted lattice graph in two dimensions, as follows. The vertices v of Γ are non-negative vertices (i,j) in the two-dimensional integer lattice $\mathbb{Z}^2$. (Below, $\mathbb{N} = \{1,2,3,\ldots\}$ denotes the natural numbers; i, j, k, m, n, and g are integers throughout.) Three sets of directed edges e come out of each vertex $v = (i,j)$: northward, northeastward, and eastward. One northeastward edge goes into $(i+1,j+1)$ with weight $X(A_i,B_j)$. For each $0 < g < m-i$, one eastward edge goes into $(i+g,j)$ with weight $-\Delta(g) < 0$; and for each $0 < g < n-i$, one northward edge goes into $(i,j+g)$ with weight $-\Delta(g) < 0$. The deterministic function $\Delta : \mathbb{N} \to (0,\infty)$ is called the "gap penalty". Database searches usually use affine gap penalties $\Delta(g) = a + bg$, more for algorithmic convenience than for biological relevance.

A *directed path* $\pi = (v_0,e_1,v_1,e_2,\ldots,e_k,v_k)$ in Γ is a finite, alternating sequence of vertices and edges that starts and ends with a vertex. For each $i = 1,2,\ldots,k$, the directed edge e_i comes out of vertex v_{i-1} and goes into vertex v_i. We say that the path π *starts* at v_0 and *ends* at v_k.

Denote subsequences of a sequence $\mathbf{A}$ by $\mathbf{A}[i,i') = A_iA_{i+1} \cdots A_{i'-1}$ for $i < i'$, with $\mathbf{A}[i,i') = \varepsilon$ if $i' \le i$ (ε is the empty sequence). Every gapped alignment of the subsequences $\mathbf{A}[i,i')$ and $\mathbf{B}[j,j')$ corresponds to exactly one directed path that starts at $v_0 = (i,j)$ and ends at $v_k = (i',j')$. The alignment's score is the "path weight" $W_\pi = \sum_{k'=1}^{k} W(e_{k'})$.

Define the "global score" $S_{i,j} = \max_\pi W_\pi$, where the maximum is taken over all paths π starting at $v_0 = (0,0)$ and ending at $v_k = (i,j)$. The paths π starting at v_0 and ending at v_k with weight $W_\pi = S_{i,j}$ are "optimal global paths" and correspond to "optimal global alignments" between the sequences $\mathbf{A}[0,i)$ and $\mathbf{B}[0,j)$.

Define also the "local score" $\hat{S}_{i,j} := \max_\pi W_\pi$, where the maximum is taken over all paths π ending at $v_k = (i,j)$, regardless of their starting point. Define the "local maximum"

$$\hat{M}_{m,n} = \max_{0 \le i \le m, 0 \le j \le n} \hat{S}_{i,j}.$$

The paths π ending at $v_k = (i,j)$ with local score $W_n = \hat{S}_{i,j} = \hat{M}_{m,n}$ are "optimal local paths" corresponding to the "optimal local alignments" between subsequences of $\mathbf{A} = \mathbf{A}[0,m)$ and $\mathbf{B} = \mathbf{B}[0,n)$. "Ungapped local alignment" is the case where $\Delta(g) = \infty$ identically, because then no optimal local path includes a northward or eastward edge, i.e., as the terminology suggests, gaps are absent from optimal ungapped local alignments.

Now, having developed notation for a particular sequence pair, we introduce randomness. Under the usual "independent letters" model, each letter in the sequences **A** and **B** is chosen independently and randomly from a fixed distribution on the alphabet Σ. Under certain conditions, the distribution of the random local maximum $\hat{M}_{m,n}$ approximates the following distribution, where $\rho_y \approx K\exp(-\lambda y)$ for large y:

$$\Pr\left[\hat{M}_{m,n} > y\right] \approx 1 - \exp(-\rho_y mn) \approx 1 - \exp(-Kmne^{-\lambda y}), \tag{3}$$

a Type 1 (Gumbel) EVD with "scale parameter" λ [1, 21]. (Note: the p-value $p = \Pr\left[\hat{M}_{m,n} \geq y\right] \approx 1 - \exp\left(-Ke^{\lambda} mne^{-\lambda y}\right)$, so K in bioinformatics usually denotes the quantity we call Ke^{λ} here; see

`http://www.ncbi.nlm.nih.gov/blast/tutorial/Altschul-3.html.`

The present definition of K is convenient for our purposes, however.) Because λ appears in a double exponent, its errors have a much greater practical impact than errors in K, so as a rule of thumb, in practice λ in Equation (3) must be known within 1% to 4%; K, within about 10%. In Equation (3), $\Pr\left[\hat{M}_{m,n} > y\right] \approx Kmne^{-\lambda y}$ if $Kmne^{-\lambda y}$ is small. The parameters λ and K in Equation (3) or its approximation can be fit from simulation, if desired.

As the next section shows, Equation (3) is unsurprising. A mathematical proof of $\rho_y \approx K\exp(-\lambda y)$ will require substantial extension to current theory, however.

6 Islands in Local Alignment

"Islands" are a useful concept in local alignment [2, 40, 43]. Consider a maximal local path π starting at the vertex $v_0 = (i', j')$ and ending at $v_k = (i, j)$ with score $\hat{S}_{i,j}$. If more than one optimal local path to v_k exists, invent an arbitrary tie-breaking rule, e.g., take $v_0 = (i', j')$ with the smallest i', and if ties are not yet broken, the smallest j'. Let us then say that v_k *belongs to* v_0, or equivalently, v_0 *owns* v_k. In addition, let the "island with root v_0" be the set of all vertices belonging to v_0. The alignment graph Γ is then partitioned into islands: every vertex v_k is contained in a unique island.

Consider now the event $\left[\hat{M}_{m,n} > y\right]$ that the maximum local score exceeds y. Thus, there exists a vertex on an island with local score exceeding y. Let us call the corresponding island a "grand island" and its root a "grand root". Let ρ_y be the expected density of grand roots in a large rectangle $[0,m] \times [0,n]$. The expected number of grand roots is $\rho_y mn$.

There is reason to believe that (in loose terms) for large m, n, and y, grand roots are only weakly dependent. Because the grand roots are weakly dependent rare events, and there are a large number of possible grand roots, the number N of grand roots in $[0,m] \times [0,n]$ is approximately Poisson distributed with mean $\mu = \rho_y mn$. (In mathematical terms, the grand roots are almost a "Poisson process" [19]. In fact, (although the corresponding limit theorem requires great mathematical depth to state precisely) if the expected number of grand islands $\rho_y mn$ is fixed at a constant value, as m and n grow, the score y grows only like $\log(mn)$. Thus, a typical grand island is a negligible proportion $y^2/(mn)$ of the rectangle $[0,m] \times [0,n]$, making dependence of grand islands very unlikely.) Because $\hat{M}_{m,n} \leq y$ if and only if $N = 0$ (no grand islands), Equation (3) is equivalent to $\Pr[N=0] \approx \exp(-\rho_y mn)$.

For large y, the equation $\rho_y \approx K\exp(-\lambda y)$ can be justified, as follows. If an island contains a vertex with a score $x+y$, a maximal path from the root to the vertex can be broken into two subpaths with scores close to x and y, corresponding to patching together two independent islands with maximal scores x and y. Thus, $\rho_{x+y} \approx K^{-1}\rho_x\rho_y$, where the extra factor K^{-1} reflects "patching". By taking logs, the only continuous, bounded, non-constant solution to the functional equation $\rho_{x+y} \approx K^{-1}\rho_x\rho_y$ is $\rho_y = K\exp(\lambda y)$, for some $\lambda > 0$.

In gapless alignment, where $\Delta(g) = \infty$ identically, the islands in the alignment matrix collapse onto diagonals, the lines containing vertices of the form $v = (i,j)$, where $i - j =$ constant. The theory of random sums then gives analytic expressions for the constants K and λ [16, 27]. (In the terminology of random sums, island roots are actually descending ladder indices in a cumulative sum along the diagonal $i - j =$ constant. [6]) The density A^{-1} of island roots is known [27]. In addition, the probability that a root yields an island with score exceeding y is about $Ce^{-\lambda y}$, where the theory of random sums yields formulas for C and λ [27]. Thus, the density of grand roots on any diagonal (or therefore, in $[0,m] \times [0,n]$) is about $A^{-1}Ce^{-\lambda y}$, i.e., $\rho_y \approx Ke^{-\lambda y}$ with $K = C/A$.

The Poisson clumping heuristic, an important general technique in applied probability due to Aldous, also yields an explanation of why the Gumbel distribution is relevant to local alignment statistics [1].

7 The Finite-Size Correction in Gapped Local Alignment

The Gumbel distribution is a good approximation in local alignment, if the sequences **A** and **B** are "infinitely long". Altschul and Gish introduced the "finite-size correction" (FSC) to improve on the Gumbel approximation in ungapped alignment [3]. In ungapped alignment, the FSC extended the Gumbel-style approximation down to short sequences (e.g., length 50 or less); in gapped alignment, its use is a necessity for typical protein sequences (e.g., length 300).

The motivation for the FSC comes from the Poisson process for the grand roots, mentioned above. At first glance, the Poisson process occurs in an area mn, but in fact not all grand roots in the area generate a grand island. To build up to a score y, a grand island requires some length of aligned sequences. Let the random variate $\hat{I}_y$ represent the length of **A** required; and similarly, $\hat{J}_y$, of **B**. Because any grand root within length $\hat{I}_y$ of the end A_{m-1} of the sequence **A** (or $\hat{J}_y$ of the end B_{n-1} of the sequence **B**) occurs "too late", the Poisson process for grand roots really has mean $\rho_y \left(m - \hat{I}_y\right)\left(n - \hat{J}_y\right)$, not $\rho_y mn$.

For large y, the substitution of the approximations $\hat{I}_y \approx \hat{J}_y \approx ay + b$ yield the original form of the FSC [3]. In fact, recognition of the randomness in $\hat{I}_y$ and $\hat{J}_y$ permits a better approximation, because the mean of the Poisson distribution is itself an expectation $\rho_y \mathbf{E}\left[\left(m - \hat{I}_y\right)^+ \left(n - \hat{J}_y\right)^+\right]$ (where the symbol $x^+ = x$ if $x > 0$ and 0 otherwise). The constants a and b can be determined from simulations, as can the expectation $\mathbf{E}\left[\left(m - \hat{I}_y\right)^+ \left(n - \hat{J}_y\right)^+\right]$. Note, however, that the ansatz (i.e., the assumed form) $\hat{I}_y \approx \hat{J}_y \approx ay$ can sometimes yield more accurate FSCs after fitting from simulation than the ansatz $\hat{I}_y \approx \hat{J}_y \approx ay + b$, simply because adding unimportant and unknown parameters to a statistical model (e.g., b) generally reduces the accuracy of estimates for the important parameters (here, a).

There are other useful approaches to FSCs, such as the addition of a correction to λ as, e.g., $\lambda + \theta n^{-1}$ if $m = n$, and then simulating to determine θ [2].

8 The Independent Diagonals Approximation

Consider the following problem. Let $\mathbf{X}$ be a PSSM of length m with columns $X_0, \ldots, X_{m-1}$. By definition, the PSSM gives any sequence $\mathbf{a} = a_0 \cdots a_{m-1}$ of length m a total score $\sum_{k=0}^{m-1} X_k(a_k)$. (By convention, empty sums are 0, e.g., if $m = 0$, so the sequence is empty, the corresponding empty sum is 0.) Now, generate a sequence $\mathbf{A} = A_0 \cdots A_{n-1}$ of length n by choosing its letters independently from the same distribution. Let $S_m^{(j)} = \sum_{k=0}^{m-1} X_k(A_{j+k})$ be the total score of the subsequence $\mathbf{A}[j, j+m)$, and let $M_n = \max_{j=0,\ldots,n-m} S_m^{(j)}$. For large y, let us attempt to approximate the p-value $\Pr[M_n > y]$ of the maximum global score M_n of a subsequence in $\mathbf{A}$.

Because the distribution of each letter A_k is the same, the sums $S_m^{(j)}$, $j = 0, \ldots, n-m$, all have the same distribution as $S_m^{(0)}$. If they were also independent, the problem is easily solved, as follows. If we simply assume independence, then our desired distribution is

$$\Pr[M_n \le y] \approx \left(\Pr\left[S_m^{(0)} \le y\right]\right)^{n-m+1}, \tag{4}$$

and Exercise 0.1 in Section 2 (Staden's method [42]) calculates the distribution of $S_m^{(0)}$.

Consider the corresponding alignment matrix (best represented with the sequence $\mathbf{A} = A_0 \cdots A_{n-1}$ left of its western edge and the columns $X_0, \ldots, X_{m-1}$ of the PSSM under its southern edge). The approximation in Equation (4) is an "independent diagonals approximation". Essentially, it assumes that the long diagonals of the alignment matrix (i.e., those diagonals with the longest length m) are independent [26]. In fact, each long diagonal shares many letters with other diagonals, introducing dependencies. The independent diagonals approximation of $\Pr[M_n > y] = 1 - \Pr[M_n \leq y]$ usually improves as the threshold y increases (which is usually what is required in practice).

The EVD and Poisson approximations can both be applied to the problem, but they are often less accurate than the independent diagonals approximation. The ease of computation is EVD > Poisson > independent diagonals, but the accuracy of approximation is often in the opposite order.

Exercise 0.4. Generalize the problem above to a sequence $\mathbf{A} = A_0, \ldots, A_{n-1}$ with independent letters drawn from different distributions.

Solution 0.4. We have

$$\Pr[M_n \leq y] \approx \Pr\left[S_m^{(0)} \leq y\right] \cdots \Pr\left[S_m^{(n-m)} \leq y\right],$$

where $\Pr\left[S_m^{(j)} \leq y\right]$ is computed by Staden's method [42].

Exercise 0.5. Consider the set-up of Exercise 0.4. Approximate the distribution of the maximum score of gapless local alignment of $\mathbf{A} = A_0, \ldots, A_{n-1}$ against the PSSM $\mathbf{X}$, i.e., the distribution of the maximum $\hat{M}_n$ of a subsequence in $\mathbf{A}$ scored against a contiguous subset of columns in the PSSM $\mathbf{X}$.

Solution 0.5. We have

$$\Pr\left[\hat{M}_n \leq y\right] \approx \Pr\left[\hat{M}_m^{(-m)} \leq y\right] \cdots \Pr\left[\hat{M}_m^{(n)} \leq y\right],$$

where the maximum local alignment score on each diagonal has the form

$$\hat{M}_m^{(j)} = \max_{0 \leq i \leq i' \leq m} \sum_{k=i}^{i'-1} X_k(A_{j+k}),$$

so the overall maximum score of gapless local alignment has the form $\hat{M}_n = \max_{j=m,\ldots,n} \hat{M}_m^{(j)}$. (The local alignment includes more diagonals than the global alignment in Exercise 0.4. To include the extra diagonals without burdening the notation, adopt the convention that $X_k(A_{j+k}) = 0$ if $j + k < 0$ or $n \leq j + k$.) Because $\hat{M}_i^{(j)}$ for each diagonal $j = -m, \ldots, n$ can be computed by dynamic programming with recursions like those in Exercise 0.2 in Section 2, Exercise 0.2 in Section 2 combined with the independent diagonals approximation gives the solution.

9 The Combinatorial Approximation

Let us consider a generalization of Exercise 0.5 to the sums of local alignment scores. Instead of just one PSSM $\mathbf{X}$, consider several PSSMs $\mathbf{X}_0, \ldots, \mathbf{X}_{b-1}$ of length $m(0), \ldots, m(b-1)$. (The PSSMs might constitute a "block model," which models a biologically functional sequence by gapless local alignment to a series of "blocks".) Generate a sequence $\mathbf{A} = A_0 \cdots A_{n-1}$ of length n by choosing its letters independently from the same distribution. Consider the maximum $\hat{M}_n = \max \sum_{a=0}^{b-1} \hat{M}_{a,n}$, where each $\hat{M}_{a,n}$ is a maximum score over local alignments whose scores have the form

$$\hat{S}_{a,m(a)}^{(j(a))} = \sum_{k=i(a)}^{i'(a)-1} X_{a,k}(A_{j(a)+k}),$$

where $0 \leq i(a) \leq i'(a) \leq m(a)$. Here, $j(a)$ is the first position of the alignment of the whole PSSM $\mathbf{X}_a$ in the sequence $\mathbf{A}$, from which the subsequence $\mathbf{A}[j(a)+i(a), j(a)+i'(a))$ is chosen. In addition, however, the local alignments contributing to $\hat{M}_n$ are subject to a "combinatorial restriction"; they occur in the sequence $\mathbf{A} = A_0 \cdots A_{n-1}$ in the same order as the PSSM order $\mathbf{X}_0, \ldots, \mathbf{X}_{b-1}$, i.e., $j(a-1)+i'(a-1) \leq j(a)+i(a)$, for $a = 1, \ldots, b-1$. In $\hat{M}_n$, therefore, the maximum is taken over all possible local alignments subject to the combinatorial restriction. For large y, the problem is to approximate the p-value $\Pr\left[\hat{M}_n > y\right]$.

For each PSSM X_a, where $a = 0, \ldots, b-1$, the corresponding local alignment yielding $\hat{M}_{a,n}$ has an island root. The combinatorial restriction implies that there are approximately

$$\binom{n}{b} = \frac{n!}{(n-b)!b!} \tag{5}$$

ways of choosing the letters in the sequence $\mathbf{A} = A_0 \cdots A_{n-1}$ to correspond to the island roots. Equation (5) neglects finite-size corrections (which can often be unimportant in database retrieval applications).

Now, the distribution of a sum of independent variates is easy to compute (with dynamic programming again, because a sum can be computed with dynamic programming). Thus, we could solve the above generalization of Exercise 0.5 if the subalignments of $\mathbf{X}_0, \ldots, \mathbf{X}_{b-1}$ against the random sequence $\mathbf{A} = A_0 \cdots A_{n-1}$ were probabilistically independent. Unfortunately, the combinatorial restriction makes the subalignments corresponding to $\hat{M}_{a,n}$, where $a = 0, \ldots, b-1$, dependent; they must be in the correct order. The combinatorial restriction forces them into an effective length j of random sequence much less than n, but it is believable that, once restricted, the subalignments are independent. What then is the effective length j?

We need to match the total number j^b of possible positions for b independent island roots to the number of possible positions for b island roots subject to the combinatorial restriction, i.e.,

$$j^b = \binom{n}{b}, \tag{6}$$

where the right side has the interpretation after Equation (5). (The idea behind Equation (6) is explained in greater detail in reference [26].) With the effective length j in hand, Exercise 0.5 in Section 8 can compute the distributions of variates $\hat{M}'_{a,j}$ to approximate the distributions of the actual $\hat{M}_{a,n}$, where $a = 0, \ldots, b-1$, contributing the maximum $\hat{M}_n$. Under the assumption that $\hat{M}'_{a,j}$ are independent, the techniques of Section 2 then compute the distribution of $\hat{M}_n = \sum_{a=0}^{b-1} \hat{M}'_{a,j}$.

Exercise 0.6. Consider the problem above, but replace the local alignments with global alignments, i.e., consider $M_n = \max \sum_{a=0}^{b-1} M_{a,n}$ where each $M_{a,n}$ is a maximum score over global alignments whose scores have the form

$$S_{a,m(a)}^{(j(a))} = \sum_{k=0}^{m(a)-1} X_{a,k}(A_{j(a)+k}).$$

For large y, approximate the p-value $\Pr[M_n > y]$.

Solution 0.6. The global alignment consumes a fixed length $m_0 = \sum_{a=0}^{b-1} m(a)$ of the sequence $\mathbf{A} = A_0 \cdots A_{n-1}$. Consider the number of the possible positions for b starts of alignments to X_a, $a = 0, \ldots, b-1$, subject to the combinatorial restriction, which is known to be the right side of Equation (7):

$$j^b = \binom{n - m_0 + b}{b}. \tag{7}$$

The left side has the same interpretation as in Equation (6). Other combinatorial restrictions lead to similar equations for effective lengths [41]. Sometimes, in practice, the right sides of equations like Equations (6) and (7) must undergo *ad hoc* changes to maintain a sensible interpretation, e.g., $n - m_0 + b$ might be replaced by $(n - m_0)^+ + b$ to avoid negative numbers in the combinatorial coefficient.

Acknowledgements The author would like to thank Drs. Yonil Park, Sergey Sheetlin, Maricel Kann, and Stephen Altschul for helpful discussions. This research was supported by the Intramural Research Program of the National Library of Medicine at the National Institutes of Health.

References

1. Aldous, D.: Probability Approximations Via the Poisson Clumping Heuristic. Springer-Verlag, New York (1989)
2. Altschul, S.F., Bundschuh, R., Olsen, R., Hwa, T.: The estimation of statistical parameters for local alignment score distributions. Nucleic Acids Research **29**, 351–361 (2001)
3. Altschul, S.F., Gish, W.: Local alignment statistics. Methods in Enzymology **266**, 460–480 (1996)
4. Altschul, S.F., Gish, W., Miller, W., Myers, E.W., Lipman, D.J.: Basic local alignment search tool. Journal of Molecular Biology **215**, 403–410 (1990)
5. Altschul, S.F., Madden, T.L., Schaffer, A.A., Zhang, J., Zhang, Z., Miller, W., Lipman, D.J.: Gapped BLAST and PSI-BLAST: A new generation of protein database search programs. Nucleic Acids Research **25**, 3389–3402 (1997)
6. Asmussen, S.: Applied Probability and Queues. Springer-Verlag, New York (2003)
7. Barbour, A.D., Holst, L., Janson, S.: Poisson Approximation. Clarendon Press, Oxford (1992)

8. Bejerano, G.: Branch and bound computation of exact p-values. Bioinformatics **22**, 2158–2159 (2006)
9. Bejerano, G., Friedman, N., Tishby, N.: Efficient exact p-value computation for small sample, sparse, and surprising categorical data. Journal of Computational Biology **11**, 867–886 (2004)
10. Bellman, R.: Dynamic Programming. Courier Dover Publications, New York (2003)
11. Bleistein, N., Handelsman, R.A.: Asymptotic Expansions of Integrals. Dover, New York (2007)
12. Carvalho, L.E., Lawrence, C.E.: Centroid estimation in discrete high-dimensional spaces with applications in biology. Proceedings of the National Academy of Sciences of the United States of America **105**, 3209–3214 (2008)
13. Daniels, H.E.: Exact saddlepoint approximations. Biometrika **67**, 59–63 (1980)
14. Daniels, H.E.: Tail probability approximations. International Statistical Review **5**, 37–48 (1987)
15. Dayhoff, M.O., Schwartz, R.M., Orcutt, B.C.: A model of evolutionary change in proteins. In: Atlas of Protein Sequence and Structure, vol. Supp 3, pp. 345–352. National Biomedical Research Foundation, Silver Spring, MD. (1978)
16. Dembo, A., Karlin, S., Zeitouni, O.: Critical phenomena for sequence matching with scoring. Annals of Probability **22**, 1993–2021 (1994)
17. Dempster, A.P., Laird, N.M., Rubin, D.B.: Maximum likelihood from incomplete data via the EM algorithm. Journal of the Royal Statistical Society **B39**, 1–39 (1977)
18. Durbin, R., Eddy, S., Krogh, A., Mitchison, G.: Biological Sequence Analysis. Cambridge Unversity Press, Cambridge (1998)
19. Dwass, M.: Probability and Statistics. W.A. Benjamin, New York (1970)
20. Feller, W.: An Introduction to Probability Theory and Its Applications, vol. 2. Wiley and Sons, New York (1971)
21. Galambos, J.: The Asymptotic Theory of Extreme Order Statistics. Wiley, New York (1978)
22. Hammersley, J.M., Handscomb, D.C.: Monte Carlo Methods. Chapman and Hall, London (1964)
23. Hartmann, A.K.: Sampling rare events: Statistics of local sequence alignments. Physical Review E **65** (2002). Art. no.-056102
24. Henikoff, S., Henikoff, J.G.: Amino acid substitution matrices from protein blocks. Proceedings of the National Academy of Sciences of the United States of America **89**, 10,915–10,919 (1992)
25. Huber, P.J.: Robust Statistics. Wiley-Interscience, New York (2003)
26. Kann, M.G., Sheetlin, S.L., Park, Y., Bryant, S.H., Spouge, J.L.: The identification of complete domains within protein sequences using accurate E-values for semi-global alignment. Nucleic Acids Research **35**, 4678–4685 (2007)
27. Karlin, S., Dembo, A.: Limit distributions of maximal segmental score among Markov-dependent partial-sums. Advances in Applied Probability **24**, 113–140 (1992)
28. Kendall, M., Stuart, A.: The Advanced Theory of Statistics, vol. 2. Griffin and Company, London (1977)
29. Liu, J.S., Chen, R.: Sequential Monte Carlo methods for dynamic systems. Journal of the American Statistical Association **93**, 1032–1044 (1998)
30. Milosavljevic, A., Jurka, J.: Discovering simple DNA sequences by the algorithmic significance method. Computer Applications in the Biosciences **9**, 407–411 (1993)
31. Nagarajan, N., Keich, U.: FAST: Fourier transform based algorithms for significance testing of ungapped multiple alignments. Bioinformatics **24**, 577–578 (2008)
32. Needleman, S.B., Wunsch, C.D.: A general method applicable to the search for similarities in the amino acid sequence of two proteins. Journal of Molecular Biology **48**, 443–453 (1970)
33. Park, Y., Sheetlin, S., Spouge, J.L.: Accelerated convergence and robust asymptotic regression of the Gumbel scale parameter for gapped sequence alignment. Journal of Physics A: Mathematical and General **38**, 97–108 (2005)
34. Rabiner, L.R.: A tutorial on Hidden Markov-Models and selected applications in speech recognition. Proceedings of the IEEE **77**, 257–286 (1989)

35. Schaffer, A.A., Aravind, L., Madden, T.L., Shavirin, S., Spouge, J.L., Wolf, Y.I., Koonin, E.V., Altschul, S.F.: Improving the accuracy of PSI-BLAST protein database searches with composition-based statistics and other refinements. Nucleic Acids Research **29**, 2994–3005 (2001)
36. Siegmund, D., Yakir, B.: Approximate p-values for local sequence alignments. Annals of Statistics **28**, 657–680 (2000)
37. Siegmund, D., Yakir, B.: Tail probabilities for the null distribution of scanning statistics. Bernoulli **6**, 191–213 (2000)
38. Siegmund, D., Yakir, B.: Approximate p-values for local sequence alignments (vol 28, pg 657, 2000). Annals of Statistics **31**, 1027–1031 (2003)
39. Smith, T.F., Waterman, M.S.: Identification of common molecular subsequences. Journal of Molecular Biology **147**, 195–197 (1981)
40. Spouge, J.L.: Path reversal, islands, and the gapped alignment of random sequences. Journal of Applied Probability **41**, 975–983 (2003)
41. Spouge, J.L., Marchler-Bauer, A., Bryant, S.H.: The combinatorics and extreme value statistics of protein threading. Annals of Combinatorics **3**, 81–93 (1999)
42. Staden, R.: Methods for calculating the probabilities of finding patterns in sequences. Computer Applications in the Biosciences **5**, 89–96 (1989)
43. Waterman, M.S., Vingron, M.: Rapid and accurate estimates of statistical significance for sequence data base searches. Proceedings of the National Academy of Sciences of the United States of America **91**, 4625–4628 (1994)
44. Webb-Robertson, B.J., McCue, L.A., Lawrence, C.E.: Measuring global credibility with application to local sequence alignment. PLoS Computational Biology **4** (2008). e1000077
45. Wilbur, W.J.: Accurate Monte Carlo estimation of very small p-values in Markov chains. Computational Statistics **13**, 153–168 (1998)
46. Yu, Y.K., Bundschuh, R., Hwa, T.: Hybrid alignment: High-performance with universal statistics. Bioinformatics **18**, 864–872 (2002)
47. Yu, Y.K., Hwa, T.: Statistical significance of probabilistic sequence alignment and related local hidden Markov models. Journal of Computational Biology **8**, 249–282 (2001)

Part II
Phylogenetics

This second part of the book collects material about evolutionary relationships including phylogenetic trees, networks, and genome wide association studies.

Practical Implications of Coalescent Theory

Paul Marjoram and Paul Joyce

Abstract The coalescent has become perhaps the most widely-used population genetics model. By modeling the ancestry of a sample, rather than the evolution of the entire population from which the sample is drawn, it provides a computationally efficient framework for data simulation. Furthermore, from a theoretical perspective, it provides the under-pinnings for many useful analysis techniques. In this chapter, we introduce the coalescent, describe some of the problems that it has been used to address, discuss practical implications that follow from the insight it provides, and summarize some of the available software.

1 Introduction — What is the coalescent?

In 1982 John Kingman took to heart the advice of Danish philosopher Søren Kierkegaard and realized that "Life can only be understood backwards, but it must be lived forwards." Applying this perspective to the world of population genetics, and generalizing an earlier result of Sewall Wright [112], led him to the development of the *coalescent*, a mathematical model for the evolution of a sample of individuals drawn from a larger population [40, 41, 42]. The coalescent has come to play a fundamental role in our understanding of population genetics and has been at the heart of a variety of widely-employed analysis methods. For this it also owes a large debt to Richard Hudson, who arguably wrote the first paper about the coalescent that the non-specialist could easily understand [33]. In this chapter, we introduce the coalescent, summarize its implications, and survey its applications.

The central intuition of the coalescent is driven by parallels with pedigree-based designs. In those studies, the shared ancestries of the sample members, as described

Paul Marjoram
University of Southern California, e-mail: pmarjora@usc.edu

Paul Joyce
University of Idaho, e-mail: joyce@uidaho.edu

L.S. Heath and N. Ramakrishnan (eds.), *Problem Solving Handbook in Computational Biology and Bioinformatics*, DOI 10.1007/978-0-387-09760-2_4,

by the pedigree, are used to inform any subsequent analysis, thereby increasing the power of that analysis. The coalescent takes this a step further by making the observation that there is no such thing as unrelated individuals. We are all related to some degree or other. In a pedigree, the relationship is made explicit. In a population-based study, the relationships are still present, albeit more distant, but the details of the pedigree are unknown. However, it remains the case that analyses of such data are likely to benefit from the presence of a model that describes those relationships. The coalescent *is* that model.

The ideas behind the coalescent are simplest to understand in the context of the original papers, in which it was developed as a model for haploid data. Consequently, while this form is inappropriate for many types of data, it is here that we begin our description. In Figure 1, we show a population evolving through a number of (discrete) generations. Arrows point from parent to offspring. We assume that a sample of individuals are drawn from the present-day population (at the bottom of the figure), and these are indicated in red. We then trace the ancestry of these individuals back through time (i.e., up the page) until a most recent common ancestor (MRCA) is reached. This ancestry, also indicated in red, is modeled by the coalescent. The behavior of the ancestry is derived by considering the probability, p_2, that any given pair of individuals share a common ancestor in the previous generation. This is a function of the population size and the variance, σ^2, of the number of offspring born to any given parent. More specifically, in the context of the commonly-used *Wright-Fisher* model [21, 112], a model in which each generation consists entirely of offspring of the the previous generation, we have $p_2 = \sigma^2/(2N)$, where $2N$ is the haploid population size. In fact, for mathematical convenience, it is more common to consider the coalescent evolving on a continuous time-scale, with time re-scaled so that p_2 is exponentially distributed with mean 1. We refer to this as the *coalescent time-scale* in what follows.

Although we have introduced the coalescent in the context of the Wright-Fisher model, it in fact works for a wide variety of models, via a change of time-scale that depends on σ^2. Of course, without mutation, life is boring because all individuals in the sample will necessarily be the same (since they all descend from the MRCA). Thus, the key parameter in the basic coalescent is $\theta = 4Nu$, where u is the mutation rate per individual per generation. This parameter controls the rate at which mutations are added to the tree, but it is important to note the confounding of the mutation rate and population size. Intuitively speaking, if we double the population size, but half the mutation rate, behavior of the coalescent, and hence of samples drawn from such a population, remains unchanged. Mutations are assumed to occur at constant rate, $\theta/2$ throughout the history of the population. Now, if we consider a situation in which there are currently k lines of ancestry in the coalescent, we can derive the key results of the coalescent. First, the time (on the coalescent time-scale) to the next event (back in time) is exponentially distributed with mean $k(k-1)/2 + k\theta/2$. Second, the probability that this event is a coalescence is $(k-1)/(\theta+k-1)$; otherwise it is a mutation. By starting with $k = n$, where n is the sample size, and iterating the Markov process described by these two results until we reach $k = 1$, we fully describe the coalescent tree of the sample.

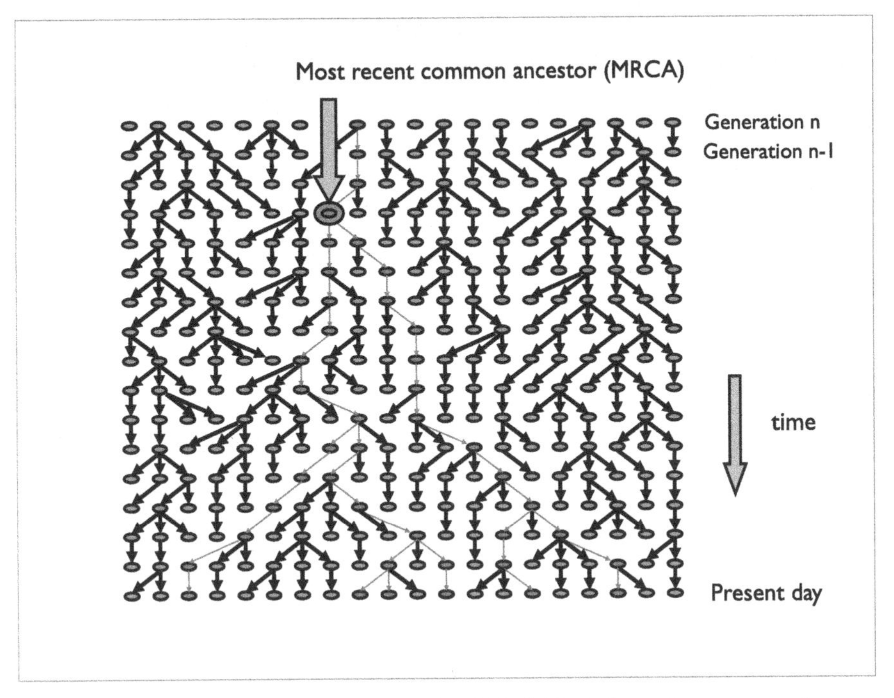

Fig. 1 An illustration of the ancestry of a sample of 6 individuals drawn from a present-day population. The ancestry is shown in red. The number of 'lines of ancestry' decreases as we move up the page, before reaching a MRCA. The process in red is formally described by the coalescent.

Of course, in most applications we need to allow for more complex features, such as recombination, population structure, selection, and changes in population size. While we avoid details here, instead referring readers to one of the many more comprehensive discussions of the coalescent (*e.g.,* [29, 66]), it is instructive to consider the changes induced to the coalescent by the addition of recombination. These were first described by Hudson [32]. When a recombination event occurs it results in a bifurcation of the ancestry as we look back in time. The region of the chromosome to the left of the recombination event was inherited from one copy of the chromosome, while the region to the right was inherited from the other copy. We illustrate this in Figure 2. This results in the coalescent being described by the relative probabilities of three events as we look back in time. The time to the next event is now distributed exponentially with mean $k(k-1)/2+k(\theta+\rho)/2$, where $\rho = 4Nr$ and r is the recombination rate per individual per generation. Furthermore, that event, when it occurs, is a coalescence with probability $(k-1)/(\theta+\rho+k-1)$, a mutation with probability $\theta/(\theta+\rho+k-1)$ and (otherwise) a recombination (with probability $\rho/(\theta+\rho+k-1)$). Once again we note the confounding, this time between the recombination rate and the population size. As a consequence of this the coalescent process is now described by a graph rather than a tree. This graph has become known as the *ancestral recombination graph* [26], and has an elegant alternative construction given by [110, 111]. As the length of the region being considered grows, the number of recombination events grows, leading to the graph assuming unmanageable proportions (see below).

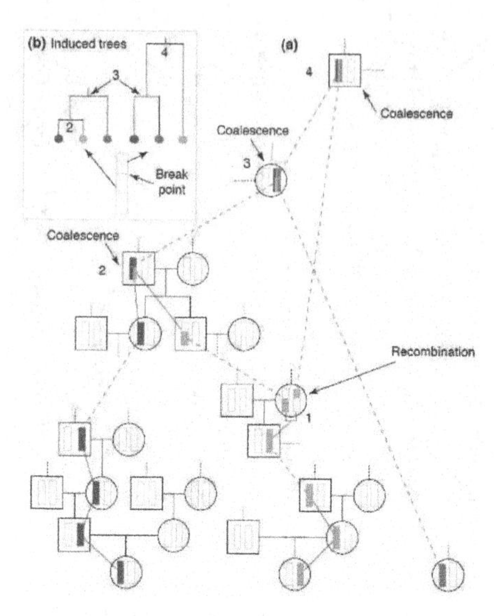

Fig. 2 The action of recombination, and its effects on ancestry. (a) We follow the ancestry of a chromosomal region indicated in color and note how recombination induces bifurcations in the ancestry as we move back through time. (b) Note that the ancestry at any specific single position in the region is still described by a tree. There are a total of 3 trees in this particular example. (Reprinted from *Trends in Genetics* **18**:83-90, M. Nordborg and S. Tavaré, "Linkage disequilibrium: what history has to tell us". Copyright 2002, with permission from Elsevier)

2 Motivating Problems

2.1 Early Human Evolution

One of the signature early applications of the coalescent was to inference regarding the early history of humans. The application was natural because, due to technological limitations, the amount of molecular variation data that was available was small and therefore required analysis using relatively complicated models, such as the coalescent. Indeed, several of the earliest data sets consisted of short regions of mitochondrial DNA (mtDNA) [5, 11, 105, 109] or Y chromosome [9, 27, 38, 108]. Since mtDNA is maternally inherited it is naturally described by the original version of the coalescent, with its reliance upon the existence of a single parent for each individual and its recombination-free nature. Similar logic applies to data from the Y chromosome. For example, Cann *et al.* [5] studied a mtDNA sample of 147 people, drawn from five geographic populations and concluded that they all descend from one woman, who they christened "Mitochondrial Eve" (mtEve) and who lived around 200Kyrs ago, likely in Africa. In a complementary analysis, the age of Y chromosome Adam, the analog of mtEve for the paternal line of ancestry, was inferred from Y chromosome data [12, 27, 93]. In another application, Slatkin and Hudson used a mtDNA data set due to Di Rienzo and Wilson [11], consisting of just 620 base-pairs, and explored properties of the distribution of *pairwise differences* [88]. The number of pairwise differences between any given pair of individuals is simply defined as the number of sites at which their nucleotides are different. The distribution of pairwise difference is found by calculating this number for *every* pair of individuals in a given sample. Di Rienzo and Wilson had observed that, for most human populations, this distribution was unimodal and relatively smooth. Slatkin and Hudson showed that such a distribution was contrary to that which would be expected for a population of constant-size, but was a consistent signature of population growth, a scenario that is distinctly plausible for human populations. It should be noted that the story can be complicated by the existence of more realistic, and more complicated features, such as population structure [50], and that some human populations, particularly those of African origin [11], exhibit pairwise difference distributions of other shapes, but the application remains interesting for its relation to the *Out of Africa* debate. This debate centers around an argument, originally made from fossil data (*e.g.*, [91]), that the MRCA for all extant humans lived in Africa. The ability to analyze mitochondrial, and then nuclear DNA, gave new impetus to this discussion, resulting in the current consensus that the Out of Africa event occurred 100K-200K years ago. However, the debate continues, particularly with regards to the likelihood of genetic admixture between the new, human lineage and existing archaic human populations (*e.g.*, [14, 15, 17, 23, 24, 25, 65, 72, 95]). Of course, the successes of the HapMap project [99], a multi-country public resource that seeks to characterize variation in human populations (see www.hapmap.org), have only accelerated work on these questions. For a review of the current state of play regarding these and other issues of human ancestry, see [73, 74, 78].

2.2 Association studies

The issue of relating phenotypic to genotypic variation is clearly of prime importance. Historically, many association methods revolved around marginal tests of relationships between each 'marker' and the phenotype. The meaning of 'marker' has changed over time to reflect the march of technology. Most recently, the commonest form of genotypic data is that of single nucleotide polymorphisms (SNPs) — single base positions at which more than one type (and usually exactly two) are observed across a population of interest. However, it is clear that the patterns of variation at nearby SNPs are related (because of shared ancestry) and this is reflected by the existence of linkage disequilibrium (LD) between such markers. Thus, understanding the nature and causes of LD, and exploiting such features within an association study, should help improve power. A parallel can be drawn with pedigree studies, in which it would clearly be nonsensical not to include the pedigree information within the analysis. In real populations the 'pedigree' is unobserved, but modeling (or inferring) it in some way would clearly have great potential utility.

A related issue arises when one tests large numbers of such loci. Correlation (as measured by LD, say) occurs between SNPs positioned near to each other along the genome. It is exactly this correlation that is modeled by versions of the coalescent that include recombination. Because SNPs are not independent, applying the most popular correction for multiple comparisons, the *Bonferroni* correction, which itself assumes independence between tests, will be conservative. The extent of the conservativeness increases as the extent of LD increases. This has, in part at least, motivated the concept of 'effective number of tests' [8, 68], which can reduce the severity of a multiple comparisons correction, although the utility of this idea continues to be the subject of discussion [62, 69, 81].

2.3 Imputation

The issue of data imputation, the inference of the (unobserved) true state of missing data, is one that is rapidly growing in importance. Modern high-throughput genotyping technologies, such as the 'SNP-chips' of Affymetrix or the bead array technology of Illumina, offer fantastic efficiencies of scale but have relatively high missing data rates. These rates can be as high as 20% [37]. Thus, there is a pressing need for algorithms than can accurately impute this missing data. Such an approach can be expected to add power to an association study, for example [49, 83]. A related application is that of next-generation sequencing technologies, such as those using polymerase colony (so-called 'polony' — www.polonator.org), 454 (www.454.com), or Solexa technologies (www.solexa.com). While such data is in short supply at the moment, there is little doubt that many future studies will collect such data. These technologies result in short regions of sequenced data, sometimes overlapping, separated by gaps in which no sequence data was obtained. Thus, there will be a need to develop algorithms that can successfully impute that missing data. Note that 'imputation' is being used in slightly different ways in these two examples. In the context of SNP-chips, for example, we wish to infer type at SNPs that have been 'sequenced' (*i.e.*, are included on the SNP-chip being used for a particular study) but for which the genotyping algorithm has not made a call. In the context of next-generation sequence data we wish to infer type at sites for which no data has been collected for that individual.

3 Implications of the Coalescent

We now stop to consider what useful insights coalescent theory and modeling have provided us. We list a series of such implications below.

Implication number one — it's a model, stupid!

The first thing to note about the coalescent is that it is a model. Therefore it is wrong. This is not a criticism specifically directed at the coalescent, but rather at models in general. If the coalescent was completely accurate it would no longer be a model — it would be reality. The key point about a model is that it should be right enough to be useful, but wrong enough (*i.e.,* simple enough) to be usable. It must capture the key features of reality, without the computational complexity inherent in many of the non-essential details. In the coalescent, the inaccuracies occur because, as one example, it is a model derived for infinite populations, and real populations are, rather obviously, finite. However, most populations are large enough that they behave much like a population of infinite size. Consequently, the coalescent has proven to be a highly useful tool for the study of all populations.

Implication number two — a sample is enough

There is a beautiful result due to Saunders *et al.* [82] showing that, in a sample of size n, drawn from a population of size N, the probability that the MRCA of the sample is the same as the MRCA of the entire population is $\frac{n-1}{n+1}\frac{N+1}{N-1}$. The striking thing here is that this probability approaches the value of 1 very quickly, meaning that one can make accurate inference regarding the MRCA of the population by studying relatively small sample sizes. (For example, a sample of size 50 has a probability of about 0.96 of having an identical MRCA to that of the population.) It is this result that enabled early investigators to study samples of human mitochondria and draw meaningful inferences about mtEve.

Implication number three — but mtDNA is a sample of size 1

The coalescent without recombination is an excellent model for the ancestry of a sample of mtDNA drawn from a human population. However, it is important to note that mtDNA, because of the lack of recombination, represents a single realization of an *extremely* variable process. We illustrate this in Figure 3, where we show four randomly sampled coalescent trees. We see how variable independent realizations of the same coalescent process can be. In reality, of course, this tree will be unobserved for any given sample, but it follows that implications regarding the mutation rate (say) drawn from the bottom-left tree would be very different from those drawn from the bottom-right tree. (Intuitively speaking, since the total length of the latter tree is longer, we would expect to estimate a higher mutation rate from data resulting from that tree than for data resulting from the first tree.) Most of the variation in tree height occurs in the final few lines of the coalescent *i.e.* the oldest few lines. Indeed, the average time taken for the last two lines of ancestry to coalesce is higher than the average height of the rest of the tree! Adding new individuals to the sample tells you little about the last few lines of ancestry, because the additional new lines are likely to coalesce with the existing tree before the last few lines are reached.

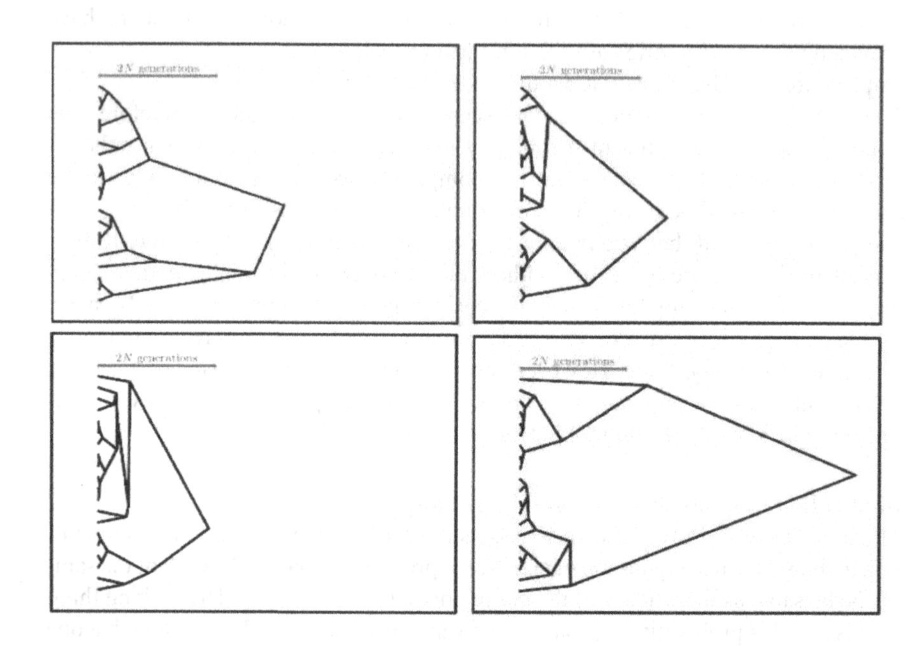

Fig. 3 A random sample of four coalescent trees for samples of size 20 (without recombination). The red line indicates a time corresponding to $2N$ generations. Note how much the shape of the tree varies from one realization to another. These trees were drawn using `coalator`, a program written by Bernhard Haubold (available at http://guanine.evolbio.mpg.de/coalator/).

These properties provide a principal motivation for wishing to look at nuclear, rather than mitochondrial (or Y chromosome) DNA. The ancestry of each chromosome provides an essentially independent realization of the evolutionary process for that population. Furthermore, because of the presence of recombination, regions that are reasonably widely-spaced within a chromosome also provide largely independent replicates. By combining analysis across multiple chromosomes, for example, we can then reduce the variance of our resulting parameter estimates. This is an approach that has proven to be of some utility in the Out of Africa debate, for example.

Implication number four- correlation matters

The coalescent provides a powerful illustration of why correlation matters. An excellent example is the distribution of pairwise differences within a sample. The number of pairwise difference between two haplotypes (short, chromosomal regions inherited from a single parent) is defined as the number of sites at which they are not identical. Under the *infinite sites* model, a model that assumes each mutation occurs at a unique new site, Fu and Li [22] showed that the probability of observing k pairwise differences when the mutation rate is θ is given by $\left(\frac{\theta}{\theta+1}\right)^k/(\theta+1)$. This distribution is, somewhat unintuitively, monotonic decreasing for all θ. An example is shown in Figure 4.

However, when we look at typical pairwise difference distributions for a sample of haplotypes simulated using a coalescent model, we see a very different picture. Four typical realizations are shown in Figure 5. This data was simulated using a constant-sized population. The bimodality in these figures is caused by the correlation induced by the coalescent. In particular, as is discussed in 'Implication number three', it follows from the fact that typical realizations of the coalescent contain a relatively long period of time during which there are exactly two lines of ancestry. Mutations will accumulate along those lines. Pairwise comparisons that involve two individuals that descend from the same one of these last two lines of ancestry will have relatively few pairwise differences; comparisons involving two individuals drawn from opposite sides of the tree will have a greater number of pairwise differences. The number of individuals that trace their ancestry to each of these last two lines is in fact uniformly distributed over the range $[1, n-1]$, where n is the sample size (another example of the surprisingly elegant results that often follow from the coalescent).

As noted earlier, when pairwise difference distributions are calculated for most human ethnic groups, they are in fact unimodal, which has been used as evidence of recent expansion of human populations sizes (*e.g.*, [88]).

Implication number five — size matters

The size of the ancestral recombination graph (ARG), the form of the coalescent that models recombination [26, 32], rapidly becomes unmanageably large as the recombination rate increases. This means that the coalescent in its exact form can only be used to model relatively short genomic regions (*e.g.*, it cannot manage entire chromosomes). This has provoked two responses. The first response is to approximate the coalescent by modeling a subgraph of the ARG. The subgraph must be such that the computational burden remains manageable when simulating entire chromosomes but must also ensure that the behavior of simulated data sets will remain essentially indistinguishable from those that would be produced by using the full ARG. This has now been successfully achieved in [7, 47, 52, 55]. It remains to be seen whether these simplified versions of the coalescent can form the basis of more efficient versions of analysis approaches that currently use the coalescent. The second response to the complexity of the coalescent when modeling long genomic regions is appropriate when the desire is to simulate data. Historically, before the appearance of the coalescent, genomic data was simulated using explicit forward-time models of the biological process of interest. However, the computational complexity of such an approach meant that it could only be applied over very short regions. But, with the rapid advances in computational hardware over the last few decades,

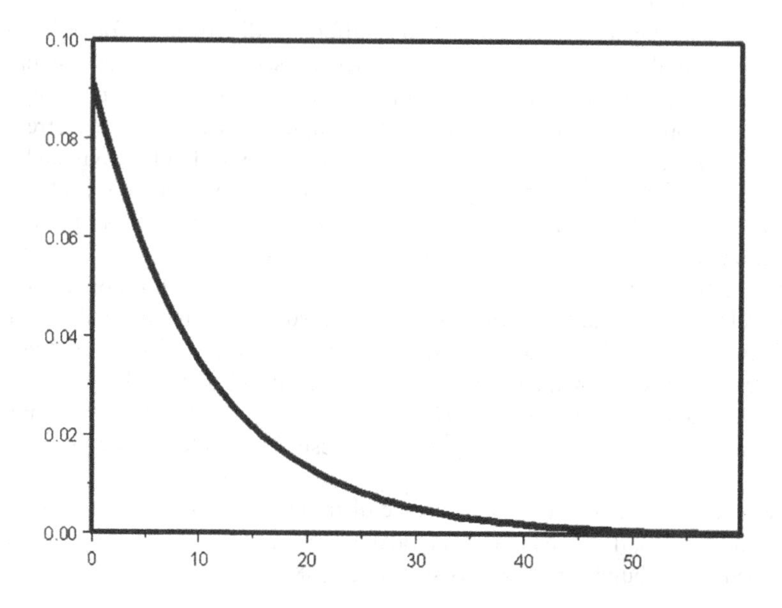

Fig. 4 The theoretical distribution for the number of pairwise differences between a pair of individuals for $\theta = 10$. The x-axis shows the number of pairwise differences, whereas the y-axis shows the probability of observing that number of pairwise differences.

forward-time simulation has again become practical, even for long regions. Many such programs have now appeared (see Section 4.1).

Implication number six
There is no implication number six.

Implication number seven — it ain't all roses
Despite the power and elegance of the coalescent model, it is not without its problems. Most significant of these is the difficulty of dealing with selection. There is an elegant pair of papers by Krone and Neuhauser [43, 64] in which a version of the coalescent is developed for contexts in which selection is present. Further publications have discussed other specific forms of, or contexts for, selection (*e.g.*, [18, 19, 36, 63, 67, 71, 85, 86, 87, 90]). However, the field is currently lacking a single, unified coalescent framework that can successfully be used for populations under a variety of forms of selection.

Implication number eight — the coalescent gives insight as to how distantly related genes are within a species.

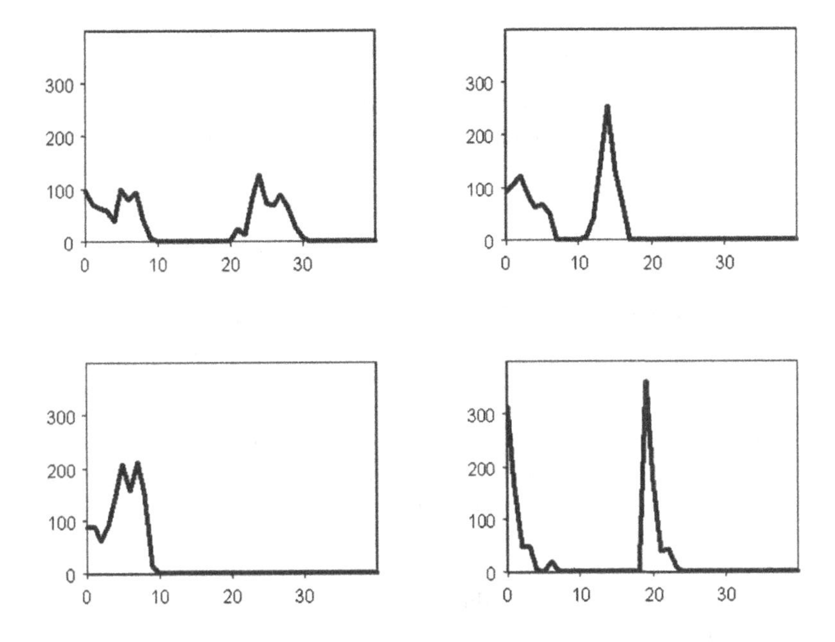

Fig. 5 Four independent realizations of the observed distribution of pairwise differences in samples of size 50, with $\theta = 10$. Again, the number of pairwise differences is shown on the x-axis, with frequency being on the y-axis. Note how correlation induces differences, such as multimodality, from the theoretical distribution shown in Figure 4.

The coalescent provides information on the history of genes within a population or species; by contrast, phylogenetic analysis studies the relationship between species. However, the former can also be used to help with the latter. Central to a phylogenetic analysis of molecular data is the assumption that all individuals within a species have coalesced to a common ancestor at a more recent time point than the time of speciation. If this assumption is met then it does not matter which homologous DNA sequence region is analyzed to infer the ancestral relationship between species. The true phylogeny should be consistently preserved regardless of the genetic locus used to infer the ancestry. If there is a discrepancy between the inferred phylogeny at one locus versus another then that discrepancy can be explained by the stochastic nature of statistical inference. However, the within species ancestry and the between species ancestry are not always on different time scales and completely separable. It is possible that a particular homologous region of DNA used to produce a phylogeny between species could produce a different phylogeny than another homologous region, and that the difference is real. One explanation of this phenomena is called lineage sorting and it occurs when the time to speciation is more recent than the time to the most recent common ancestor (MRCA) of the gene. This makes it appear as though two subpopulations from the same species are more distantly related than two distinct species.

However, the coalescent model can actually help determine if lineage sorting is plausible. For example, if based on external evidence (possibly fossil evidence), the time to speciation is at least v generations into the past, then it is reasonable to ask how likely is it that a population has not reached a common ancestor by time v. Converting from generations to the coalescent time scale, define $t = v/2N$. If T is the time it takes a population to reach a common ancestor, then

$$e^{-t} \leq P(T > t) \leq 3e^{-t}.$$

This quick calculation can determine if lineage sorting is a reasonable explanation. If $3e^{-t}$ is small, then the coalescent time scale and the phylogenetic time scale are likely to be different and lineage sorting is likely not to be the appropriate explanation.

When lineage sorting is plausible, then it becomes a central issue for inference of species phylogeny. This becomes particularly important when considering phylogenetic analysis at the genomic level. A coherent likelihood framework for combining gene trees and determining how each informs the likelihood of a particular species tree, requires a combination of coalescent and phylogenetic likelihood analysis. A multitude of papers have investigated this phenomenon using the coalescent framework, *e.g.*, [10, 48, 77].

A related point is that molecular variation data cannot inform you about times before the time to most recent common ancestor of the sample [97]. The data we see today is a consequence of the type of the MRCA and interplay of the genealogy, and mutations that have occurred on that genealogy, since that time. There is no further information about evolutionary history before the MRCA. Thus another implication of coalescent theory is that the it provides appropriate insight as to how distantly related genes are within a species, which can help resolve issues in phylogenetic analysis.

Implication number nine — robustness means lack of power

A collection of non-recombining homologous DNA sequences from a single population may look like a considerable amount of data, but in fact it represents a multivariate sample of size one. This point was made clear earlier. A rigorous way to quantify the inferential limitations on the coalescent is to consider the properties of estimates of the mutation rate parameter θ. Since this is the parameter that distinguishes one neutral coalescent model from the next, inference on θ can be very informative with regard to the power of the coalescent to resolve competing hypothesis. An under-appreciated result by Fu and Li [22] produces the optimal variance of any estimator for θ. No matter how hard one toils to make one's estimation algorithm use the data most efficiently, one will never get an unbiased estimator for θ with variance smaller than

$$\frac{\theta}{\sum_{j=1}^{n-1} 1/(\theta + j)}.$$

Note that $\sum_{j=1}^{n-1} 1/(\theta + j) \approx \ln(\theta + n)$. So as the sample size n becomes large, the variance of the estimator decreases at a very slow rate. The above is called the Cramér-Rao lower bound on the variance. It shows that among unbiased estimators the best one can do is this lower bound.

The estimator is based on the following idealized situation. Suppose that we assume that every mutation that separates all individuals at a particular locus in the population is revealed, and the full ancestry is resolved. Further assume that the number of mutations between each coalescent event is observable. Define Y_j to be the number of mutations that occur during the time the sample has j distinct ancestors. Therefore, $P(Y_j = y_j)$ is the probability that y_j mutations occur before a coalescence. This is analogous to flipping an (unfair) coin and asking what is the probability of getting y_j tails before a heads. This produces the well known geometric distribution given by

$$P(Y_j = y_j) = \left(\frac{\theta}{j-1+\theta}\right)^{y_j} \left(\frac{j-1}{j-1+\theta}\right).$$

Because of independence we can write,

$$\begin{aligned} f(y_2, y_3, \ldots, y_n; \theta) &= P(Y_2 = y_2, Y_2 = y_3, \cdots, Y_n = y_n; \theta) \\ &= \prod_{j=2}^{n} P(Y_j = y_j) \\ &= \prod_{j=2}^{n} \left(\frac{\theta}{j-1+\theta}\right)^{y_j} \left(\frac{j-1}{j-1+\theta}\right) \end{aligned}$$

For notational convenience we will denote the likelihood by

$$L_n(\theta) = f(Y_2, Y_3, \ldots, Y_n; \theta)$$

Statistical theory guarantees that the variance of any unbiased estimator is bounded below by $-E\left(\frac{\partial^2}{\partial \theta^2} \log L_n\right)$, which can be calculated to be

$$\frac{\theta}{\sum_{j=1}^{n-1} 1/(\theta + j)}.$$

Implication number ten — the answer is probably 2

Finally, on a rather less serious note, we offer the following implication of coalescent theory: if you are asked a question involving coalescences and coalescent times, the answer is likely to be 2. For example:

Question: On the coalescent time-scale, how long on average does it take a population to coalesce to a common ancestor?

Question: What is the average length of the external branches in a coalescent? (An

external branch is defined as a branch that connects directly to a tip of the tree.)
Question: How many ancestral lines are there during the (on average) longest coalescence time?
Question: When coalescence occurs it involves how many individuals?
Question: When there are n individuals in a sample, the rate of coalescence is n choose what?
Answer: 2.

4 Software Review

4.1 Simulation software

One of the main uses of the coalescent is as a method for efficient simulation of data sets. As such, it can be used as a tool in power studies, or for evaluating the efficiency of methods that estimate parameters from genetic data. In this section, we introduce just some of the software available. We begin with programs that simulate the full coalescent model. However, as discussed in implication number five, there has been a recent trend to develop algorithms that approximate the coalescent in order to improve computational efficiency in contexts that had previously been intractable (such as for genome-wide data), so we go on to include examples of this trend, before closing with algorithms that conduct explicit, forward-time simulations of populations (rather than taking the genealogical perspective of the coalescent). For a more thorough review of this field, see Excoffier and Heckel [16].

We first list the coalescent-based simulators:

- By far the most popular coalescent simulation software is `ms`, due to Richard Hudson [35]. This allows simulation of the coalescent for a variety of differing demographic scenarios. More recently, the software has been broadened to include recombination and gene conversion hot spots, in the form of the `msHot` software of Hellenthal and Stephens [30]. Both are available at http://home.uchicago.edu/~rhudson1/source/mksamples.html.
- The `SelSim` software of Spencer and Coop [90] allows for coalescent-based simulation of populations experiencing natural selection and recombination. Available at http://www.stats.ox.ac.uk/mathgen/software.html.
- Users wishing to simulate more complex demographic settings might make use of `SIMCOAL 2.0`, a package due to Laval and Excoffier [45], which allows for arbitrary patterns of migration within complex demographic scenarios. Available at http://cmpg.unibe.ch/software/simcoal2/.
- The `GENOMEPOP` software of Cavajal-Rodriguez [6] also allows for complex demographic scenarios but is aimed at simulating coding regions. It is available at http://darwin.uvigo.es/.
- In [46], Li and Stephens introduced an urn-model that approximates the coalescent. The goal is to produce data that will closely approximate that resulting from

the coalescent, but at much greater computational efficiency. While no software is available, this elegant construction has been used to simulate data for power studies (*e.g.,* [13]), and forms the back-bone for some data imputation schemes [49, 83].

- Other approximations to the coalescent were introduced by McVean and Cardin [55] and Marjoram and Wall [52]. An update to the latter program, `MACS` [7], allows for a far wider range of demographic scenarios, and is available at http://hsc.usc.edu/ garykche

We now list a couple of the forward-simulation algorithms:

- `simuPOP` is a program due to Peng and Kimmel [70] that provides much flexibility via user-written Python scripts. It is available at http://simupop .sourceforge.
- The `FREGENE` software of Hoggart et al. [31] uses a re-scaling of population size to provide extremely efficient forward simulation of large data sets. It is available at http://www.ebi.ac.uk/projects/BARGEN.

4.2 Parameter Estimation

The coalescent is also widely used as a foundation for model-based analysis, for example in parameter estimation. An early approach centered around rejection methods, where data are simulated under a variety of parameter values and then the parameter value that generated each particular instance of those data sets is *accepted* if the data matches that seen in an observed data set of interest; otherwise the generating parameter is *rejected.* Taking a Bayesian perspective, the set of accepted parameter values then forms an empirical estimate of the posterior distribution of the parameter conditional on the data. [79]. However, in practical applications the probability of simulating data identical to the observed data is vanishingly small, even if the correct parameter value is used. This has provoked a move towards so-called *Approximate Bayesian Computation*, in which the requirement for an exact match is relaxed. There has been widespread interest in this development in recent years (*e.g.,* [3, 76, 84, 107]), but here, as in most examples discussed in this section, there is little off-the-shelf software. For most applications users must write their own code!

A related methodology is that of Markov chain Monte Carlo, Metropolis-Hastings sampling [28, 56]. Here, at least, there is custom software in the form of the comprehensive `LAMARC` package of Joe Felsenstein's group [44]. This is available from http://evolution.gs.washington.edu/lamarc/ and can be used to estimate a variety of population demographics parameters, such as mutation, recombination and migration rates. There are also a large number of importance sampling algorithms in existence, which again estimate a variety of population demographics parameters. A good example is the `GENETREE` software of Bob Griffiths (*e.g.,* [2]) which can be found at http://www.stats.ox.ac.uk/~griff/software.html.

Finally, we would like to recall an old result of Geoff Watterson [106] in which a simple, but remarkably accurate, moment-based estimate of mutation rate was derived by considering the number of polymorphic sites in the sample. This result pre-dates the coalescent, but can (retrospectively) be very simply derived from insights given by the coalescent. There is also a range of excellent existing software packages for estimating the other most popular demographic parameter, the recombination rate, *e.g.*, [1, 20, 34, 46, 54, 89, 104].

4.3 Association Mapping

There is an abundance of association mapping software that does not exploit the coalescent in any form. We exclude those from discussion in the present context, except to point out a family of approaches that attempt to avoid the computational complexity inherent in using the coalescent over anything other than short genomic regions but instead consider coarser approximations of the ancestry, often via the use of ideas drawn from cluster analysis. These methods focus on analysis of haplotypes rather than marginal tests of each SNP. Some examples are [13, 53, 58, 59, 75, 96, 98, 103], resulting in software such as

- `TREESCAN` [75], available at http://darwin.uvigo.es/software/treescan.html,
- `VORONOI` [58], available from john.molitor@imperial.ac.uk,
- `GENEPBM` [60], available from amorris@well.ox.ac.uk, and
- `BEAGLE` [4], available at http://www.stat.auckland.ac.nz/~browning/beagle/beagle.html.

More recently, a variety of methods have appeared in which the ancestry of the haplotypes is modeled explicitly using the coalescent or an approximation thereof. Such methods tend to be computationally very intensive but do add power. Examples include

- `COLDMAP` [61] — available from amorris@well.ox.ac.uk,
- `MARGARITA` [57] available at http://www.sanger.ac.uk/Software/analysis/margarita,
- A related effort is the `BIM-BAM` software [83] which uses an approximation to the coalescent in an imputation step. Available at http://stephenslab.uchicago.edu/software.html, and
- The `LAMARC` software package also contains a coalescent-based association mapping tool — available at http://evolution.gs.washington.edu/lamarc/
- Most recently, the CAMP software of Kimmel *et al.* [39] is available at http://www2.icsi.berkeley.edu/kimmel/software/camp.

For a comprehensive selection of cutting-edge tools for association studies we refer the reader to the "Genome-wide Association Study Software" that was used in the analysis of the Wellcome Trust Case-Control Consortium Data [100]. This is available from http://www.stats.ox.ac.uk/~marchini/software/gwas/gwas.html.

4.4 Imputation

Imputation algorithms fall into two broad classes: those that are based upon a coalescent model, or an approximation thereof, and those that are not. In the former category, we have the IMPUTE package of Marchini et al. [49], available at http://www.stats.ox.ac.uk/~marchini/#software. In the latter category, we have the MACH software package, available at http://www.sph.umich.edu/csg/abecasis/MaCH/index.html, and the NPUTE software of Roberts et al. [80] which is available at http://compgen.unc.edu/software. We also note the imputation step within the BIM-BAM package of [83] and the error detection algorithm EDUT of Toleno et al. [101]. (It is relatively straightforward to re-cast error detection algorithms as data imputation algorithms.)

5 Further Reading

For a more formal treatment of the coalescent, we refer readers to the books by Tavaré and Zeitouni [94] or Wakeley [102]. For a somewhat less formal book-length treatment, see Hein *et al.* [29]. Useful article-length reviews of the coalescent can also be found, *e.g.*, [33, 66, 92]. For a longer discussion of model-based analyses using the coalescent, see [51].

6 Exercises for the reader

1. In Section 1, we gave a formulation of the coalescent. By considering the expected time for which the coalescent has k lines, for $k = 2, 3, \ldots, 10$, and summing these terms, derive the mean and variance of the height of the coalescent for a sample of size 10.
2. Repeat the analysis of Exercise 1 for a sample of size n, showing that the expected height of the tree can be written as $2[1 - 1/n]$. Note that the expected height of the tree never exceeds 2, *regardless of sample size.*
3. Under the infinite sites model, derive the probability that two individuals are identical under the coalescent model. Extend this to k individuals. Derive the probability that a sample of n individuals are identical.
4. In a similar manner to the previous exercise, derive the expected number of mutations that exist between two individuals. Extend this to find the expected number of mutations in a sample of size n.
5. On page 1 we indicate that the mutation parameter $\theta = 4Nu$ reflects a confounding between u, the mutation probability per individual per generation, and the haploid population size $2N$. Why are these parameters confounded? (Hint: consider a pair of individuals in a population with discrete generations and calculate (a) the probability that the pair coalesce in the previous generation, and (b) the

probability that at least one of them experiences a mutation in the previous generation.)

6. As discussed on Page 3, the theory for the coalescent is derived by assuming that the population being modeled is of infinite size. This means that the possibility of some events can be ignored. What events are these? How would the behavior of the coalescent change if the theory were derived for a population of finite size?
7. In Implication Number Three, we discuss the fact that the last few lines of the coalescent explain most of the total variance of the evolutionary process. As an illustration of this, consider the variance of the tree height. Using the results of Section 1, derive the variance of the time for which the standard coalescent (*i.e.* without recombination) has k lines, for $k = 1, 2, \ldots, 10$, and plot this as a function of k.
8. As discussed on Page 3, unimodal distributions of pairwise differences are a signature of population growth. Population growth typically results in trees that resemble so-called *star phylogenies*, (trees in which all lines coalesce at a single point — the root). Why should such a tree topology result in unimodal pairwise difference distributions?

Acknowledgements Marjoram was funded by grants from the NIH (GM069890 and MH084678). Joyce was supported by grants from the NIH(P20 RR16448, NIH R01 GM076040-01) and the NSF (NSF-DEB-0515738). The authors thank Monty Python for suggesting implication number six, and Simon Tavaré and Bernhard Haubold for providing useful comments on earlier versions of the manuscript.

References

1. Auton, A., McVean, G.: Recombination rate estimation in the presence of hotspots. Genome Research **17**, 1219–1227 (2007)
2. Bahlo, M., Griffiths, R.: Coalescence time for two genes from a subdivided population. J. Math. Biol. **43**, 397–410 (2001)
3. Beaumont, M.A., Zhang, W., Balding, D.J.: Approximate Bayesian computation in population genetics. Genetics **162**, 2025–2035 (2002)
4. Browning, B., Browning, S.: Efficient multilocus association testing for whole genome association studies using localized haplotype clustering. Genet Epidemiol **31**, 365–375 (2007)
5. Cann, R., Stoneking, M., Wilson, A.: Mitochondrial DNA and human evolution. Nature **325**, 31–36 (1987)
6. Carvajal-Rodriguez, A.: GENOMEPOP: A program to simulate genomes in populations. BMC Bioinformatics **9**, 223 (2008)
7. Chen, G., Marjoram, P., Wall, J.: Fast and flexible simulation of DNA sequence data. Genome Res. **19**, 136–142 (2009)
8. Cheverud, J.: A simple correction for multiple comparisons in interval mapping genome scans. Heredity **87**, 52–58 (2001)
9. Cooper, G., Amos, W., Hoffman, D., Rubinsztein, D.: Network analysis of human Y microsatellite haplotypes. Hum. Mol. Genet. **5**, 1759–1766 (1996)
10. Degnan, J., Salter, L.: Gene tree distributions under the coalescent process. Evolution **59**, 24–37 (2005)

11. Di Rienzo, A., Wilson, A.C.: Branching pattern in the evolutionary tree for human mitochondrial DNA. Proc. Nat. Acad. Sci. **88**, 1597–1601 (1991)
12. Dorit, R.L., Akashi, H., Gilbert, W.: Absense of polymorphism at the ZFY locus on the human Y chromosome. Science **268**, 1183–1185 (1995)
13. Durrant, C., Zondervan, K.T., Cardon, L.R., Hunt, S., Deloukas, P., Morris, A.P.: Linkage disequilibrium mapping via cladistic analysis of single-nucleotide polymorphism haplotypes. Am. J. Hum. Genet. **75**, 35–43 (2004)
14. Eswaran, V., Harpending, H., Rogers, A.: Genomics refutes an exclusively African origin of humans. Journal of Human Evolution **49**, 1–18 (2005)
15. Excoffier, L.: Human demographic history: Refining the recent African origin model. Current Opinion in Genetics & Development **12**, 675–682 (2002)
16. Excoffier, L., Heckel, G.: Computer programs for population genetics data analysis: A survival guide. Nat. Rev. Genet. **7**, 745–758 (2006)
17. Fagundes, N.J.R., Ray, N., Beaumont, M., Neuenschwander, S., Salzano, F.M., Bonatto, S.L., Excoffier, L.: Statistical evaluation of alternative models of human evolution. Proc. Natl. Acad. Sci. **104**, 17,614–17,619 (2007)
18. Fearnhead, P.: Perfect simulation from non-neutral population genetic models: Variable population size and population sub-division. Genetics **174**, 1397–1406 (2006)
19. Fearnhead, P.: The stationary distribution of allele frequencies when selection acts at unlinked loci. Theor. Pop. Biol. **70**, 376–386 (2006)
20. Fearnhead, P., Donnelly, P.: Estimating recombination rates from population genetic data. Genetics **159**, 1299–1318 (2001)
21. Fisher, R.A.: The Genetical Theory of Natural Selection. Clarendon Press (1930)
22. Fu, Y.X., Li, W.H.: Maximum likelihood estimation of population parameters. Genetics **134**, 1261–1270 (1993)
23. Garrigan, D., Hammer, M.: Reconstructing human origins in the genomic era. Nat. Rev. Genet. **7**, 669–680 (2006)
24. Garrigan, D., Hammer, M.: Ancient lineages in the genome: a response to Fagundes et al. Proc. Natl. Acad. Sci. **105**, E3 (2008)
25. Green, R., Krause, J., Ptak, S., Briggs, A., Ronan, M.: Analysis of one million base pairs of Neanderthal DNA. Nature **444**, 330–336 (2006)
26. Griffiths, R.C., Marjoram, P.: An ancestral recombination graph. In: P. Donnelly, S. Tavaré (eds.) Progress in Population Genetics and Human Evolution, *IMA Volumes in Mathematics and its Applications*, vol. 87, pp. 100–117. Springer Verlag (1997)
27. Hammer, M.: A recent common ancestry for the human Y chromosome. Nature **378**, 376–378 (1995)
28. Hastings, W.K.: Monte Carlo sampling methods using Markov chains and their applications. Biometrika **57**, 97–109 (1970)
29. Hein, J., Schierup, M.H., Wiuf, C.: Gene Genealogies, Variation and Evolution. Oxford University Press, Oxford (2005)
30. Hellenthal, G., Stephens, M.: msHOT: Modifying Hudson's ms simulator to incorporate crossover and gene conversion hotspots. Bioinformatics **23**, 520–521 (2007)
31. Hoggart, C.J., Chadeau-Hyam, M., Clark, T.G., Lampariello, R., Whittaker, J.C., Iorio, M.D., Balding, D.J.: Sequence-level population simulations over large genomic regions. Genetics **177**, 1725–1731 (2007)
32. Hudson, R.R.: Properties of a neutral allele model with intragenic recombination. Theor. Popn. Biol. **23**, 183–201 (1983)
33. Hudson, R.R.: Gene genealogies and the coalescent process. In: D. Futuyma, J. Antonovics (eds.) Oxford Surveys in Evolutionary Biology, vol. 7, pp. 1–44. Oxford University Press (1990)
34. Hudson, R.R.: Two-locus sampling distributions and their application. Genetics **159**, 1805–1817 (2001)
35. Hudson, R.R.: Generating samples under a Wright-Fisher neutral model. Bioinformatics **18**, 337–338 (2002)

36. Hudson, R.R., Kaplan, N.L.: The coalescent process in models with selection and recombination. Genetics **120**, 831–840 (1988)
37. Huentelman, M., Craig, D., Shieh, A., Corneveaux, J.: SNiPer: improved SNP genotype calling for Affymetrix 10K GeneChip microarray data. BMC Genomics **6**, 149 (2005)
38. Jobling, M., Tyler-Smith, C.: Fathers and sons: The Y chromosome and human evolution. Trends in Genetics **11**, 449–456 (1995)
39. Kimmel, G., Karp, R., Jordan, M., Halperin, E.: Association mapping and significance estimation via the coalescent. Am. J. Hum. Genet. **83**, 675–683 (2008)
40. Kingman, J.F.C.: The coalescent. Stoch. Proc. Applns. **13**, 235–248 (1982)
41. Kingman, J.F.C.: Exchangeability and the evolution of large populations. In: G. Koch, F. Spizzichino (eds.) Exchangeability in probability and statistics, pp. 97–112. North-Holland Publishing Company (1982)
42. Kingman, J.F.C.: On the genealogy of large populations. J. Appl. Prob. **19A**, 27–43 (1982)
43. Krone, S.M., Neuhauser, C.: Ancestral processes with selection. Theor. Popn. Biol. **51**, 210–237 (1997)
44. Kuhner, M.K.: LAMARC 2.0: maximum likelihood and Bayesian estimation of population parameters. Bioinformatics **22**, 768–770 (2006)
45. Laval, G., Excoffier, L.: SIMCOAL 2.0: A program to simulate genomic diversity over large recombining regions in a subdivided population with a complex history. Bioinformatics **20**, 2485–2487 (2004)
46. Li, N., Stephens, M.: Modelling linkage disequilibrium, and identifying recombination hotspots using SNP data. Genetics **165**, 2213–2233 (2003)
47. Liang, L., Zollner, S., Abecasis, G.R.: GENOME: A rapid coalescent-based whole genome simulator. Bioinformatics **23**, 1565–1567 (2007)
48. Liu, L., Pearl, D.: Species trees from gene trees: reconstructing Bayes posterior distributions of a species phylogeny using estimated gene tree distributions. Mathematical Biosciences Institute Tech. Report (2006)
49. Marchini, J., Howie, B., Myers, S., McVean, G., Donnelly, P.: A new multipoint method for genome-wide association studies by imputation of genotypes. Nat Genet **39**, 906–913 (2007)
50. Marjoram, P., Donnelly, P.: Pairwise comparison of mitochondrial DNA sequences in subdivided populations and implications for early human evolution. Genetics **136**, 673–683 (1994)
51. Marjoram, P., Tavaré, S.: Modern computational approaches for analysing molecular genetic variation data. Nat. Rev. Genet. **7**, 759–770 (2006)
52. Marjoram, P., Wall, J.D.: Fast "coalescent" simulation. BMC Genetics **7:16** (2006)
53. McPeek, M.S., Strahs, A.: Assessment of linkage disequilibrium by the decay of haplotype sharing, with application to fine-scale genetic mapping. Am. J. Hum. Genet. **65**, 858–875 (1999)
54. McVean, G., Myers, S., Hunt, S., Deloukas, P.: The fine-scale structure of recombination rate variation in the human genome. Science **304**, 581–584 (2004)
55. McVean, G.A.T., Cardin, N.J.: Approximating the coalescent with recombination. Phil. Trans. R. Soc. B **360**, 1387–1393 (2005)
56. Metropolis, N., Rosenbluth, A.W., Rosenbluth, M.N., Teller, A.H., Teller, E.: Equations of state calculations by fast computing machines. J. Chem. Phys. **21**, 1087–1091 (1953)
57. Minichiello, M., Durbin, R.: Mapping trait loci by use of inferred ancestral recombination graphs. Am. J. Hum. Genet. **79**, 910–922 (2006)
58. Molitor, J., Marjoram, P., Thomas, D.: Application of Bayesian clustering via Voronoi tesselations to the analysis of haplotype risk and gene mapping. Am. J. Hum. Genet. **73**, 1368–1384 (2003)
59. Molitor, J., Marjoram, P., Thomas, D.: Application of Bayesian spatial statistical methods to the analysis of haplotype effects and gene mapping. Gen. Epi. **25**, 95–105 (2003)
60. Morris, A.: Direct analysis of unphased SNP genotype data in population-based association studies via Bayesian partition modelling of haplotypes. Genet Epidemiol **29**, 91–107 (2005)

61. Morris, A.P., Whittaker, J.C., Balding, D.J.: Fine-scale mapping of disease loci via shattered coalescent modeling of genealogies. Am J Hum Genet **70**, 686–707 (2002). DOI 10.1086/339271
62. Moskvina, V., Schmidt, K.M.: On multiple-testing correction in genome-wide association studies. Genet Epidemiol **32**, 567–573 (2008)
63. Navarro, A., Barton, N.H.: The effects of multilocus balancing selection on neutral variability. Genetics **161**, 849–63 (2002)
64. Neuhauser, C., Krone, S.M.: The genealogy of samples in models with selection. Genetics **145**, 519–534 (1997)
65. Noonan, J., Coop, G., Kudaravalli, S., Smith, D.: Sequencing and analysis of Neanderthal genomic DNA. Science **314**, 1113–1118 (2006)
66. Nordborg, M.: Coalescent theory. In: D.J. Balding, M.J. Bishop, C. Cannings (eds.) Handbook of Statistical Genetics, pp. 179–208. John Wiley & Sons, Inc., New York (2001)
67. Nordborg, M., Innan, H.: The genealogy of sequences containing multiple sites subject to strong selection in a subdivided population. Genetics **163**, 1201–1213 (2003)
68. Nyholt, D.: A simple correction for multiple testing for SNPs in linkage disequilibrium with each other. Am. J. Hum. Genet. **74**, 765–769 (2004)
69. Nyholt, D.: Evaluation of Nyholt's procedure for multiple testing correction - author's reply. Hum. Hered. **60**, 61–62 (2005)
70. Peng, B., Kimmel, M.: simuPOP: A forward-time population genetics simulation environment. Bioinformatics **21**, 3686–3687 (2005)
71. Pfaffelhuber, P., Haubold, B., Wakolbinger, A.: Approximate genealogies under genetic hitchhiking. Genetics **174**, 1995–2008 (2006)
72. Plagnol, V., Wall, J.: Possible ancestral structure in human populations. PLoS Genet **2**(e105) (2006)
73. Portin, P.: Evolution of man in the light of molecular genetics: A review. Part I. Our evolutionary history and genomics. Hereditas **144**, 80–95 (2007)
74. Portin, P.: Evolution of man in the light of molecular genetics: A review. Part II. Regulation of gene function, evolution of speech and of brains. Hereditas **145**, 113–125 (2008)
75. Posada, D., Maxwell, T., Templeton, A.: TreeScan: A bioinformatic application to search for genotype/phenotype associations using haplotype trees. Bioinformatics **21**, 2130–2132 (2005)
76. Pritchard, J.K., Seielstad, M.T., Perez-Lezaun, A., Feldman, M.W.: Population growth of human Y chromosomes: A study of Y chromosome microsatellites. Mol. Biol. Evol. **16**, 1791–1798 (1999)
77. Rannala, B., Yang, Z.: Bayes estimation of species divergence times and ancestral population sizes using DNA sequences from multiple loci. Genetics **164**, 1645–1656 (2003)
78. Relethford, J.H.: Genetic evidence and the modern human origins debate. Heredity **100**, 555–563 (2008)
79. Ripley, B.D.: Stochastic simulation. John Wiley & Sons, Inc., New York (1982)
80. Roberts, A., McMillan, L., Wang, W., Parker, J., Rusyn, I., Threadgill, D.: Inferring missing genotypes in large SNP panels using fast nearest-neighbor searches over sliding windows. Bioinformatics **23**, i401–i407 (2007)
81. Salyakina, D., Seaman, S.R., Browning, B.L., Dudbridge, F., Müller-Myhsok, B.: Evaluation of Nyholt's procedure for multiple testing correction. Hum. Hered. **60**, 19–25 (2005)
82. Saunders, I.W., Tavaré, S., Watterson, G.A.: On the genealogy of nested subsamples from a haploid population. Adv. Appl. Prob. **16**, 471–491 (1984)
83. Servin, B., Stephens, M.: Imputation-based analysis of association studies: Candidate regions and quantitative traits. PLoS Genet. **3**, e114 (2007)
84. Siegmund, K., Marjoram, P., Shibata, D.: Modeling DNA methylation in a population of cancer cells. Statistical Applications in Genetics and Molecular Biology **7**, a18 (2008)
85. Slade, P.F.: Simulation of 'hitch-hiking' genealogies. J. Math. Biol. **42**, 41–70 (2001)
86. Slade, P.F.: The structured ancestral selection graph and the many-demes limit. Genetics **169**, 1117–1131 (2005)

87. Slatkin, M.: Simulating genealogies of selected alleles in a population of variable size. Genetics Research **78**, 49–57 (2001)
88. Slatkin, M., Hudson, R.R.: Pairwise comparisons of mitochondrial DNA sequences in stable and exponentially growing populations. Genetics **129**, 555–562 (1991)
89. Smith, N.G.C., Fearnhead, P.: A comparison of three estimators of the population-scaled recombination rate: accuracy and robustness. Genetics **171**, 2051–2062 (2005)
90. Spencer, C.C.A.: SelSim: A program to simulate population genetic data with natural selection and recombination. Bioinformatics **20**, 3673–3675 (2004)
91. Stringer, C., Andrews, P.: Genetic and fossil evidence for the origin of modern humans. Science **239**, 1263–1268 (1988)
92. Tavaré, S.: Line-of-descent and genealogical processes, and their applications in population genetics models. Theor. Popn. Biol. **26**, 119–164 (1984)
93. Tavaré, S., Balding, D.J., Griffiths, R.C., Donnelly, P.: Inferring coalescence times for molecular sequence data. Genetics **145**, 505–518 (1997)
94. Tavaré, S., Zeitouni, O.: Lectures on Probability Theory and Statistics. Springer-Verlag (2001)
95. Templeton, A.: Genetics and recent human evolution. Evolution **61**, 1507–1519 (2007)
96. Templeton, A.R.: A cladistic analysis of phenotypic associations with haplotypes inferred from restriction endonuclease mapping or DNA sequencing. V. Analysis of case/control sampling designs: Alzheimer's disease and the Apoprotein E locus. Genetics **140**, 403–409 (1995)
97. Templeton, A.R.: Haplotype trees and modern human origins. Yrbk Phys Anthropol **48**, 33–59 (2005)
98. Templeton, A.R., Maxwell, T., Posada, D., Stengard, J.H., Boerwinkle, E., Sing, C.F.: Tree scanning: A method for using haplotype trees in phenotype/genotype association studies. Genetics **169**, 441–453 (2005)
99. The International HapMap Consortium: A haplotype map of the human genome. Nature **437**, 1299–1320 (2005)
100. The Wellcome Trust Case Control Consortium: Genome-wide association study of 14,000 cases of seven common diseases and 3,000 shared controls. Nature **447**, 661–678 (2007)
101. Toleno, D., Morrell, P., Clegg, M.: Error detection in SNP data by considering the likelihood of recombinational history implied by three-site combinations. Bioinformatics **23**, 1807–1814 (2007)
102. Wakeley, J.: Coalescent Theory: An Introduction. Roberts & Company (2008)
103. Waldron, E., Whittaker, J., Balding, D.: Fine mapping of disease genes via haplotype clustering. Genet Epidemiol **30**, 170–179 (2006)
104. Wall, J.D.: A comparison of estimators of the population recombination rate. Mol. Biol. Evol. **17**, 156–163 (2000)
105. Wallace, D.: 1994 Willliam Alan Award Address - Mitochondrial DNA variation in human evolution, degenerative disease, and aging. Am. J. Hum. Genet. **57**, 201–223 (1995)
106. Watterson, G.A.: On the number of segregating sites in genetical models without recombination. Theor. Popn. Biol. **7**, 256–276 (1975)
107. Weiss, G., von Haeseler, A.: Inference of population history using a likelihood approach. Genetics **149**, 1539–1546 (1998)
108. Whitfield, L.S., Sulston, J.E., Goodfellow, P.N.: Sequence variation of the human Y chromosome. Nature **378**, 379–380 (1995)
109. Wills, C.: When did Eve live? An evolutionary detective story. Evolution **49**, 593–607 (1995)
110. Wiuf, C., Hein, J.: The ancestry of a sample of sequences subject to recombination. Genetics **151**, 1217–1228 (1999)
111. Wiuf, C., Hein, J.: Recombination as a point process along sequences. Theor. Popul. Biol. **55**, 248–259 (1999)
112. Wright, S.: Evolution in Mendelian populations. Genetics **16**, 97–159 (1931)

Graph Model of Coalescence with Recombinations

Laxmi Parida

1 Introduction

One of the primary genetic events shaping an autosomal chromosome is *recombination*. This is a process that occurs during meiosis, in eukaryotes, that results in the offsprings having different combinations of (homologous) genes, or chromosomal segments, of the two parents. The presence of these recombination events in the evolutionary history of each chromosome complicates the genetic landscape of a population, and understanding the manifestations of these genetic exchanges in the chromosome sequences has been a subject of intense curiosity (see [Hud83, Gri99, HSW05] and citations therein).

The most important mathematical object in this context is the Ancestral Recombinations Graph (ARG) introduced by Griffith and Marjoram [GM97]. Utilizing the assumption of coalescent theory that *there is no such thing as unrelated individuals*, ARG, in a similar spirit captures all the historical recombination events that relate the individuals. The resulting topology of coalescence is a tree and that of ARG is a general network. If the flow of genetic material is given a direction, say from ancestors to descendants, both the structures are acyclic, i.e., without directed cycles (why?) but only the former is without any undirected cycles as well.

Taking a reconstruction perspective, techniques from phylogenetic tree estimation can be used for coalescent trees, such as the structures based on only mtDNA or only on non-recombining Y (NRY) chromosomes. However, uncovering the recombinational palimpsest of the participating chromosomes is far from straightforward. One of the approaches has been to study the network as a sort of perturbation of an underlying tree. The effect of recombinations on the traditional phylogenetic tree reconstruction [SH00], on combinatorial complexity, in terms of deviations and error bounds [WH99b], and on the overall effect on ancestral relationships [WH99a, WH99b, Gri99, DSL$^+$07] has been studied in literature. The

Laxmi Parida
IBM T J Watson Research, e-mail: parida@us.ibm.com

L.S. Heath and N. Ramakrishnan (eds.), *Problem Solving Handbook in Computational Biology and Bioinformatics*, DOI 10.1007/978-0-387-09760-2_5,

other approach has been to explicitly model the recombination events in the phylogeny [GBB+07, PMC+08] and the reader is directed to [PJM+09] for algorithmic approaches to estimate the ARG.

Yet another area that will gain immensely from this understanding is that of population simulation. The evolution of the statistical properties of an ideal population, with genealogical relationship between sequences in a diploid population, can be understood through simulations. The underlying ideas of ARG have been used in simulation algorithms [GSN$^+$02], Schaffner05), with migrations, populations subdivision and other influencing factors layered in, to simulate human population evolution. Some understanding of the ARG has been applied to the study of genetic variations in human populations [MJC+09].

Population stratification (or subclustering) is of importance to GWAS (genome-wide association studies), disease studies, medical informatics and related areas. Needless to mention, these subtle substructures that exist in the populations can be unraveled by a having a good understanding of the effect of the genetic exchange events on genetic composition of populations.

In this chapter we discuss a random graphs framework to study pedigree history in an ideal population. A graph is a natural *outil de choix*, since each individual or unit in a population can be represented by a vertex (v) and the binary parent-child relationship is naturally captured by an edge (e) in a graph. An ancestor is related to a descendant by a simple directed path. Again, the edges are introduced at random (details in the next section) to reflect generic evolution of populations. The topological entity of interest in this graph is the graph-theoretic LCA (least common ancestor) which is also of tremendous interest to population geneticists for at least two reasons. The first is to naturally cap the structure of variations by a "root", thus defining a finite structure (perhaps with infinite details) for study, analysis and hypothesis-generation. The second is a somewhat whimsical curiosity about a hypothetical mitochondrial Eve or Y-chromosomal Adam (or Y-chromosomal Aaron) in the prehistoric past of humans. In any case, we define and study the LCA in the random graph setting. With some simple derivations and observations, this framework integrates the underlying mathematical objects in pedigree graph, mtDNA or NRY tree, ARG, HUD etc. used in population genetics literature, into a single unified random graph framework. The random graphs framework gives an alternative parametrization of the ARG that does not use the recombination rate ρ and instead uses a parameter M based on the (estimate of) the number of non-mixing segments in the extant units. This seems more natural in a setting that attempts to tease apart the population dynamics from the biology of the units. This framework also gives a purely topological definition of GMRCA, analogous to MRCA on trees (which has a purely topological description i.e., it is a root, graph-theoretically speaking, of a tree). Also, with a natural extension of the ideas from random-graphs it is possible to design sampling (simulation) algorithms to construct random instances of ARG/unilinear transmission graph with *uniform* sampling of the space of ARG instances.

2 Random Graph Framework: Pedigree Graph

The Wright Fisher Population Model. The ideal population or Wright Fisher Model assumes three properties of the evolving population: (1) constant size, (2) non-overlapping generations, and (3) *panmictic* with random mating and no selection. While the first two properties appear non-realistic at first glance, these assumptions are reasonable for the purposes of the study of the genetic variations at the population level. In fact, models with varying population size and/or overlapping generations can be reparameterized for an equivalent Wright-Fisher Model (see texts such as [HSW05, JHTS04, BÖ0]). Panmictic means that there is no substructuring of the population due to mating restrictions caused by mate selection, geography or any other such factors. Thus the model assumes equal sex ratio and equal fecundity.

Here we discuss a rather straightforward random graph model for studying the evolution of population over generations. As discussed in an earlier chapter, coalescent theory works on individuals from the present-day (we call them *extant units*). This random graph model uses two parameters: K is the number of extant units and $2N$ is the size of the population at each generation.

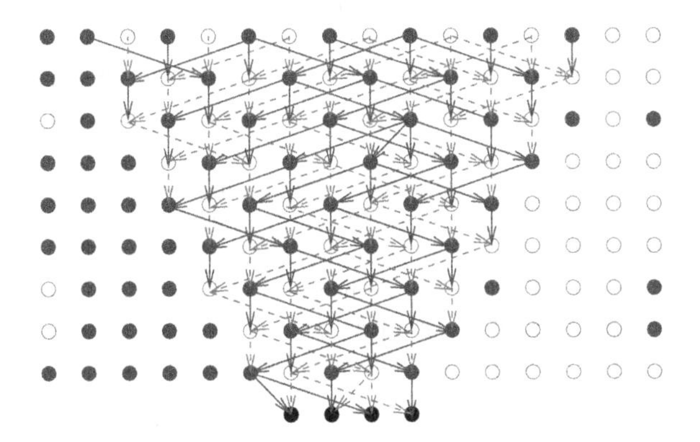

Fig. 1 The first 10 generations of an instance of a relevant pedigree graph $G_{PG}(K,N)$ with $K = 4$ and $N = 8$. The solid (blue) dots represent one gender, say males and the hollow (red) dots represent the other gender (females). Each row is a generation with the direction on edges indicating the flow of the genetic material and the four extant units are at the bottom row, i.e., row 0. Under the Wright Fisher Population model there are equal number of males and females in each row and the two distinct parents, one male and one female from the immediately preceding generation are randomly chosen.

The random graph is defined as follows. Let V be the set of vertices and E the set of edges in a directed graph $G(V,E)$. Each vertex v corresponds to an individual or a unit in the population. The edges denote the flow of genetic material between the units. The characteristics of these two sets are as follows.

Vertices: The vertex set V is a countably infinite set. We suppose that the vertices are organized in rows, each of fixed size. Each row represents a generation and is numbered as $0,1,2,3,\ldots$. Row 0 has K vertices and each row > 1 has $2N$ ($N \geq 1$) vertices, N of which are colored blue and N are colored red, denoting the gender of the units. The fixed size of $2N$ per row (or generation) is due to the constant population size model. The panmictic nature of the WF population dictates that the number of blue and red vertices be equal.

The vertex set in row 0 has K elements whose color is immaterial: these K nodes are also called the *extant* vertices.

The N vertices of each color in each row g can be labeled by a pair (g, j) where $1 \leq j \leq N$. A graph instance is *vertex-labeled* if this label is associated with every vertex v of the graph instance.

Edges: All the edges in E are directed and are only between vertices of adjacent rows. Also, the direction of the edge is from the vertex at row $g+1$ to the vertex at row g. Let $(v,u) \in E$ be a directed edge from v to u. The direction indicates the flow of genetic information and v is called the *parent* of u and similarly u is the *child* of v. When a node u at row g has two incoming edges (v_1,u) and (v_2,u), then v_1 and v_2 (of row $g+1$) must have different colors.

The parent vertex (vertices) is chosen at random reflecting the panmictic nature of the WF population. A parent and child are in adjacent rows due to the non-overlapping generations in the model. However this can be easily relaxed to have overlapping generations and the essence of each discussion below still holds.

Note that one needs to distinguish between a specific *instance* or *realization* (sometimes called *replica* in the simulation parlance) of the random graph from the entity random graph itself which is a probability measure on the space of infinite directed graphs with a countably infinite set of vertices. An instance of the random graph is obtained after executing the edge construction procedure as below.

> Repeat for each row g: For each vertex u in row i, pick exactly one blue vertex v_1 and exactly one red vertex v_2 *at random* from row $g+1$. The two directed edges are v_1u and v_2u.

Then,

1. Every instance of $G_{PG}(K,N)$ is a directed acyclic graph (DAG). This follows from the fact that no vertex can be an ancestor of itself.
2. An instance of $G_{PG}(K,N)$ corresponds to the entity termed *pedigree graph* in literature [SH06].

The number of vertices in row g is denoted by k_g. Note that when $g = 0$, $k_g = K$.

Forbidden structures. Can the pedigree graph be monochromatic? It turns out, that if the color is not taken into account, then the graph has certain forbidden structures (see Fig 2). These are topologies where the parents of a set of vertices cannot be colored satisfying the condition that the two parents of a vertex must be of different colors. Due to the forbidden structures, we retain the colors of the vertices in the pedigree graph.

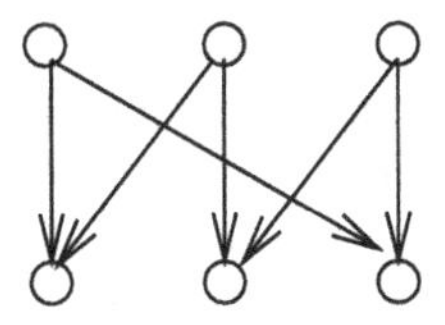

Fig. 2 Can the pedigree graph be monochromatic? *Forbidden structure* in an instance of the pedigree graph. There exists no consistent assignment of red and blue colors (different genders) to the parents of the three vertices in the bottom row.

2.1 Least Common Ancestor (LCA)

The vertex set of an instance of the pedigree graph can be trimmed by focussing only on the flow of genetic information to the extant vertices. This is termed the *relevant* pedigree graph. A vertex v_a is an *ancestor* of vertex v if there exists a directed path from v_a to v. In graph-theoretic terms, it means that any vertex on the relevant pedigree must be an ancestor of at least one extant vertex. However, a relevant pedigree graph is also an infinite object. In the rest of the discussion, a pedigree graph is always a *relevant pedigree graph*.

A common ancestor v_a of vertices $v_1, v_2, \ldots, v_k$ is called the *least common ancestor* (LCA) if no descendant of v_a is also a common ancestor (see texts such as [CLR90]). In the rest of the paper, an LCA always refers to an LCA of all the K extant vertices in the pedigree graph.

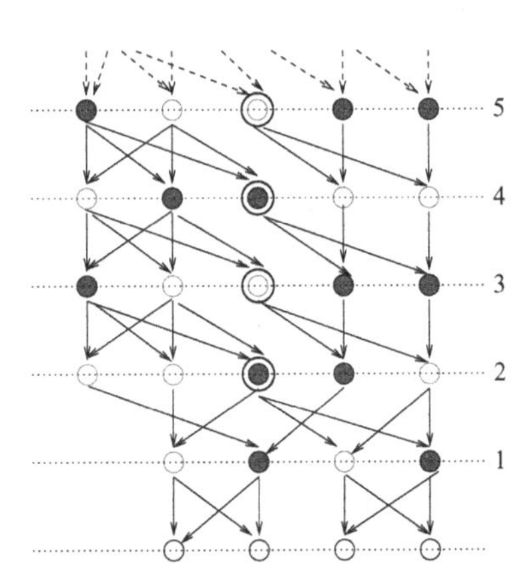

Fig. 3 An instance of the pedigree graph $G_{PG}(4,3)$ (i.e., 4 extant vertices and population of size 3 for each gender at every generation) with an infinite number of LCAs. A possible coloring (or gender assignment) is shown for rows 1 and above; the colors of vertices of row 0 are immaterial. Only the first six rows are shown, marked from 1 to 5 (bottom row is 0). The LCAs are shown with an extra concentric ring. Rows 2 or higher: The *block-vertices* are the two leftmost vertices; the *path-vertices* are the two vertices in the right; the *LCA-vertex* is in the center. The same pattern of edges can be followed for all rows to define an infinite number of LCAs.

Why are we interested in the topological entity LCA? It turns out that in population genomics, the most common recent ancestor (TMRCA) [JHTS04, HSW05], is exactly the same as LCA. Hence in the following we study the behavior of LCA in the pedigree graph.

Note that even though in every instance of the pedigree graph the indegree and outdegree of every vertex is bounded: indegree by 2 and outdegree by $2N$ and each row is bounded by $2N$ vertices, the number of possible LCA's might not be finite. Let $Z(K,N)$ be the random variable denoting the number of LCAs in $G_{PG}(K,N)$. Then

$$0 \leq Z(K,N) \leq \infty.$$

Fact 1 *1. There exist instances where $Z(K,N)$ attains the value* 0.
2. There exist instances where $Z(K,N)$ attains the value ∞.

The reader is directed to [Par09] for the proofs of these statements. However, for illustrative purposes in Fig 3 we create an instance of the random graph $G_{PG}(4,3)$ that has an infinite number of LCAs, i.e., with $Z(K,N) = \infty$. The construction is as follows. Row 0 has the four extant vertices. The outgoing edges from vertices in Row 0, 1 and 2 are constructed as shown in the figure. The vertices in row 3 and higher are of three categories: (1) Two vertices of different colors called the *blocked-vertices* (two left vertices in the figure); (2) one vertex, called the *LCA-vertex* of any color (the middle vertex in the figure); (3) two vertices of the same color, but different from the color of the LCA-vertex called the *path-vertices* (two right vertices in the figure). The edge constructions follow a simple pattern as shown in the figure. Under this construction, the following can be verified: (1) the instance of the pedigree graph is valid i.e., the color of the two parents of a node are of different colors, (2) every LCA-vertex of row 2 and higher is indeed the LCA of all the extant vertices.
As is common in population genetics, we make the assumption that any ancestor of an LCA is not of consequence and can be *excised* from the relevant pedigree graph.

Is the pedigree graph after excising all the ancestors of all the LCAs finite? The answer to the question of finiteness of the excised pedigree graph with fixed parameters $K(>1)$ and $N(>2)$ is summarized here:

1. It is possible that an instance of the (excised) pedigree graph is infinite.
2. Further, there are infinitely many such instances.

3 Pedigree Subgraphs

It is perhaps not very surprising to note that a pedigree graph may have multiple LCAs. However, it is rather surprising to note that even in a finite population model, the number of LCAs could be infinite (Fact 1). This counter intuitive characteristic of the pedigree graph can be addressed by exploiting the biparental mode and coupling genetic exchange information with the topology of the pedigree graph.

Population dynamics vs biology. Usually, a mutation rate θ is associated with a population and different populations can be Wright Fisher populations with different mutation rates. Note that different values of θ do not (and should not) affect the population dynamics under such population models. Also, θ is not a direct observable: it is inferred from the observed mutations or allele values. In fact mutations, reflected as allele frequencies, can be even viewed as external markers (say like *Lagrangian markers* in fluid dynamics) to study the evolution of the statistical properties of the ideal population. Ideally, θ does not affect the topology of the unilinear transmission trees: it only affects the sequences that each unit represents. Since θ is 'external' to the population, this additional parameter does not affect the modeling (or understanding) of the dynamics of the population.

Then, how about the parameter recombination rate ρ in a biparental model? Analogous to mutations (and other duplication-model genetic events), this should not affect the population dynamics but only the sequences of the units. Again, it is (by current biotechnologies) not a direct observable but can be inferred from the sequences in the population. However, just as in the unilinear transmission model, the vertices that have no paths to an extant unit is not relevant for the study, so in the biparental model, vertices that have no genetic material ancestral to any in the extant units are not of relevance. The definition of the *relevant pedigree graph* is now extended to exclude those vertices that do not carry any genetic material to the extant units (although there may be a path in the graph to an extant vertex). It now seems more natural to annotate the vertices of the biparental graph with the non-mixing segments (i.e. a segment that is inherited completely from the mother or the father but not mixed by the two parents) of genetic materials. Thus instead of ρ, a more natural parameter seems to be M, the number of nonmixing units in the extant population. The parameter M models the biparental mode as a natural extension of the (well-accepted) unilinear transmission mode and ρ continues to be 'external' to the population.

We identify two classes of subgraphs of the pedigree graph $G_{PG}(K,N)$ as our objects of study (Fig 4 shows the different subgraph models of the pedigree graph): **(1) Unilinear Transmission:** A Monochromatic Subgraph $G_{PT}(K,N)$ is induced on the vertices of one color (either only blue or only red). Thus each vertex has exactly one parent. The biological interpretation of a monochromatic subgraph is as follows. The genetic material that is transmitted only through the blue vertices (father) is the NRY chromosome. Similarly, the genetic material that is transmitted only through the red vertices (mother) is the mitochondrial DNA. These subgraphs of the pedigree graph actually represent the duplications-only model. Fact 3 has an interesting consequence: *All* the genetic material in the extant sequences can be traced back to a unique vertex in the pedigree graph. Topologically, this vertex is the LCA and is called the TMRCA (the most recent common ancestor).

(2) Genetic Exchange Model: In this model genetic material is additionally associated with the vertices. M is an upper bound on the number of non-mixing (or genetic exchange) segments in the extant units, which is used as an additional parameter. A Mixed Subgraph $G_{PGE}(K,N,M)$ is induced on some blue and some red vertices. Thus the vertices have may have either one or two parents. Fact 5 has an interesting consequence: *All* the genetic material in the extant sequences can be traced back to a unique vertex in the pedigree graph. Topologically, this vertex is the LCAA (defined in Section 6). This is called the GMRCA (the grand most recent common ancestor).

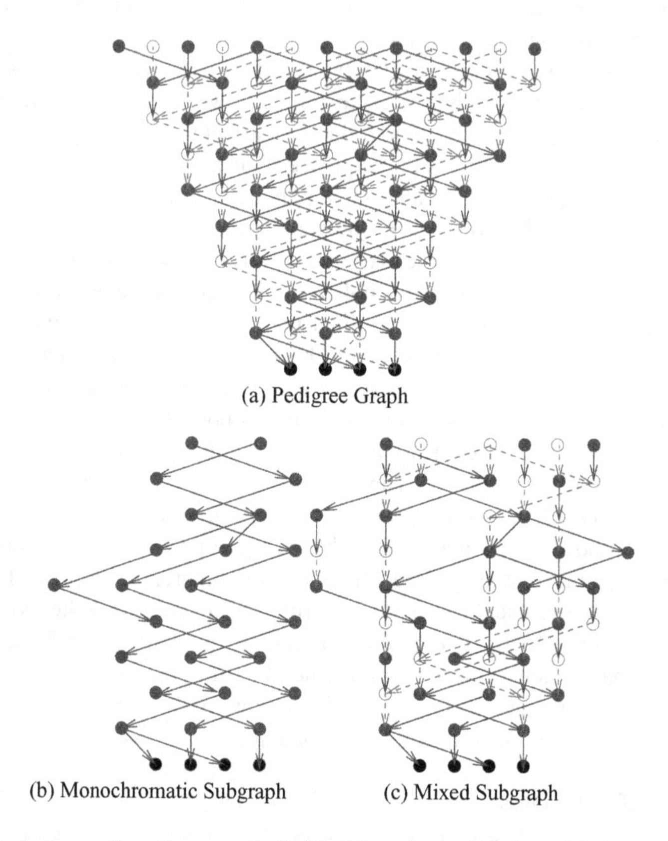

(a) Pedigree Graph

(b) Monochromatic Subgraph (c) Mixed Subgraph

Fig. 4 (a) An instance of a pedigree graph. (b) The Monochromatic Subgraph induced on the blue nodes (each node has exactly one parent) (b) A Mixed Subgraph induced on a subset of the blue and the red nodes (each node has one or two parents).

4 Unilinear Transmission: Monochromatic Subgraphs

Mutation events or genetic events such as the ones leading to Short Tandem Repeat (STR) polymorphisms are modeled as *duplication* events or simply *non-recombining* events [HSW05, JHTS04]. Hence this is also called the duplications-only model and each vertex has only one parent. Since we do not model any gender-specific characteristics, the duplications-only model is equivalent to the monochromatic (all vertices of the same color) model in our general setting. The following is easily verified.

Fact 2 *The monochromatic subgraph is a forest (tree), i.e., the graph has no closed paths.*

Hence the monochromatic subgraph is written as $G_{PT}(K,N)$.

Fact 3 *Given an instance of a monochromatic subgraph $G_{PT}(K,N)$:*

1. *The number of vertices can neither increase with depth nor be zero at any row, i.e.,*
$$K \geq k_1 \geq k_2 \geq k_3 \geq, ..., (\geq 1).$$
2. *The number of LCAs is at most 1.*
3. *$(k_g = 1) \Leftrightarrow$ The vertex at row g is the LCA.*

Proof Sketch: 1. This follows from the fact that the graph is a tree. 2. Assume the contrary that an instance has $l > 1$ LCAs. Let v_1 and v_2 be two distinct LCAs. Then there must exist vertices u_1 and u_2 (possibly with $u_1 = u_2$) where u_2 is an extant vertex and there is a path from v_1 to u_1, a path from v_2 to u_1 and a path from u_1 to u_2. Further let u_1 be such that there is no other vertex u' with a path from v_1 to u', v_2 to u' and u' to u_1 (if such is the case then we call u' as u_1). Observe that each vertex of the monochromatic subgraph has at most one parent. However, u_1 must have at least two parents (each on the two distinct paths to the two LCAs v_1 and v_2) contradicting this fact. Hence the assumption must be wrong and the number of LCAs $l \leq 1$. 3. This follows from 2. □

5 Genetic Exchange Model: Mixed Subgraph

Given K extant sequences, the most recent common ancestor (MRCA) is a sequence S from some generation such that the genealogy of *every* segment (nucleotide) of every extant sequence can be traced back to S. Further, amongst all such common ancestors, S is the most recent one. In population genetics, usually the term MRCA is used for the duplications-only model, and the term grand MRCA (GMRCA) is used when the genetic events include rearrangements of the sequence, such as recombinations (see texts such as [HSW05]). In the following we use M, an upper bound on the number of mixing segments in the extant units, to parameterize a general genetic exchange model. A Mixed Subgraph $G_{PGE}(K,N,M)$ is defined as follows. For each instance G of the mixed subgraph:

1. Each vertex in G is annotated with M nonmixing segments and must have genetic material that flows to at least one of the extant vertices.
   This implies that a vertex may have only one parent (if the other parent has no genetic material flowing to an extant unit).

2. Each genetic mixing event is equally likely to occur.
   This is only a simplifying assumption and in the same spirit as random mating or panmictic condition of the Wright Fisher population.

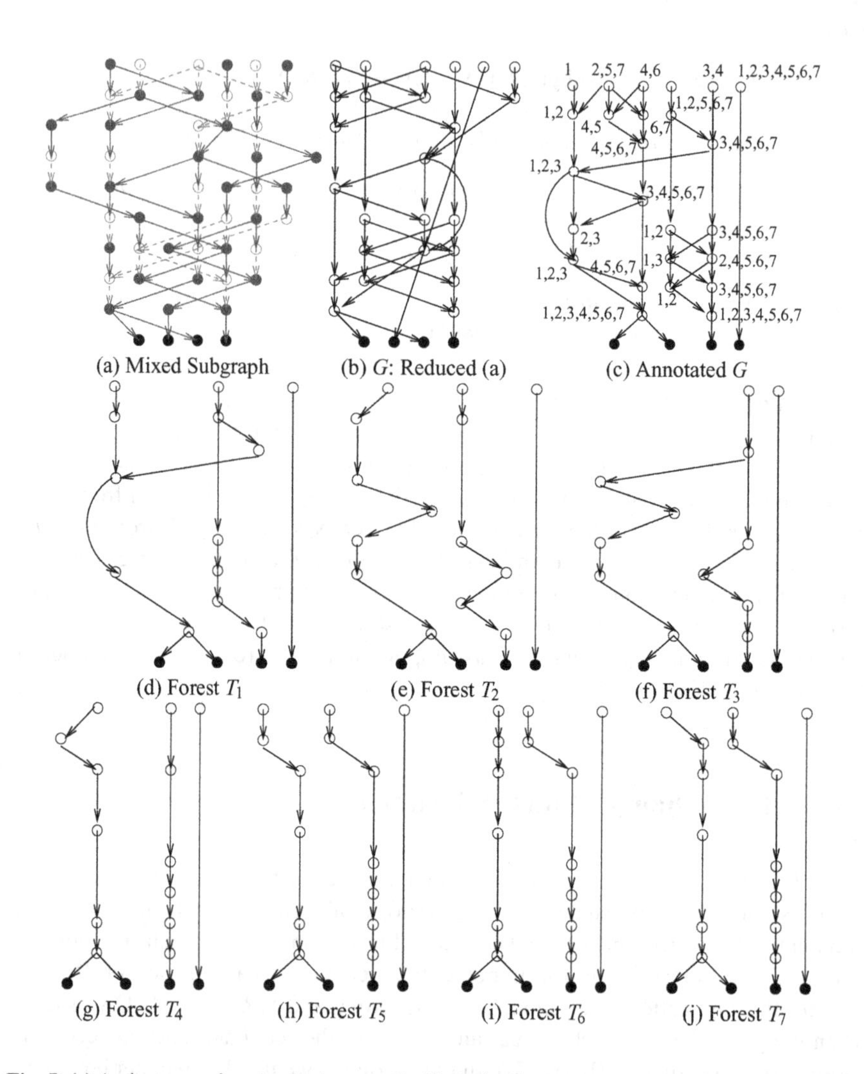

Fig. 5 (a) An instance of a mixed subgraph $G_{\text{GE}}(4,N,7)$. (b) For clarity, the chain-paths have been replaced by a single edge. (c) G with possible annotation of genetic units $\{1,2,3,4,5,6,7\}$. (d)-(j) The 7 distinct trees (forests), induced by each nonmixing unit i, embedded in G. In other words, $G = \bigcup_{i=1}^{7} T_i$.

Genetic Material gm **Notation.** The genetic material of a unit v is $gm(v)$ and the flow of the genetic material through an edge e is $gm(e)$. The M nonmixing units are written as $\{1,2,\ldots,M\}$and is associated with each extant vertex v. The genetic material may have nonconsecutive segments, say, 2,3, 7, i.e. $gm(v) = \{2,3,7\}$. Thus for all v and e of the subgraph instance

$$gm(v), gm(e) \subset \{1,\ldots,M\}.$$

The flow of the genetic material $gm(e)$ through an edge e and the genetic material $gm(v)$ of a vertex v are not independent and are related by the two rules.

1. Rule 1: Let u be a vertex with d ascendant (incoming) edges $e_i, i = 1..d$ (the valid values of d are 1 or 2).

$$gm(u) = \begin{cases} gm(e_1), & \text{if } d = 1, \\ gm(e_1) \bigsqcup gm(e_2). & \text{if } d = 2 \end{cases}.$$

   (Note that $S = S_1 \bigsqcup S_2$ denotes that S is the disjoint union of S_1 and S_2, i.e., $S_1 \cap S_2 = \emptyset$.)
2. Rule 2: Let v be a vertex with d descendant (outgoing) edges $e_i, i = 1..d$.

$$gm(v) = \bigcup_{i=1}^{d} gm(e_i).$$

Let m graphs $G_i(V_i, E_i)$ with vertex set V_i and edge st E_i be defined on (labeled) vertices, $1 \leq i \leq m$. Then the induced graph on vertices $V = \bigcup_{i=1}^{m} V_i$ (with edges $E = \bigcup_{i=1}^{m} E_i$) is written as

$$G = G_1 \cup G_2 \cup \ldots \cup G_m.$$

Fact 4 *Given G, an instance of a mixed subgraph* $G_{PGE}(K,N,M)$, *the following hold.*

1. *For each vertex v and each edge e of G,*

$$gm(v), gm(e) \neq \emptyset.$$

2. *Let* $V_m = \{v \mid m \in gm(v)\}$, *for a nonmixing unit* $1 \leq m \leq M$ *with induced graph* T_m *on* V_m.

   a. T_m *is a forest for all* $1 \leq m \leq M$.
   b. $G = T_1 \cup T_2 \cup \ldots \cup T_M$.

3. *Let the set of vertices at depth g be* V_g. *The following holds for each depth g.*

   a. $|V_g| \leq KM$.
   b. *For each nonmixing unit* $1 \leq m \leq M$, $\left|\{v \in V_g \mid m \in gm(v)\}\right| \leq K$.

Proof Sketch: (1) This follows from the definition of the mixed subgraph (each node must have genetic material that flows to at least one extant vertex). (2a) Assume that the result is not true: For some m, T_m has a closed path. By the nature of the direction of the edges in the pedigree graph, then there exists a vertex v with two distinct paths P_1 and P_2 to u. Without loss of generality, let the two paths be nonintersecting, except at v and u. Clearly, by Rule 1, the two incoming edges e_1 and e_2 on u cannot be such that $m \in gm(e_1), gm(e_2)$. This leads to a contradiction and the assumption is false. (2b) Let the vertex set of G be V and the edge set be E. Then

$$V_1 \cup V_2 \cup \ldots \cup V_M \subseteq V, \qquad \text{(by definition of } V_m)$$
$$V \subseteq V_1 \cup V_2 \cup \ldots \cup V_M.$$
$$\text{(by (1) each } v \text{ of } G \text{ must belong to at lease one of } T_m)$$

Thus

$$V = V_1 \cup V_2 \cup \ldots \cup V_M,$$
$$E = E_1 \cup E_2 \cup \ldots \cup E_M. \qquad \text{(by similar arguments)}$$

Hence the result. (3a) and (3b) In each T_m the number of vertices per row does not exceed K. Also, each vertex in T_m must be annotated with the genetic element m. Thus by 2(b), the number of vertices in G cannot exceed KM and the number of nodes with nonmixing element m cannot exceed K. □

An example is shown in Fig 5 of the embedded trees in an instance of a mixed subgraph. An instance of a graph where each vertex v has genetic material $gm(v)$ defined is said to be *gm*-annotated. Note that in a monochromatic subgraph, for any two distinct vertices v_1 and v_2, $gm(v_1) = gm(v_2)$. Thus a monochromatic subgraph can be considered to be always *gm*-annotated.

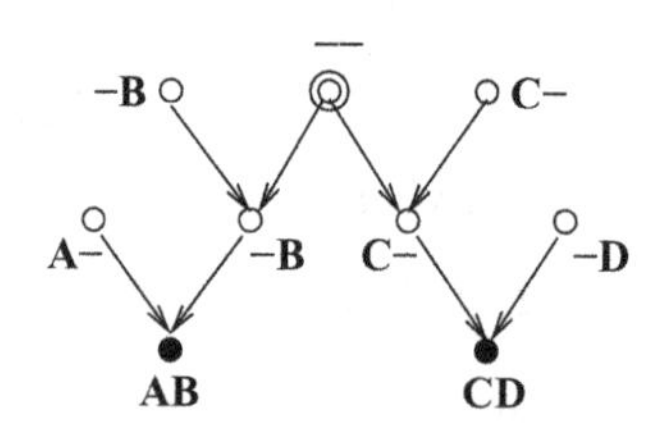

Fig. 6 Ancestor without ancestry: Each node is annotated with the genetic material labeled as a combination of **A, B, C, D**, or -. The symbol '-' denotes material that is not ancestral to any in the extant units. The LCA (shown with an extra concentric ring) has no ancestral material although it is the LCA of the two extant vertices with genetic material **AB** and **CD** respectively. Thus this LCA is an ancestor without any ancestry.

6 Topological definition of GMRCA: Least Common Ancestor with Ancestry (LCAA)

Conceptually the term LCA is equivalent to MRCA. However, LCA is *not* equivalent to GMRCA, which additionally is also ancestral to the genetic material in the extant vertices. This is due to the following fact: If node v_a is an ancestor of some node v in $G_{\text{PG}}(K,N)$, then it is possible that not *all* the genetic material of v_a is ancestral to the genetic material of v. It is also possible that v_a is ancestral to no genetic material of v. In the latter case a topological ancestor is not a 'genetic' ancestor. For example see Fig 6. A natural question is if there exists a purely topology-based definition that captures the notion of a GMRCA. We call this the LCA with ancestry or LCAA, a graph-theory based term for the GMRCA (it is defined in the fact below). Given G an instance of a mixed subgraph $G_{\text{PGE}}(K,N,M)$, by Fact 4, let

$$G = T_1 \cup T_2 \cup \ldots \cup T_M.$$

Fact 5 *1. The following two definitions of LCAA are equivalent:*

(a) (population genetics based) the least, or most recent, common ancestor of the K extant units that is also ancestral to all the genetic material in the K units.
(b) (graph-theory based) the LCA of the LCA's of $T_1, T_2, \ldots, T_M$.

2. The number of LCAAs is at most 1.
3. $(k_g = 1) \Leftrightarrow$ The vertex at row g is the LCAA.

Proof Sketch: We prove the second statement first. 2. Let every genetic unit (say a nucleotide) be tagged by a two tuple, its position in the chromosome and the label of extant vertex. Thus assuming there are c nucleotides and K extant units, there are cK distinct tuples. Next the genetic flow from vertex to vertex through the edges is marked by the tuples. It is easy to see that a path marked with a specific tuple is a chain (that does not branch). Thus if vertex v is a GMRCA, then by definition, v is on all the cK paths. Thus there cannot exist more than one GMRCA since all the marked paths are chains. 1. Let v'' be the LCAA by definition (a). Note that in tree, T_m, $1 \le m \le M$, the LCA of T_m, say vertex v_m, is also the LCAA of the extant units, corresponding to genetic material m. Further let v' be the LCA of $v_1, v_2, \ldots, v_M$. Then, clearly, v' is a CAA (common ancestor with ancestry) of the K extant units. Case 1: If v'' is an ancestor of v', it contradicts the definition of LCAA and $v' = v''$. Case 2: If v' is an ancestor of v'', it contradicts the definition of LCA of $v_1, v_2, \ldots, v_M$ and $v' = v''$. Case 3: There is no path between v' and v''. Then both are LCAAs but this contradicts 2. and $v' = v''$. Next, let v'' be the LCAA by definition (b). Then v'' is also a CAA of the K extant units. Let v' be an LCAA. By considering the three cases as before, we show that $v' = v''$. 3. This follows from 1. and 2. □

The most interesting and a fundamental observation is that the mixed subgraph is indeed the union of M trees (or forests). This provides a possible conduit for extending the results from coalescent theory to this more general setting.

7 Discussion

One of the consequences of studying recombinational population genomics through the lens of a random graph model is the observation that GMRCA indeed has a purely topological definition: it is the LCA of the LCA's of the M trees of

$G_{\mathrm{PGE}}(K,N,M)$. Thus it is possible to compute the expected depth of a GMRCA using results from coalescence theory and TMRCA (we leave this as food for thought for the reader).

Also, the mixed subgraph (or the ARG) is topologically nothing but the union of M trees. This simple view is very important to understanding how recombinations shape the autosomal chromosome. This also paves the way for a natural continuity with coalescent theory and all its implications, making it possible to apply many results from this theory, such as time to the most recent common ancestor or the probability of the GMRCA of the sample being that of the population, and so on. Most importantly, it provides a basis for reconstructability of the ancestral recombinations graph from a set of haplotypes, just as the reconstruction of mtDNA or NRY trees from mitochondrial or NRY DNA samples.

Finally, this model also opens up the possibility of population sampling algorithms that do not use a recombination "rate" parameter, which is somewhat difficult to characterize for a population.

It is perhaps not too far-fetched to say that recombinations is nature's use of combinatorics to bring about genetic diversity in populations. And, the combinatorists' view presented in the chapter, hopefully, will synergize with population geneticists and statisticians to produce new insights into the rather complicated phenomenon of genetic exchange shaping populations.

Exercises

1. Construct an instance of $G_{\mathrm{PG}}(4,3)$ with no LCA's.
   What is the probability of an instance of $G_{\mathrm{PG}}(4,3)$ having no LCA's? (Hint: see [Par09] for the definition of a natural probability measure)
2. Construct at least two instances of $G_{\mathrm{PG}}(4,3)$ with an infinite number of LCA's. Argue that it is possible to construct infinitely many such instances. (Hint: study the figure below)

   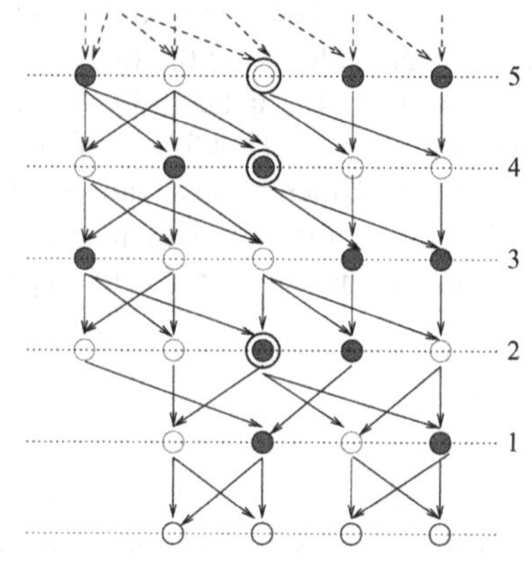

   What is the probability of an instance of $G_{\mathrm{PG}}(4,3)$ having an infinite LCA's?

3. For fixed parameters, $K > 1$ and $N > 2$, argue that
   a. there are infinite number of instances of $G_{PG}(K,N)$ each with no LCAs.
   b. there are infinite number of instances of $G_{PG}(K,N)$ each with an infinite number of LCAs.
4. Show that the monochromatic subgraph is topologically a forest (tree).
5. Show that for fixed parameters, $K > 1$ and $N \geq 1$, there are infinitely many monochromatic subgraphs with no LCAs.
6. Consider the Mixed Subgraph of Figure5(a). Is it possible to have an annotation with 5 genetic elements? Why?
   Is it possible to have an annotation with 7 genetic units distinct from the one in Figure5(c), other than relabelings? Why?
7. Argue the following: *The effective value of M for a pedigree graph, with N blue or red vertices and K extant units, at some depth is:*

$$M = NK^{-1}.$$

References

[BÖ0] R Bürger. *The mathematical theory of selection, recombination, and mutation.* New York, Wiley, 2000.

[CLR90] T. H. Cormen, C. E. Leiserson, and R. L. Rivest. *Introduction to Algorithms.* The MIT Press, Cambridge, Massachusetts, 1990.

[DSL+07] Joanna L. Davies, Frantiek Simank, Rune Lyngs, Thomas Mailund, and Jotun Hein. On recombination-induced multiple and simultaneous coalescent events. *Genetics*, 177:2151–2160, December 2007.

[Føl55] Erling Følner. On groups with full Banach mean value. *Mathematica Scandinavica*, 3: 243254, December 1955.

[GM97] R. C. Griffiths and P. Marjoram. An ancestral recombinations graph. *Progress in Population Genetics and Human Evolution (P Donnelly and S Tavare Eds) IMA vols in Mathematics and its Applications*, 87:257–270, 1997.

[Gri99] R. C. Griffiths. The time to the ancestor along sequences with recombination. *Theoretical Population Biology*, 55(2):137–144, April 1999.

[GSN+02] Stacey B. Gabriel, Stephen F. Schaffner, Huy Nguyen, Jamie M. Moore, Jessica Roy, Brendan Blumenstiel, John Higgins, Matthew DeFelice, Amy Lochner, Maura Faggart, Shau Neen Liu-Cordero, Charles Rotimi, Adebowale Adeyemo, Richard Cooper, Ryk Ward, Eric S. Lander, Mark J. Daly, and David Altshuler. The structure of haplotype blocks in the human genome. *Science*, 296(5576):2225 – 2229, 2002.

[GBB+07] Dan Gusfield, Vikas Bansal, Vineet Bafna and Yun S. Song. A decomposition theory for phylogenetic networks and incompatible characters. *Journal of Computational Biology*, 14(10): 1247–1272, 2007.

[HSW05] Jotun Hein, Mikkel H. Schierup, and Carsten Wiuf. *Gene Genealogies, Variation and Evolution: A Primer in Coalescent Theory.* Oxford Press, 2005.

[Hud83] R. R. Hudson. Properties of a neutral allele model with intragenic recombination. *Theoretical Population Biology*, 23(2):183–201, April 1983.

[Hud90] R. R. Hudson. *Gene genealogies and the coalescent process.* Oxford Surveys in Evolutionary Biology. Oxford University Press, Oxford, 1990.

[JHTS04] M.A. Jobling, M. Hurles, and C. Tyler-Smith. *Human Evolutionary Genetics: Origins, Peoples and Disease*. Mathematical and Computaional Biology Series. Garland Publishing, 2004.

[KC64] Motoo Kimura and James F. Crow. The number of alleles that can be maintained in a finite population. *Genetics*, 49(4):725–738, 1964.

[Kim69] Motoo Kimura. The number of heterozygous nucleotide sites maintained in a finite population due to steady flux of mutations. *Genetics*, 61(4):893–903, 1969.

[Kin82] J. F. C. Kingman. On the genealogy of large populations. *Journal of Applied Probability*, 19A:2743, 1982.

[MJC+09] Marta Melé, Asif Javed, Francesc Calafell, Laxmi Parida, Jaume Bertranpetit, Genographic Consortium. Recombination-based genomics: a genetic variation analysis in human populations. *under submission*, 2009.

[Par07] Laxmi Parida. *Pattern Discovery in Bioinformatics: Theory and Algorithms*. Chapman Hall Press, 2007.

[Par09] Laxmi Parida. Ancestral Recombinations Graph: A Reconstructability Perspective using Random-Graphs Framework. under submission, 2009.

[PMC+08] Laxmi Parida, Marta Melé, Francesc Calafell, Jaume Bertranpetit, Genographic Consortium. Estimating the Ancestral Recombinations Graph (ARG) as Compatible Networks of SNP Patterns. *Journal of Computational Biology*, 15(9):1–22, 2008.

[PJM+09] Laxmi Parida, Marta Melé, Francesc Calafell, Jaume Bertranpetit, Genographic Consortium. Minimizing recombinations in consensus networks for phylogeographic studies. *BMC Bioinformatics*, 10(1):S72, DOI = 10.1186/1471-2105-10-S1-S72, ISSN = 1471-2105, 2009.

[SFG+05] Stephen F. Schaffner, Catherine Foo, Stacey Gabriel, David Reich, Mark J. Daly and David Altshuler. Calibrating a coalescent simulation of human genome sequence variation. *Genome Res.*, 15:1576 - 1583, 2005.

[SH00] Mikkel H. Schierup and Jotun Hein. Consequences of recombination on traditional phylogenetic analysis. *Genetics*, 156:879–891, October 2000.

[SH06] Mike Steel and Jotun Hein. Reconstructing pedigrees: A combinatorial perspective. *Journal of Theoretical Biology*, 240(3):360–367, 2006.

[WH99a] Carsten Wiuf and Jotun Hein. Recombination as a point process along sequences. *Theoretical Population Biology*, 55:248–259, 1999.

[WH99b] Carsten Wiuf and Jotun Hein. The ancestry of a sample of sequences subject to recombination. *Genetics*, 151:1217–1228, March 1999.

Phylogenetic Trees From Sequences

Paul Ryvkin and Li-San Wang

Abstract In this chapter, we review important concepts and approaches for phylogeny reconstruction from sequence data. We first cover some basic definitions and properties of phylogenetics, and briefly explain how scientists model sequence evolution and measure sequence divergence. We then discuss three major approaches for phylogenetic reconstruction: distance-based phylogenetic reconstruction, maximum parsimony, and maximum likelihood. In the third part of the chapter, we review how multiple phylogenies are compared by consensus methods and how to assess confidence using bootstrapping. At the end of the chapter are two sections that list popular software packages and additional reading.

1 Basics

1.1 Definition

We assume the reader knows the basics in graph theory, for which many good textbooks are available [3, 48]. Recall that an unrooted tree is a connected, undirected, acyclic graph, and that a rooted tree has in addition a significant node called the root. A phylogeny T is a tree where the leaves of T, $L(T)$, are labeled and no internal node has degree 2 except for the root. As with trees, phylogenies may be rooted or unrooted. Each leaf of the tree, i.e., each external node, is also called a *taxon* (plural form *taxa*). Any edge in the phylogeny that is incident to a leaf is an *external edge*. Internal nodes and edges are defined accordingly, i.e., they are not external.

Paul Ryvkin
Genomics and Computational Biology Graduate Program, University of Pennsylvania, e-mail: pry@mail.med.upenn.edu

Li-San Wang
Penn Center for Bioinformatics, e-mail: lswang@mail.med.upenn.edu

L.S. Heath and N. Ramakrishnan (eds.), *Problem Solving Handbook in Computational Biology and Bioinformatics*, DOI 10.1007/978-0-387-09760-2_6,

A phylogeny reflects the evolutionary history and relationship among taxa. In a rooted phylogeny, an internal node corresponds to the most recent ancestral species of all taxa in the subtree rooted by the node: all such taxa are descendants of this species, and this node corresponds to the state of the most common recent ancestor from which these species diverged. A complete rooted subtree in a phylogeny is sometimes called a *clade*, and the taxa in a clade are called a *cluster*. It is useful to think of a rooted phylogeny as a collection of compatible clusters (the concept of compatibility will be further discussed in Section 1.2). The clusters in a phylogeny are organized hierarchically by the topology of the phylogeny: for any two distinct clusters $A,B \subseteq L(T)$, exactly one of the following relations takes place: $A \subset B$, $B \subset A$, or $A \cap B = \emptyset$. The first two scenarios imply the relation between their most recent common ancestors: if $A \subset B$, then the most recent common ancestor of taxa in A is a descendant of the most recent common ancestor of taxa in B. Moreover, every phylogeny with the same set of taxa will have the following *trivial* clusters: the cluster of all taxa, and the single-taxon clusters, one for each taxon – these clusters are trivial because they are in every phylogeny and do not provide any information on the evolutionary relationships.

While a rooted phylogeny details the evolutionary hierarchy by its clusters, an unrooted phylogeny does so without the "directionality" information. Instead the unrooted phylogeny tells which subsets of species are closer evolutionarily than the other species through *bipartitions*. Given an unrooted phylogeny, for each edge we can define a bipartition on the taxa – the two subsets of the bipartition are the taxa on the two sides of the edge. Again, any bipartition induced by an external edge is trivial because it has exactly one taxon at one side and is in every phylogeny with the same taxon set. An unrooted phylogeny can be regarded as a collection of "compatible" bipartitions, i.e. partitions that can co-exist in a phylogeny.

Clusters and bipartitions are very useful concepts: they carry all the information that is available in a phylogeny and they allow us to define phylogenies without explicitly constructing them – this provides great mathematical convenience as the following examples demonstrate.

1. Two phylogenies are equivalent (or isomorphic in graph theoretical terminology) if and only if they have the same set of clusters or bipartitions. The proof is simple: we can identify edges (and nodes, accordingly) shared by the two trees by matching their induced clusters or bipartitions.
2. A *subtree* is different from a clade: whereas a clade consists of all descendant taxa of an ancestral species, a subtree is the result of restricting a rooted phylogeny T to a subset of taxa X. The restriction of T to X is the phylogeny T' such that $L(T') = X$ and the set of clusters of T' is the set of unique clusters obtained by restricting each cluster of T to X (the unrooted version is defined similarly by replacing clusters with bipartitions). Though one can also define a subtree by describing an algorithm that computes the subtree, the definition is more cumbersome due to the many special cases and details the algorithm has to deal with.

Please refer to Figure 1 for some examples.

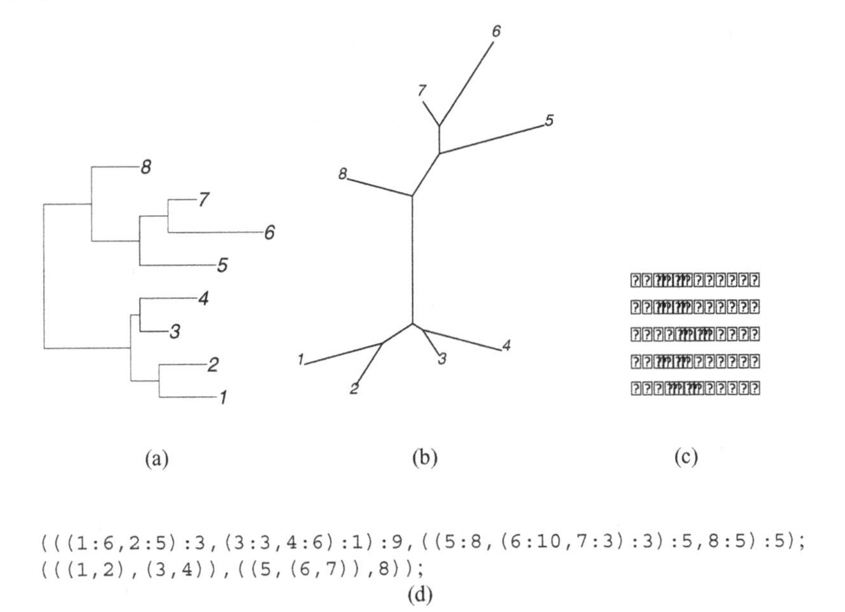

Fig. 1 An example phylogeny with eight taxa: (a) drawn as a rooted phylogeny, (b) drawn as an unrooted phylogeny, (c) list of nontrivial bipartitions, (d) its parenthesized representation with and without edge lengths.

1.2 Combinatorics of phylogenetics; compatibility

We describe some additional properties of phylogenies without proof due to space limitations.

Let T be an unrooted phylogeny with n taxa. Let $L(T)$ and $N(T)$ be the set of taxa and internal nodes of T, respectively. The number of external edges in T is $|L(T)|$, the number of elements in set $L(T)$, and the number of internal edges in T is $|N(T)|-1$.

T is *binary*, or *(fully) resolved*, if every internal node has degree 3; moreover, T is binary if and only if $N(T) = L(T) - 2$. An internal node with degree > 3 is *unresolved*.

The number of distinct phylogenies with n taxa,

$$(2n-5)!! = (2n-5)(2n-7)\cdots 1,$$

grows super-exponentially as the number of taxa increases: when $n = 4$ there are only 3 different binary phylogenies, but for $n = 53$ the number becomes 2.75×10^{80}, which is more than the estimated number of atoms in the whole universe (you will hear this often if you go to computational phylogenetics seminars). This combinatorial explosion means brute-force approaches are intractable and has important implications for the development of phylogenetic algorithms.

Not all clusters or bipartitions can coexist in the same phylogeny, i.e., are compatible. The following theorem is given by Buneman [4].

Theorem 0.1. *Given a set $\mathscr{B}$ of nontrivial, distinct bipartitions on taxon set X, the following two properties are equivalent:*

1. *($\mathscr{B}$ is* setwise compatible*) There exists a phylogeny such that its set of nontrivial bipartitions is $\mathscr{B}$;*
2. *($\mathscr{B}$ is* pairwise compatible*) For any two bipartitions $\{A|A'\}$ and $\{B|B'\}$ in $\mathscr{B}$, $A' = X - A$, $B' = X - B$, exactly one of the following four intersections is empty: $A \cap B$, $A \cap B'$, $A' \cap B$, $A' \cap B'$.*

1.3 Text representation of a phylogeny

The most common text representation of a phylogeny is the parenthesized format (commonly referred to as the Newick format [34]). The representation is for a rooted phylogeny (for unrooted phylogenies, simply root them arbitrarily) and recursive: (1) each taxon is represented by its label, (2) the representation of each clade in the phylogeny is the parenthesized list of representations of all its immediate subclades, separated by commas, (3) the representation usually ends with a semicolon, though it is straightforward to check if the representation is complete by the balance of parentheses, and (4) one can optionally add edge lengths by appending each edge length after the clade right below the edge, separated by a colon. Refer to Figure 1 for an example.

1.4 Operations that change the topology of a phylogeny

The following operations take an unrooted binary phylogeny as input and generate a new phylogeny by changing its topology. Illustrations are given in Figure 2.

1. A *nearest-neighbor-interchange* (NNI) operation first picks an internal edge (x,y). Let the other two nodes adjacent to them be n_1, n_2, and, n_3, n_4. Pick one of n_1 or n_2, and pick one of n_3 or n_4; say n_1 and n_3 are picked. Remove edges (x,n_1), (y,n_3) from the phylogeny, and add edges (x,n_3) and (y,n_1). In other words, we obtain the new phylogeny by swapping the two clades rooted at n_1 and n_3.
2. A *subtree-pruning-regrafting* (SPR) operation picks two edges (x,y), and (u,v). The edge (u,v) is bisected to create edges (u,w) and (w,v). Pick one of the end points for edge (x,y), say x. The edge (x,y) is first removed from the phylogeny, and the edge (w,y) added to the phylogeny. This makes x a degree-2 node, which has to be suppressed: let the two nodes adjacent to x be a, b; remove edges (x,a) and (x,b), remove node x, then add edge (a,b). This operation detaches the clade rooted at y and reattaches it to the edge (u,v).

3. A *tree-bisecting-reconnecting* (TBR) operation removes an edge (x,y), then suppresses the two degree-2 nodes x and y. This creates two disconnected subtrees; choose one edge from each of the two trees, say (a,b) and (c,d). Bisect the two edges by adding nodes u and v, and add edge (u,v) to reconnect the two subtrees.

Note in SPR and TBR, (x,y) and (u,v) do not have to be internal edges, though the resulting operation might be degenerate; for example, if x is a leaf in a TBR operation, then one of the two subtrees broken by detaching (x,y) is a single-leaf tree (x), so u must be x and the operation becomes an SPR operation.

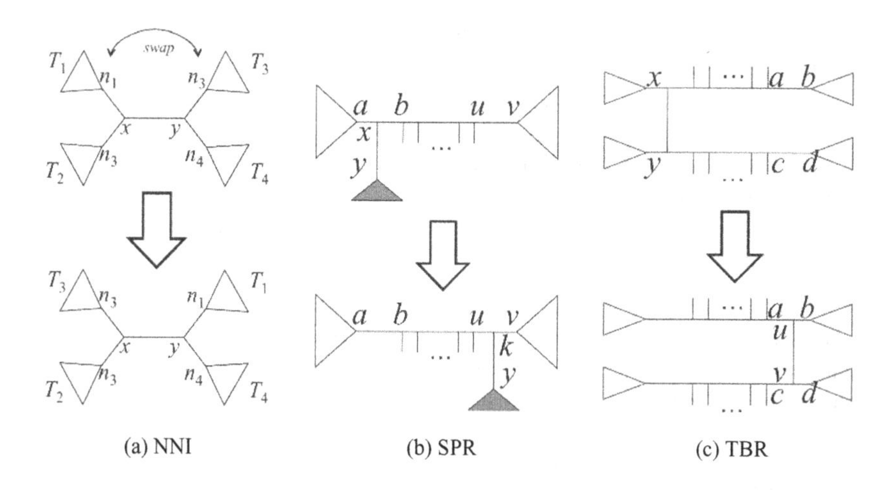

Fig. 2 Tree-topology changing operations: (a) nearest neighbor interchange (NNI), (b) subtree pruning regrafting (SPR), (c) tree bisection reconnection (TBR).

We encourage the reader to try some examples, and find out the relationship between the three operations: NNI is a special case of SPR, which in turn is a special case of TBR. Let n be the number of taxa in the phylogeny; the number of distinct NNI, SPR, and TBR operations are $O(n)$, $O(n^2)$, and $O(n^3)$, respectively (meaning they are bounded by an, bn^2, and cn^3 for some constants a,b,c). Moreover, it is provable that NNI (and as a result SPR, TBR) operations are enough to modify any one phylogeny to any other. The importance of these operations will be revealed when we discuss phylogeny reconstruction algorithms.

2 Sequence evolution

2.1 Jukes-Cantor model

A popular class of models for sequence evolution is the continuous-time discrete Markov process. In such models, each position in a sequence or site is considered a random variable whose state changes with time; in the case of DNA sequences, the state space corresponds to the nucleotides A,C,G,T. The Jukes-Cantor model[26] is the simplest of these models. It makes the following assumptions:

1. The sequence at the root of the phylogeny is drawn uniformly from the four nucleotides.
2. No insertions or deletions take place.
3. Each site evolves independently and identically.
4. For each site, the amount of time between any two consecutive mutation events follows an exponential distribution.
5. When a mutation takes place, the probability of substituting the current state (nucleotide) with any of the three other nucleotides is the same (Figure 3).

Note that properties (4) and (5) imply that the stochastic process of sequence substitution is a discrete Markov process, which is a mathematical necessity if the process satisfies the *memoryless* property, i.e., the distribution of the outcome only depends on the current state, but not the past history (many textbooks on Markov Chains discuss this). Second, we assume each site is independent and identically distributed (the *i.i.d.* assumption). Finally, the substitution model has the same distribution regardless of the sequence – it does not depend on the fitness of the current state and is selectively *neutral.*

The distribution of the outcome for a particular site after time t and with mutation rate μ (number of substitutions per unit time per site), can be determined using the following transition probability matrix:

$$\begin{pmatrix} P_{AA} & P_{AG} & P_{AC} & P_{AT} \\ P_{GA} & P_{GG} & P_{GC} & P_{GT} \\ P_{CA} & P_{CG} & P_{CC} & P_{CT} \\ P_{TA} & P_{TG} & P_{TC} & P_{TT} \end{pmatrix} = \frac{1}{4} \begin{pmatrix} 1+3e^{-\frac{4}{3}\mu t} & 1-e^{-\frac{4}{3}\mu t} & 1-e^{-\frac{4}{3}\mu t} & 1-e^{-\frac{4}{3}\mu t} \\ 1-e^{-\frac{4}{3}\mu t} & 1+3e^{-\frac{4}{3}\mu t} & 1-e^{-\frac{4}{3}\mu t} & 1-e^{-\frac{4}{3}\mu t} \\ 1-e^{-\frac{4}{3}\mu t} & 1-e^{-\frac{4}{3}\mu t} & 1+3e^{-\frac{4}{3}\mu t} & 1-e^{-\frac{4}{3}\mu t} \\ 1-e^{-\frac{4}{3}\mu t} & 1-e^{-\frac{4}{3}\mu t} & 1-e^{-\frac{4}{3}\mu t} & 1+3e^{\frac{4}{3}-\mu t} \end{pmatrix}$$

The model has symmetry across the four nucleotides – we can permute the nucleotides arbitrarily and still obtain the same matrix. When $t = 0$, only the diagonal is nonzero and no mutation has occurred. When $t \to \infty$ every entry in the matrix becomes $1/4$, so the resulting sequence is totally random.

Moreover, the matrix is symmetric (it is equal to its transpose); together with the fact the starting distribution of the nucleotides is uniform, it is provable that the stochastic process is equivalent if you reverse the starting sequence and the outcome. This *time-reversible* property implies that, no matter how you root the phylogeny, the distribution of the sequences at the leaves will always be the same. This is one reason why researchers focus on unrooted phylogenies. Under time-reversible models, it is impossible to locate the root without additional information. One common approach is to add the sequence of an additional species, called an *outgroup*, that is known *a priori* to be evolutionarily more distant than all the input sequences; then the external edge that is incident to the outgroup can be regarded as where the root is.

Given a phylogeny with edge lengths and sequence length k, the Jukes-Cantor model generates the sequence as follows. First the root sequence is determined by randomly choosing k nucleotides. Proceeding from the root downwards, the sequence of each child node can be obtained by applying the above transitional matrix (with t equal to the edge length; μ is the substitution rate) to each nucleotide in the parent node.

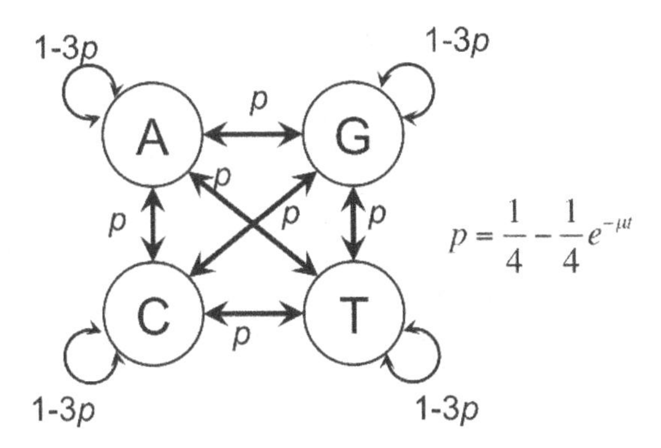

Fig. 3 Transition graph of the Jukes-Cantor model.

2.2 Distance correction

From the discussion above, the difference between a sequence and its mutated outcome will be more drastic as the amount of mutation μt increases. As we will see in the next section, if we can estimate μt correctly between any two input sequences, we can reconstruct the phylogeny correctly. It turns out that for the Jukes-Cantor model we can estimate it fairly reliably using the Hamming distance (sometimes called the *p-distance*), i.e., the proportion of mismatched sites between two sequences.

For simplicity, assume $\mu = 1$ (that is, the time is normalized by the mutation rate so that $t = 1$ amounts to one mutation per site on average). The expected Hamming distance after time t is the probability that the outcome is different from the starting nucleotide:

$$1-\frac{1}{4}\left(1+\frac{3}{4}e^{-\frac{4}{3}t}\right)=\frac{3}{4}-\frac{3}{4}e^{-\frac{4}{3}t}.$$

As $t \longrightarrow \infty$, this value becomes $3/4$, which is intuitively the expected Hamming distance between any two random sequences. Setting the above to be equal to the observed Hamming distance h and solving for t gives us the following formula:

$$\hat{t}=\frac{3}{4}\ln\left(1-\frac{4}{3}h\right).$$

This is the *distance correction* formula for the Jukes-Cantor model; Please see Figure 4 for an illustration.

What are the implications of this formula? The randomness and the limited state space (4 nucleotides) mean that, for many sites, the actual number of mutations will probably be greater than the observed number of mutations. Remember only the final result of the evolutionary process is observable, but not the intermediate states, so in the scenario of mutations like $A \longrightarrow G \longrightarrow T$ only the mutation $A \longrightarrow T$ is observable, and in $A \longrightarrow G \longrightarrow A$ it seems no mutation ever took place at all! This phenomenon is negligible when t is small: to a first approximation there is at most one mutation per site, so the distance correction curve is almost linear. But the sites are independently and identically distributed observations (*i.i.d.* replicates); with a large enough number of sites k, the average behavior (the Hamming distance) will be sufficiently informative. The variance of the distance estimate $\hat{t}$ can be shown to be inversely proportional to $\sqrt{k}$.

The formula does not work when $h \geq \frac{3}{4}$, and special rules and heuristics have been suggested to handle this singularity; for example, a very large but finite value will be used for $\hat{t}$.

2.3 Other models

It is obvious that the Jukes-Cantor model may be too simple. Many other models for DNA sequence have been suggested; here we discuss four such models. The Kimura two-parameter (K2P) model [27] assumes transitions (substitutions between purines or between pyrimidines, $A \leftrightarrow G$, or $C \leftrightarrow T$) and transversions (substitutions from purines to pyrimidines or vice versa) have different rates, but the nucleotide frequencies in the root sequence are equal. The transition probability matrix is

$$\begin{pmatrix} P_{AA} & P_{AG} & P_{AC} & P_{AT} \\ P_{GA} & P_{GG} & P_{GC} & P_{GT} \\ P_{CA} & P_{CG} & P_{CC} & P_{CT} \\ P_{TA} & P_{TG} & P_{TC} & P_{TT} \end{pmatrix} = \frac{1}{4}\begin{pmatrix} 1+a_t+2b_t & 1-a_t & 1-b_t & 1-b_t \\ 1-a_t & 1+a_t+2b_t & 1-b_t & 1-b_t \\ 1-b_t & 1-b_t & 1+a_t+2b_t & 1-a_t \\ 1-b_t & 1-b_t & 1-a_t & 1+a_t+2b_t \end{pmatrix}$$

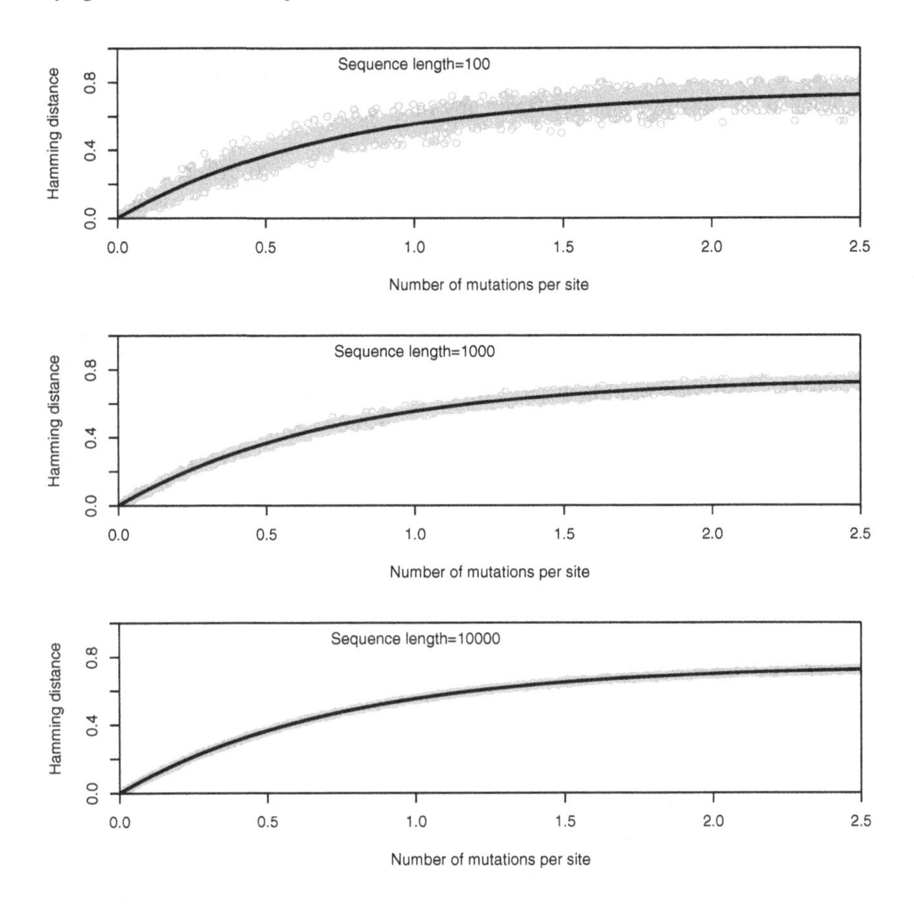

Fig. 4 Under the Jukes-Cantor model, scatter plot of the Hamming distance as a function of t (amount of evolution). The distance correction formula is simply the inverse of the curve (by flipping the x and y axes).

where $a_t = e^{-\alpha t}$ and $b_t = e^{-\beta t}$. The proportion of sites observed with transitions and transversions is $\frac{1}{4}(1-a_t)$ and $\frac{1}{2}(1-b_t)$. Let h and h' be the proportion of sites having a transition and transversion, respectively. The distance correction formula can be shown to be

$$-\frac{1}{4}\left(\ln(1-2h') + 2\ln(1-h-2h')\right)$$

The ratio $R = \alpha/(2\beta)$, called the *transition/transversion ratio* , is also estimable. Note that the Jukes-Cantor model is a special case where $\alpha = \beta$. The HKY (Hasegawa, Kishino and Yano)[22] model, likes the K2P model, assumes transition and transversion rates are unequal, and in addition assumes that the nucleotide frequencies in the root sequence may be unequal. The general-time-reversible (GTR) model is a further extension by assuming the nucleotide frequencies do not have to be uniform, the transition probability matrix is symmetric (time-reversibility), and the six pairwise substitution rates (from any nucleotide to any other nucleotide) are free parameters.

For protein sequences, the evolutionary model will have 20 different states, one for each amino acid. Some standard transition probability matrices (such as the PAM [8] matrix) have been established by fitting to experimental data.

All models above assume the sites evolve independently, identically, and neutrally, which does not necessarily happen in reality. One common modification is to introduce rate variation across sites. For example, some chosen sites may be fixed (i.e. no mutation occurs), and the proportion of these fixed sites is a model parameter. Another common modification is to assign scaling factors to the evolutionary rates for each site, where the scaling factors are usually drawn from a gamma distribution (with the shape parameter of the gamma distribution as a model parameter) [49]. Intuitively, under this model, each site scales the original phylogeny by its own rate factor, but the relative ratios between any two edge lengths as well as the topology are fixed.

3 Distance-based phylogeny reconstruction

Given an unrooted phylogeny T with edge lengths, the distance matrix D_T between taxa can be computed easily: the distance between any two taxa is the sum of the lengths of edges on the path connecting the two taxa. The distance-based approach for phylogeny reconstruction reverses this process – given an input distance matrix D on the set of taxa, our goal is to find a phylogeny with edge lengths such that its induced distance matrix is most similar to D. In this section we discuss important concepts, properties, and algorithms for distance-based phylogeny reconstruction.

3.1 *Ultrametricity and Additivity*

We first define two special types of distance matrices.

1. A distance matrix is *ultrametric* if $D[x,y] \leq \max\{D[x,z], D[y,z]\}$ for any three taxa x,y,z (this is also called the *three-point condition*).
2. A distance matrix is *additive* if for any four taxa x,y,u,v, the two larger values of the following three are equal: $\{D[x,y]+D[u,v], D[x,u]+D[y,v], D[x,v]+D[y,u]\}$ (this is also called the *four-point condition*).

Similarly, a rooted phylogeny with edge lengths is *ultrametric* if the distance between the root and any taxon is the same (i.e. the "height" of the tree is uniform). An unrooted phylogeny with edge lengths is *ultrametric* if we can root it (and bisect an edge and its length into two edges if necessary) so that the resulting rooted phylogeny is ultrametric. We have the following properties (the reader is encouraged to find the corresponding phylogeny reconstruction algorithms):

Theorem 0.2. *1. [21] A phylogeny is ultrametric if and only if its induced pairwise distance matrix is ultrametric.*
2. [47] Given a distance matrix D, there exists a phylogeny with edge lengths such that its induced distance matrix equals D if and only if D is additive. Moreover, the phylogeny is unique if all its edge lengths are strictly greater than 0.

The theorem also implies any ultrametric distance is immediately additive, though the converse does not hold. Please refer to Figure 5 for some examples.

3.2 Distance-based phylogeny reconstruction algorithms and the minimum evolution principle

Let T be the true phylogeny for the input set of species, and let the edge lengths be the amount of evolution (e.g., μt in the Jukes-Cantor model) along the edge. The additive distance matrix induced by T thus reflects the amount of evolution between any two species. If we can estimate this amount correctly from sequence data, then by the property of additivity we can reconstruct the phylogeny correctly. If the evolutionary rate is constant over the phylogeny (has the *molecular clock* property), then the length of each edge will be proportional to the amount of time and the phylogeny will be ultrametric; otherwise the edge length will be the product of the evolutionary rate and the amount of time (i.e. an edge will be longer if the evolutionary rate along it is greater).

The accuracy of additive distance matrix estimation is, of course, limited by many factors in reality; noise due to the randomness in the evolutionary process, limited amounts of data, and violations of assumptions in distance estimation are all contributing factors. Distance correction (see previous section) leads to better estimates of the actual amount of evolution and is almost always used. When the input distance matrix does not agree with the additive distance matrix induced by the actual phylogeny, there may not be a phylogeny that induces the input exactly, and we need a criterion to define how well a phylogeny matches the input. For example, the *least-squares criterion* is often used: given a phylogeny T, the distance matrix closest to input D is D_T that minimizes

$$\min_{\text{edge lengths for } T} \sum_{x,y \in L(T)} (D_T[x,y] - D[x,y])^2,$$

over all ways of assigning edge lengths in T. The least squares criterion determines the optimal edge lengths for a fixed phylogeny. If we let the length of a phylogeny be the sum of all edge lengths by the least-squares criterion, the phylogeny with smallest length corresponds to the scenario that requires the least amount of evolution – this is called the *minimum-evolution* criterion.

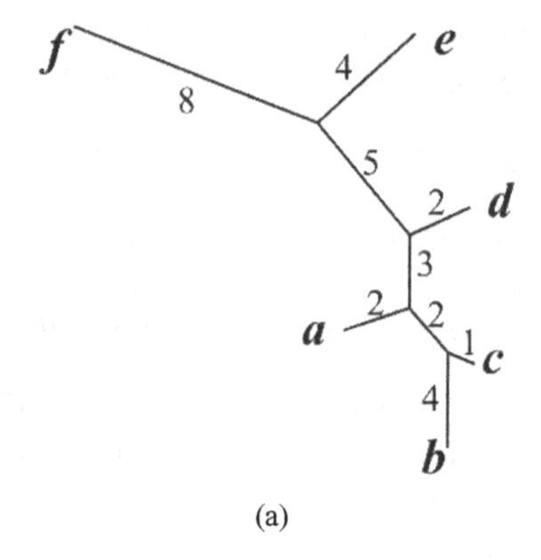

(a)

	a	b	c	d	e	f
a	0	8	5	7	14	18
b	8	0	5	11	18	22
c	5	5	0	8	15	19
d	7	11	8	0	11	15
e	14	18	15	11	0	12
f	18	22	19	15	12	0

(b)

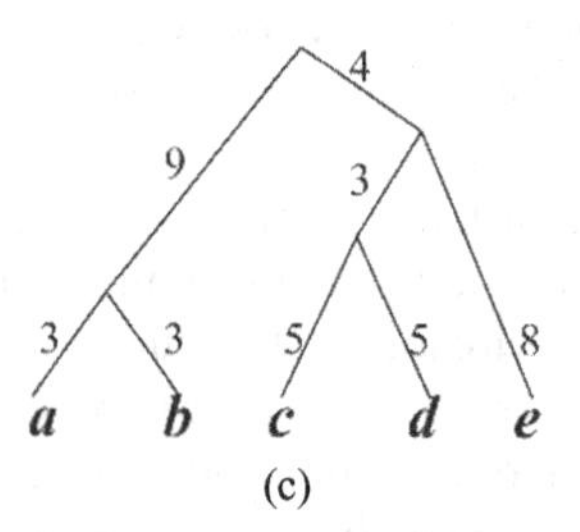

(c)

Fig. 5 Illustration and examples of distance-based phylogeny reconstruction. (a) A phylogeny with edge length, and (b) its induced distance matrix between every pair of taxa. (c) An ultrametric tree, rooted to show the distance from any taxon to the root is identical.

Minimum evolution is an NP-hard problem [7], meaning finding the optimal phylogeny is at least as difficult as other NP-complete problems such as the celebrated Traveling Salesperson Problem, and most likely no efficient algorithm (running time polynomial in the number of taxa) exists. The following algorithms are often used as heuristics for minimum evolution.

The UPGMA algorithm [42] is essentially average-linkage hierarchical clustering: starting with each taxon as a single-leaf tree, the UPGMA algorithm iteratively merges two trees that have the smallest pairwise distance between their roots, and updates the distance between the root of the new tree with other trees. Each iteration reduces the number of trees by one, and the procedure is repeated until only one tree is left. It can be shown that UPGMA can reconstruct the phylogeny correctly for ultrametric distance matrices, but not for general additive distance matrices.

Neighbor joining (NJ) [39] is another distance-based phylogeny algorithm that is widely used [1] and regarded as a gold standard that is often compared to in phylogeny algorithmic research. Like UPGMA, NJ also iteratively merges taxa until one tree is left. The difference lies in which two trees are selected for merger and how distances between trees are updated. Let $D[i,j]$ be the distance between subtrees i and j at the beginning of a particular iteration with m subtrees. Then

1. Compute the quantity $R[i]$ for each subtree i:

$$R[i] = \sum_{1 \le j \le m, j \ne i} D[i,j].$$

2. Pick two subtrees x and y such that $(m-2)D[i,j] - R[i] - R[j]$ is minimized. Merge the two subtrees into a new subtree z.
3. The lengths for edges (z,x) and (z,y) are

$$\begin{aligned} l[z,x] &= \frac{1}{2}D[x,y] + \frac{1}{2(m-2)}(R(x) - R(y)) \\ l[z,y] &= D[x,y] - l[z,x] \end{aligned}$$

4. For the new subtree z, compute $D[z,w]$ for any subtree w other than x or y:

$$D[z,w] = \frac{1}{2}(D[w,x] + D[w,y] - D[x,y]).$$

The original NJ paper [39] showed how the iterative algorithm can be seen as a greedy heuristic for the minimum evolution criterion. Moreover, NJ reconstructs the phylogeny correctly if the input matrix is additive.

A slightly different approach is to introduce perturbations to trees to see if the tree length can be improved. Since minimum evolution is NP-hard, greedy hill-climbing heuristics are often used. For example, FastME [9] starts by building the BioNJ [15] phylogeny, a variant of neighbor joining, then perturbs the phylogeny via NNI operations to reduce the length of the phylogeny until a local optimum is reached. FastME is relatively new, but the algorithm is very fast (running time in practice is proportional to n^2 for n taxa, whereas both UPGMA and NJ have running time proportional to n^3), and shows good performance in their simulation study.

[1] Cited more than 7000 times in the academic literature [16], it is probably the second most popular bioinformatics/computational biology algorithm after BLAST.

4 Maximum parsimony

In this section and the next, we discuss two approaches that, instead of aggregating the differences between sequences into a single distance matrix, directly utilize the sequence patterns at individual sites.

Given a phylogeny, we can assign sequences for internal nodes such that the overall number of mismatches (mutations) over all edges is minimized; the *parsimony score* of the phylogeny is this minimum number of mismatches. The score can be computed with a dynamic programming algorithm by Fitch [14]. First root the phylogeny arbitrarily by bisecting an edge, then compute the score for each site in two passes.

1. **Bottom-up:** each internal node x is assigned a set of possible candidate values $S(x)$ by comparing the candidate sets for its two child nodes x_1, x_2:
   If $S(x_1) \cap S(x_2) \neq \emptyset$ then $S(x) = S(x_1) \cap S(x_2)$, otherwise $S(x) = S(x_1) \cup S(x_2)$.
2. Arbitrarily choose a letter from $S(r)$ for root r.
3. **Top-down:** for each internal node x, if the chosen letter for its parent node x_p is in $S(x)$, choose the letter for x. Otherwise choose any letter from $S(x)$.

Please see Figure 6 for an example of the algorithm. For DNA sequences, the running time is proportional to nm, the size of the input sequence matrix.

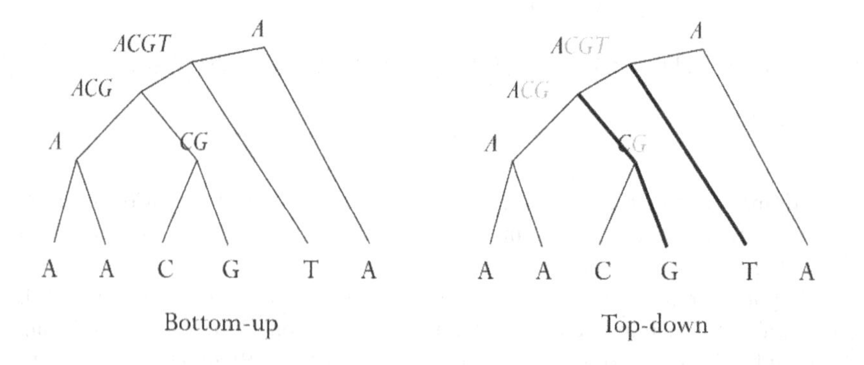

Fig. 6 An example of the Fitch algorithm for maximum parsimony. Left: bottom-up stage; letters in italic are the candidate nucleotides for the internal nodes. Right: top-down stage; chosen letters for internal nodes are in black, and edges with mutations are highlighted as wide lines. The parsimony score of the phylogeny is 3.

Given a definition for the parsimony score (e.g., as above), the goal of the maximum parsimony method is to find the phylogeny (or phylogenies) with the lowest parsimony score. This can be seen as an instance of the Occam's razor principle [33]: if multiple models explain the same observation equally well, the one with fewest assumptions or has the least complexity (in this case, the parsimony score, which reflects the number of mutations) is preferred. While computing the parsimony score of a phylogeny is straightforward at least for DNA, finding the best tree turns out to be difficult: the problem is known to be NP-hard [7]. Only for very small datasets (with fewer than 12 taxa) is brute force computation feasible, and for larger datasets hill-climbing heuristics are usually utilized. The starting trees can be random trees, trees computed by distance-based methods, or trees generated by sequential addition of taxa (adding taxa such that each additional taxon minimizes the tree's parsimony score). We can then use hill-climbing optimization by perturbing the current candidate trees to see if the parsimony score is lowered, and update the candidate tree set accordingly until some local optimum is reached. The three topological operations (NNI, SPR, TBR) are ideal for perturbing the phylogeny, because it can be proven that between any two trees that are one TBR apart (or NNI or SPR), the parsimony score will differ by at most 1 for each site (M. Steel, personal communication).

The parsimony score defined above is time-reversible: no matter how the phylogeny is rooted, its parsimony score is exactly the same. This particular parsimony is called *Wagner parsimony*. There are other kinds of parsimony scores. For example, one can set weights for different nucleotide or amino acid substitutions (for amino acid substitutions many standard matrices are available). For presence-absence (0/1) data, one can also use either *Dollo* and *Camin-Sokal* parsimony. Dollo parsimony assumes that the gain of a particular feature is highly unlikely to occur twice independently; given that the ancestral state is 0 for each character/site, the Dollo parsimony score is the number of $1 \longrightarrow 0$ (backward) changes when at most one $0 \longrightarrow 1$ (forward) change is allowed. The Camin-Sokal parsimony score is the number of $0 \longrightarrow 1$ (forward) changes when $1 \longrightarrow 0$ (backward) changes are disallowed. Both parsimony scores are easy to compute, and both are time-irreversible: the score of a phylogeny may be different when it is rooted differently.

5 Maximum likelihood

Maximum likelihood is a statistical framework that allows one to directly apply probabilistic sequence evolution models to phylogeny estimation. While in the past it was considered too computationally taxing to be practical, its use is now widespread. It can be thought of as a refinement of parsimony - instead of counting changes along a tree, it evaluates the probability of observing the sequence data given a particular tree and evolution model. The *likelihood* of a tree T with edge lengths $\mathbf{t}$ and evolution model θ describing one site is

$$L(T,\mathbf{t},\theta) = P(X|T,\mathbf{t},\theta),$$

where X is the observed state (sequence pattern) of all taxa at the site. The goal is to find a tree T which maximizes this likelihood over the entire sequence. Although there is a model defined by θ (Jukes-Cantor, for example) which gives us the probability of a transition from one state to another, states at internal nodes of the tree are unknown. Therefore we must evaluate the likelihood for all possible internal node states. Felsenstein's algorithm [11] can be used to recursively compute the likelihood at one site over the state space of nucleotides {A,C,G,T} (or amino acids in the case of proteins). Let n be the number of taxa with $x_1,\ldots,x_n$ being the observed leaf states at this site and $x_{n+1},\ldots,x_{2n-1}$ the states at internal nodes. The likelihood can be computed by post-traversal of the tree; for node i with state a and daughter nodes p and q,

$$L_i(T,\mathbf{t},\theta|a) = \begin{cases} D(i,a) & \text{if node } i \text{ is a leaf} \\ \left[\sum_{b\in\{A,C,G,T\}} P(x_p=b|x_i=a,t_{pi},\theta)L_p(T,\mathbf{t},\theta|b)\right] \times \\ \quad \left[\sum_{c\in\{A,C,G,T\}} P(x_q=c|x_i=a,t_{qi},\theta)L_q(T,\mathbf{t},\theta|c)\right] & \text{otherwise} \end{cases}$$

where

$$D(i,a) = \begin{cases} 1 \text{ if } x_i = a \\ 0 \text{ otherwise} \end{cases}$$

The likelihood of the entire tree for this site is defined by starting at the root:

$$L(T,\mathbf{t},\theta) = \sum_{a\in\{A,C,G,T\}} \pi_a L_{2n-1}(T,\mathbf{t},\theta|a).$$

Here π_a is the frequency of nucleotide a and node $2n-1$ is the root. These frequencies are defined by the the evolution model and for some models are free parameters. Assuming all sites evolve independently, one can compute the likelihood of a tree over the entirety of a sequence by taking the product of the likelihoods over all sites i:

$$L(T,\mathbf{t},\theta) = \prod_{i=1}^{N} P(X_i|T,\mathbf{t},\theta),$$

or the sum of the log likelihoods

$$\ln L(T,\mathbf{t},\theta) = \sum_{i=1}^{N} \ln P(X_i|T,\mathbf{t},\theta),$$

where N is the sequence length. Note that edge lengths $\mathbf{t}$ are usually not known, and must be treated as additional parameters to be optimized. This, combined with the fact that we must explore the entire tree space to find a tree that satisfies the optimality criterion, makes the problem intractable. There are several heuristics available for searching the tree space, however; for example, PhyML [20] employs a common hill-climbing algorithm. Taking a different approach, NJML [35] starts with a distance-based-tree, then perturbs the edges shown to be unreliable via bootstrapping while maximizing the likelihood.

Bayesian inference is a related statistical method which, as a phylogenetic reconstruction approach, is relatively more recent. In general Bayesian inference, an initial *prior distribution* of model parameters is suggested, and the parameters are iteratively adjusted to maximize the probability that the model is correct by taking into account evidence (observed data). The updated model (*posterior* distribution of the parameters) is given by Bayes' formula:

$$p(\theta|X) = \frac{p(X|\theta)p(\theta)}{p(X)}$$

where $p(X|\theta)$ is the likelihood, $p(\theta)$ is the prior distribution of the model parameters and $p(X)$ is the marginal probability of observing the data. Since the marginal probability is integrated over all possible trees and parameter values, the posterior is difficult to compute exactly. As a result, most implementations resort to sampling to approximate the posterior distribution, weighting the exploration of samples such that the algorithm converges on likely candidate models (those with high posterior probabilities). A popular stochastic sampling method is Markov Chain Monte Carlo (MCMC) [19, 23, 32], which is used by MrBayes [25]. One notable difference between Bayesian and ML methods is the need for prior distributions of parameters in Bayesian inference; there is much debate over what makes a reasonable prior and under which circumstances. Another difference is that the Bayesian method can yield a large number of trees which may be summarized by any number of methods. For example, one may simply selects the tree with the highest posterior probability averaged over all parameter values (also called the *maximum a posteriori* (MAP) criterion [29]). Alternatively, one may combine the trees with high posterior probabilities using consensus methods, for example.

6 Multiple phylogenies: comparison, consensus, and confidence

There are many situations where one may have to compare and summarize multiple phylogenies, whether they are equally optimal phylogenies from the same algorithm, results of different algorithms on the same dataset, or phylogenies for different datasets from overlapping taxon sets. In this section we discuss related problems, mathematical concepts, and common approaches.

6.1 Distances between trees

If we regard a phylogeny as a collection of bipartitions, a natural comparison between two phylogenies is to find out which bipartitions are shared and which are distinct. The *Robinson-Foulds distance* is the number of bipartitions that are in one phylogeny but not the other:

$$D_{\mathrm{RF}}(T_A, T_B) = |\mathscr{B}(T_A) \Delta \mathscr{B}(T_B)|$$
$$= |\{b : b \in \mathscr{B}(T_A) - \mathscr{B}(T_B) \text{ or } b \in \mathscr{B}(T_B) - \mathscr{B}(T_A)\}|$$

Here $\mathscr{B}(T_A)\Delta\mathscr{B}(T_B)$ is the symmetric difference set between the sets of bipartitions for trees A and B. It is straightforward to prove that the distance D_{RF} is a metric: it is nonnegative, it is zero if and only if the two phylogenies are equal, it is symmetric, and it satisfies the triangle inequality $D_{\mathrm{RF}}(T_A, T_B) \leq D_{\mathrm{RF}}(T_A, T_C) + D_{\mathrm{RF}}(T_C, T_B)$ for any phylogenies T_A, T_B, T_C. Note that the distance is defined for phylogenies sharing the same set of taxa.

Of course, no single distance measure can capture the complexity that entities like phylogenies convey. The Robinson-Foulds distance can be misleading in the *rogue-taxon* scenario, when a small number of taxa are located at opposing places in two phylogenies that are otherwise the same (see Figure 8).

Although the RF distance is widely used, there are other types of distance between phylogenies. One approach is to compute the minimum number of topology-changing operations (such as NNI, SPR, and TBR) required to change one phylogeny to the other. However, it has been shown that computing the NNI, SPR, and TBR distances are all NP-hard [1, 6, 24].

6.2 Consensus trees

There are many consensus tree methods; here we discuss three methods that are often employed. The *strict consensus* of an input set of phylogenies (with the same taxon set) is the phylogeny such that its every bipartition is in every input phylogeny (hence the name strict). Algorithmically, we can compute the strict consensus as follows: pick any input phylogeny, mark every edge whose bipartition is missing in at least one input phylogeny, and contract it by "gluing" its two endpoints together.

A relaxation of the strict consensus tree is the p-consensus [31]: it is the phylogeny whose bipartitions are in proportion $\geq pN$ of all the N input phylogenies. It can be shown that if $p > 0.5$ (by Buneman's Theorem) then such a phylogeny exists and is unique, but this is not necessarily so when $p \leq 0.5$. Strict consensus is simply the case where $p = 1$. When we require every bipartition in the consensus to be in $> N/2$ input trees (i.e., $p = (N/2+1)/N$ for even N and $p = 1/2$ for odd N), this is called the *majority consensus*. Computing the p-consensus is somewhat more complicated, because the consensus may have bipartitions from different input phylogenies, though efficient algorithms exist [2], and there are many programs available for this task. Also of note is the connection between the Robinson-Foulds distance and the strict and majority consensus trees: they are all bipartition-based. For example, it is straightforward that the RF distance between any binary input phylogeny and the strict consensus is the number of internal edges the input tree loses to become the strict consensus.

The third type of consensus tree is the *maximum agreement subtree* (MAST) [13]: the goal is to find a largest subset of input taxa such that the input phylogenies all have the same topology when we restrict them to this subset. An advantage of this method is that it is robust against the rogue taxon situation. The maximum agreement subtree is not necessarily unique, and in the worst-case scenario the maximum agreement subtree will still have at least three taxa.

Please see Figure 7 for examples.

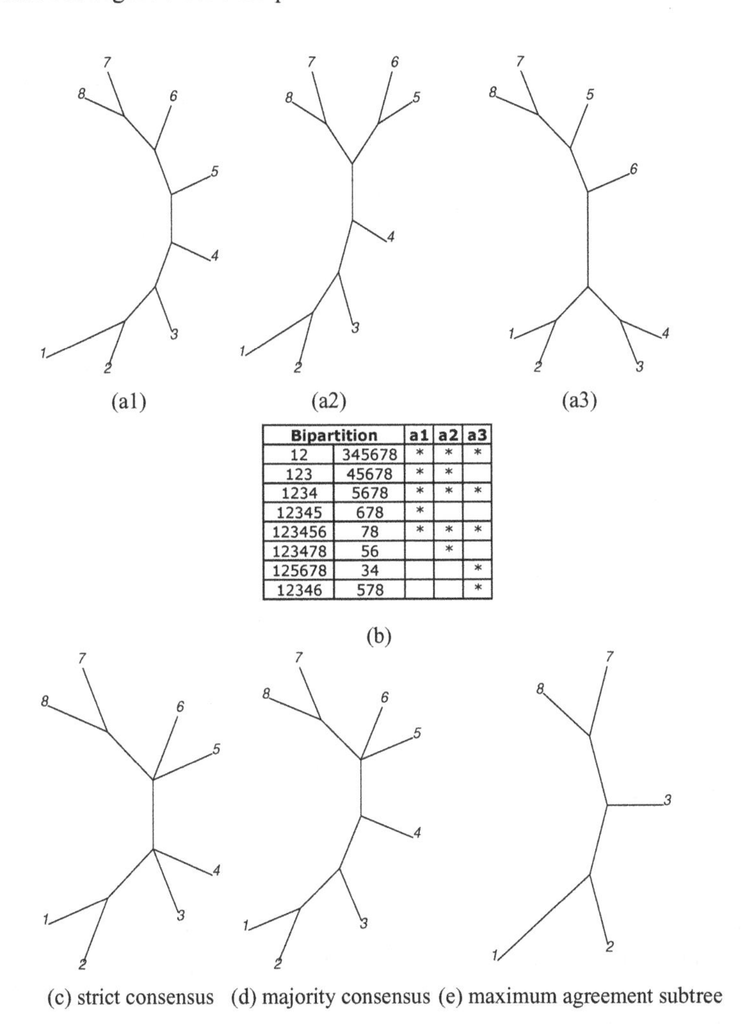

Bipartition		a1	a2	a3
12	345678	*	*	*
123	45678	*	*	
1234	5678	*	*	*
12345	678	*		
123456	78	*	*	*
123478	56		*	
125678	34			*
12346	578			*

Fig. 7 Consensus trees.(a1)-(a3) the three input trees; (b) table of nontrivial bipartitions; (c) the strict consensus; (d) the majority consensus; (e) the maximum agreement subtree.

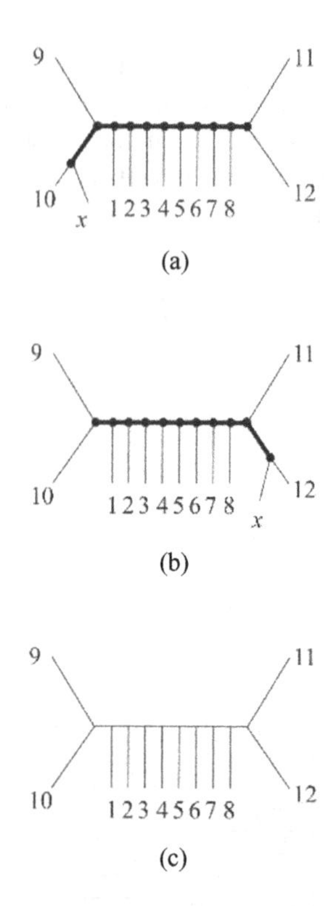

Fig. 8 A rogue taxon example: (a) and (b) are input trees; only the placement of taxon x disagrees. The bipartitions of the highlighted edges are affected by x, the RF distance between the two trees is maximal, and the strict consensus will not have any internal edge as a result. (c) The maximum agreement subtree, obtained by removing taxon x.

6.3 Bootstrapping

The *bootstrapping* procedure is adapted from bootstrapping in statistics and is widely used to report the confidence of bipartitions in phylogenetics literature. Assuming the sequence matrix is properly aligned (all sites are regarded to evolve independently and are from the same distribution in the phylogeny reconstruction procedure), bootstrapping starts by creating B *bootstrap datasets* (B is usually between 100 and 1000), each of which is the result of resampling the columns *with replacement*. We then apply the same phylogeny reconstruction algorithm on these bootstrap datasets to obtain B bootstrap phylogenies. For each internal edge of the phylogeny using the original dataset, the confidence score is the percentage of all bootstrap trees having the same bipartition as this edge. In [10] it is shown that the bootstrap scores can be interpreted as the posterior probability of bipartitions with a uniform prior for the topology of the phylogeny.

7 Software

For a survey of phylogenetics software, we recommend the online catalog compiled by Felsenstein[2].

The PHYLIP software, written by Joe Felsenstein[3], is a free collection of programs that perform many important tasks in phylogenetics. PHYLIP runs on most major platforms, and its source codes are available. PAUP [46], written by David Swofford, is another popular program for phylogenetics. The software is in beta stage but is functionally complete; though PAUP is not free, it is widely used by researchers as it comes with a programming language interface for scripting, is versatile and computationally very efficient, and is available on many platforms. For readers who are familiar with R [37], the package APE [36] is available. Mesquite [30], by Wayne Maddison, is written in Java and provides both command-line and graphical user interface for phylogenetics. The CIPRES (Cyberinfrastructure for Phylogenetic Research) project [5] is a five-year NSF-sponsored research on computational phylogenetics, which culminates in the CIPRES workbench and is available for researchers [38].

There are many special-purpose phylogenetic software packages available. For a distance-based method, we recommend FastME [9]. The TNT [17] software package provides highly efficient maximum parsimony computation. RAxML [43] and PhyML [20] are two popular software packages for maximum likelihood phylogeny reconstruction. For Bayesian methods (which we covered only briefly), we recommend MrBayes [25]. PAML [50] implements many likelihood-based hypothesis tests for phylogenetics. Seq-gen [18] is widely-used for generating simulated sequence data; for alternatives the readers may try r8s [40] and Rose [44].

8 Further reading

We recommend the following books and chapters for readers looking for in-depth knowledge on phylogenetics. The book chapter by Hillis and Swofford [45] provides a good survey on important concepts for phylogenetics and models of sequence evolution. The book *Inferring Phylogenies* by Felsenstein [12] is very nicely written and covers many important topics in phylogenetics; it can be used as a graduate-level text book or a great desk reference. For more advanced concepts, the book *Phylogenetics* written by Steel and Semple [41] is a good reference for readers who are mathematically oriented and interested in a mathematical treatise on properties of phylogenetics, and the book *Fundamentals of Molecular Evolution* by Li and Graur [28] has in-depth discussion regarding the statistical aspects of molecular systematics.

---

[2] `http://evolution.genetics.washington.edu/phylip/software.html.`

[3] `http://evolution.gs.washington.edu/phylip.html.`

9 Exercise

The following is an exercise that goes through steps for the reconstruction of a molecular phylogeny.

1. We will use programs from the PHYLIP software for this exercise. PHYLIP can be obtained from the PHYLIP website. Installation steps may vary depending on the operating system you use; please refer to the PHYLIP website for details.
2. Download the exercise dataset from the support website. This is a reformatted file of the partial 12S rRNA gene sequences from 36 mammals from the EMBL Nucleotide Sequence Database. Description of the PHYLIP sequence format can be found at the PHYLIP website Change the input file to `input.phy`.
3. We will use seqboot to generate 100 bootstrap sequences. PHYLIP has an interactive command-line user interface. The program will ask for input files since the default input file `inputfile` is missing. Enter `input.phy`. If the default output file is found, the program will prompt for how to deal with it; choose R for replacement. Then a menu will show up; enter the key for each menu item to make changes. For now we can accept the default settings to generate 100 bootstrap samples. Enter Y to accept, then the program will ask for random seed; enter any odd integer. The output sequence will be stored in the file "outfile". Change it to `input.boot`.
4. Use the program **dnadist** to compute pairwise distances. Use `input.boot` when asked for the input file. When the menu shows up, enter M to ask the program to handle multiple datasets; choose D for dataset (not weight), then enter 100 for 100 bootstrap samples. Enter Y to accept. Change the output file `outfile` to `input.boot.dist`.
5. Use the program **neighbor** to compute the neighbor joining tree. Set the input file to `input.boot.dist`, and enter M to specify 100 datasets. Enter Y to accept. Change the output file `outtree` to `input.boot.nj`; this file contains 100 neighbor joining trees.
6. We will now use the program **consense** to compute the consensus of the 100 neighbor joining trees. Set the input file to `input.boot.nj`. The default setting uses the (extended) Majority Consensus; accept the default settings by entering Y. The consensus tree will be stored in the file `outtree`, where the length of each edge is the number of input trees having the corresponding bipartition. The file `outfile` has more information, including a text-based tree plot, and a table of bipartition frequencies. How many internal edges have more than 50% frequency in the majority consensus?
7. PHYLIP comes with many phylogeny reconstruction programs. Use **dnaml** and **dnapars** to compute the maximum likelihood and maximum parsimony trees. Maximum likelihood is much slower, so be forewarned.
8. Use **treedist** to compute the pairwise Robinson-Foulds distance between the neighbor joining, maximum parsimony, and maximum likelihood bootstrap trees. Concatenate the tree trees into a single input file (you can either use the "cat" command in linux, "type" command in windows command prompt, or use a text

editor). Change the Distance Type to Symmetric Difference, and set it to compute distances between all possible pairs in the file. Which two trees are closer?

References

1. Allen, B., Steel, M.: Subtree transfer operations and their induced metrics on evolutionary trees. Annals of Combinatorics **5**, 1–15 (2001)
2. Amenta, N., Clarke, F., St. John, K.: A linear-time majority tree algorithm. In: Algorithms in Bioinformatics, *Lecture Notes in Computer Science*, vol. 2812, pp. 216–227. Springer, Heidelberg (2003)
3. Bollobas, B.: Modern Graph Theory. Springer (2002)
4. Buneman, P.: A note on the metric property of trees. J. Combin. Theory Ser. B **17**, 48–50 (1974)
5. CIPRES: Cipres. URL http://www.phylo.org/
6. DasGupta, B., He, X., Jiang, T., Li, M., Tromp, J., Zhang, L.: On computing the nearest neighbor interchange distance. Proc. DIMACS Workshop on Discrete Problems with Medical Applications **55**, 125–143 (2000)
7. Day, W.: Computational complexity of inferring phylogenies from dissimilarity matrices. Bull. Math. Biol. **49**, 461–467 (1987)
8. Dayhoff, M., Schwartz, R., Orcutt, B.: A model of evolutionary change in proteins. In: Atlas of protein sequence and structure, vol. 5, pp. 345–358. Nat. Biomed. Res. Found. (1978)
9. Desper, R., Gascuel, O.: Fast and accurate phylogeny reconstruction algorithms based on the minimum-evolution principle. J Comput Biol **9**(5), 687–705 (2002)
10. Efron, B., Halloran, E., Holmes, S.: Bootstrap confidence levels for phylogenetic trees. PNAS **93**(23), 13,429–13,429 (1996)
11. Felsenstein, J.: Evolutionary trees from DNA sequences: a maximum likelihood approach. J Mol Evol **17**(6), 368–376 (1981)
12. Felsenstein, J.: Inferring phylogenies. Sinauer Associates (2004)
13. Finden, C.R., Gordon, A.D.: Obtaining common pruned trees. Journal of Classification **2**(1), 255–276 (1985)
14. Fitch, W.: Toward defining the course of evolution: minimum change for a specific tree topology. Syst Zool **20**(4) (1971)
15. Gascuel, O.: BIONJ: an improved version of the NJ algorithm based on a simple model of sequence data. Mol Biol Evol **14**(7), 685–695 (1997)
16. Gascuel, O., Steel, M.: Neighbor-joining revealed. Mol Biol Evol **23**(11), 1997–2000 (2006)
17. Goloboff, P.: Analyzing large data sets in reasonable times: Solutions for composite optima. Cladistics **15**(4), 415–428 (1999)
18. Grassly, N., Adachj, J., Rambaut, A.: PSeq-Gen: an application for the Monte Carlo simulation of protein sequence evolution along phylogenetic trees (1997)
19. Green, P.: Reversible jump Markov chain Monte Carlo computation and Bayesian model determination. Biometrika **82**(4), 711–732 (1995)
20. Guindon, S., Gascuel, O.: A simple, fast, and accurate algorithm to estimate large phylogenies by maximum likelihood. Syst Biol **52**(5), 696–704 (2003)
21. Gusfield, D.: Algorithms on Strings, Trees, and Sequences. Cambridge University Press (1997)
22. Hasegawa, M., Kishino, H., Yano, T.: Dating of the human-ape splitting by a molecular clock of mitochondrial DNA. Journal of Molecular Evolution **22**(2), 160–174 (1985)
23. Hastings, W.: Monte Carlo sampling methods using Markov chains and their applications. Biometrika **57**(1), 97–109 (1970)
24. Hickey, G., Dehne, F., Rau-Chaplin, A., Blouin, C.: SPR distance computation for unrooted trees. Evolutionary Bioinformatics **4**, 17–27 (2008)

25. Huelsenbeck, J.P., Ronquist, F.: MRBAYES: Bayesian inference of phylogenetic trees. Bioinformatics **17**(8), 754–755 (2001)
26. Jukes, T., Cantor, C.: Evolution of protein molecules in HN Munro, ed. Mammalian protein metabolism, pp. 21–132. Academic Press, New York (1969)
27. Kimura, M.: A simple method for estimating evolutionary rates of base substitutions through comparative studies of nucleotide sequences. Journal of Molecular Evolution **16**(2), 111–120 (1980)
28. Li, W., Graur, D.: Fundamentals of Molecular Evolution. Sinauer Associates, Sunderland, Massachusetts (1991)
29. Lunter, G., Mikls, I., Drummond, A., Jensen, J.L., Hein, J.: Bayesian coestimation of phylogeny and sequence alignment. BMC Bioinfo **6**, 83 (2005)
30. Maddison, W.P., Maddison., D.: Mesquite: a modular system for evolutionary analysis. Version 2.5 (2008). URL http://mesquiteproject.org
31. Margush, T., McMorris, F.R.: Consensus n-trees. Bulletin of Mathematical Biology **43**(2), 239–244 (1981)
32. Metropolis, N., Rosenbluth, A., Rosenbluth, M., Teller, A., Teller, E.: Equation of state calculations by fast computing machines. The Journal of Chemical Physics **21**(6), 1087 (1953)
33. Moret, B., Bader, D., Warnow, T.: High-performance algorithm engineering for computational phylogenetics. The Journal of Supercomputing **22**, 99–111 (2002)
34. Olsen, G.: Gary Olsen's interpretation of the "Newick's 8:45" tree format standard. URL http://evolution.genetics.washington.edu/phylip/newick_doc.html
35. Ota, S., Li, W.H.: NJML: a hybrid algorithm for the neighbor-joining and maximum-likelihood methods. Mol Biol Evol **17**(9), 1401–1409 (2000)
36. Paradis, E., Claude, J., Strimmer, K.: APE: Analyses of phylogenetics and evolution in r language. Bioinformatics **20**(2), 289–290 (2004)
37. R Development Core Team: R: A Language and Environment for Statistical Computing. R Foundation for Statistical Computing, Vienna, Austria (2008). URL http://www.R-project.org
38. Rifaieh, R., Unwin, R., Carver, J., Miller, M.: SWAMI: Integrating biological databases and analysis tools within user friendly environment. Data Integration in the Life Sciences 4th International Workshop, DILS 2007, Philadelphia, PA, USA, June 27-29, 2007: Proceedings (2007)
39. Saitou, N., Nei, M.: The neighbor-joining method: a new method for reconstructing phylogenetic trees. Mol Biol Evol **4**(4), 406–425 (1987)
40. Sanderson, M.: r8s: inferring absolute rates of molecular evolution and divergence times in the absence of a molecular clock. Bioinformatics **19**(2), 301–302 (2003)
41. Semple, C., Steel, M.: Phylogenetics, Oxford Lecture Series in Mathematics and its Applications. Oxford University Press, Oxford (2003)
42. Sneath, P.: Numerical taxonomy. WH Freeman and Co., San Francisco (1973)
43. Stamatakis, A., Ludwig, T., Meier, H.: RAxML-III: a fast program for maximum likelihood-based inference of large phylogenetic trees. Bioinformatics **21**(4), 456–463 (2005)
44. Stoye, J.: Rose: generating sequence families. Bioinformatics **14**(2), 157–163 (1998)
45. Swofford, D., Olsen, G., Waddell, P., Hillis, D.: Phylogenetic inference. In: Molecular Systematics, vol. 2, pp. 407–514. Sunderland (1996)
46. Swofford, D.L.: PAUP*: Phylogenetic Analysis Using Parsimony (and Other Methods). Sinauer Associates (2003). Version 4
47. Waterman, M.S., Smith, T.F., Singh, M., Beyer, W.A.: Additive evolutionary trees. J Theor Biol **64**(2), 199–213 (1977)
48. West, D.B.: Introduction to Graph Theory, 2nd edn. Prentice Hall (2000)
49. Yang, Z.: Maximum likelihood phylogenetic estimation from DNA sequences with variable rates over sites: Approximate methods. Journal of Molecular Evolution **39**(3), 306–314 (1994)
50. Yang, Z.: PAML: a program package for phylogenetic analysis by maximum likelihood. Comput Appl Biosci **13**(5), 555–556 (1997)

Evolutionary Phylogenetic Networks: Models and Issues

Luay Nakhleh

Abstract Phylogenetic networks are special graphs that generalize phylogenetic trees to allow for modeling of non-treelike evolutionary histories. The ability to sequence multiple genetic markers from a set of organisms and the conflicting evolutionary signals that these markers provide in many cases, have propelled research and interest in phylogenetic networks to the forefront in computational phylogenetics. Nonetheless, the term 'phylogenetic network' has been generically used to refer to a class of models whose core shared property is tree generalization. Several excellent surveys of the different flavors of phylogenetic networks and methods for their reconstruction have been written recently. However, unlike these surveys, this chapter focuses specifically on one type of phylogenetic networks, namely *evolutionary* phylogenetic networks, which explicitly model reticulate evolutionary events. Further, this chapter focuses less on surveying existing tools, and addresses in more detail issues that are central to the accurate reconstruction of phylogenetic networks.

1 Introduction

In Charles Darwin's *Origin of Species* [17], the depiction of an evolutionary history of species took the shape of a tree. Ever since, trees, in a variety of forms, have been the mainstream of phylogenetics. Such a tree, also referred to as a *phylogeny*, is taken to model the ancestor-descendant evolutionary relationship of a group of species from their most recent common ancestor. Though appropriate for several groups of taxa, a phylogenetic tree may be inadequate for other groups. For example, evidence shows that bacteria may obtain a large proportion of their genetic diversity through the acquisition of sequences from distantly related organisms, via horizontal gene transfer (HGT) [20, 21, 39, 58, 67, 74, 81, 103]. Furthermore, additional evidence of widespread HGT in plants has emerged recently [5, 6, 73].

Luay Nakhleh
Rice University, e-mail: nakhleh@cs.rice.edu

L.S. Heath and N. Ramakrishnan (eds.), *Problem Solving Handbook in Computational Biology and Bioinformatics*, DOI 10.1007/978-0-387-09760-2_7,

Interspecific recombination is believed to be ubiquitous among viruses [82, 83], and hybrid speciation is a major evolutionary mechanism in plants and groups of fish and frogs [23, 60, 65, 66, 80, 85, 86].

These processes are collectively referred to as *reticulate* evolutionary events and occur at different evolutionary scales: the individual, the population, and the species.

1. Reticulation Between Chromosome Pairs: Meiotic Recombination. During each round of sexual reproduction, the total number of chromosomes must be halved to produce the gametes. The process is called *meiosis*, and during one phase of it the chromosome pairs (sister chromatids) exchange pieces in a precise fashion known as *meiotic recombination*. The net result is chromatids that have two or more evolutionary histories on them. Blocks of chromosomes that share a single evolutionary history are referred to as *haplotype blocks*.

2. Reticulation Within a Lineage: Sexual Recombination. For sexually reproducing organisms, there is recombination of nuclear genomes during each bout of reproduction. Each parent contributes half of its original nuclear genome—one sister chromatid from each chromosome—and each of these chromosomes have themselves undergone meiotic recombination during the process of producing the haploid gametes (sex cells). Because different parts of each parent's contribution to the genome of the next generation may have a different evolutionary history from that of the other parent's contribution, sexual recombination is a form of population-level reticulation. Organellar genomes (mitochondria and chloroplasts) are usually inherited uniparentally so they do not usually undergo any sort of sexual recombination.

3. Reticulation Among Lineages: Horizontal Gene Transfer and Hybrid Speciation. In horizontal (also called lateral) gene transfer (HGT for short), genetic material is transferred from one lineage to another. In an evolutionary scenario involving horizontal transfer, certain sites (specified by a specific substring within the DNA sequence of the species into which the horizontally transferred DNA was inserted) are inherited through horizontal transfer from another species, while all others are inherited from the parent.

In hybrid speciation, which is a form of horizontal transfer, two lineages recombine to create a new species. The new species may have the same number of chromosomes as its parent (*diploid hybridization*) or the sum of the numbers of chromosomes of its parents (*polyploid hybridization*). In a diploid hybridization event, the hybrid inherits one of the two homologs for each chromosome from each of its two parents. Since homologs assort at random into the gametes (sex cells), each has an equal probability of ending up in the hybrid. In polyploid hybridization, both homologs from both parents are contributed to the hybrid. Prior to the hybridization event, each site on the homolog has evolved in a tree-like fashion, although due to meiotic recombination (exchanges between the parental homologs during production of the gametes), different strings of sites may have different histories. Thus, each site in the homologs of the parents of the hybrid evolved in a tree-like fashion on one of the trees contained inside the network representing the hybridization event.

Looking through a macroevolutionary lens (evolution among lineages), only reticulate events at the species level fail to be modeled by a tree. However, looking through a microevolutionary lens (evolution within a lineage), sexual and meiotic recombination fail to be modeled by a bifurcating tree. Since phylogenies are usually constructed at either the population or the species level, meiotic recombination does not cause a species-level reticulate evolutionary history, but it can confound species-level inference of reticulation by producing patterns that have the appearance of species-level reticulation (more on this in Section 4).

In effect, when reticulation occurs, two or more independent evolutionary lineages are combined at some level of biological organization, thus resulting in complex evolutionary relationships that cannot be adequately modeled with trees; instead, *phylogenetic networks* become the appropriate model. Phylogenetic networks are a special class of graphs that allows for multiple paths between pairs of taxa in the phylogeny, and as such provide an extension of phylogenetic trees, in which a unique path exists between any two taxa. Phylogenetic networks come in various flavors, and a variety of methods for reconstructing them have been designed recently. There have been several recent detailed surveys of phylogenetic reconstruction methods [32, 47, 48, 61, 64, 72], some of which identify their similarities and differences. Further, Gambette has created an excellent online resource for documenting all work related to phylogenetic networks [31].

In this chapter, we focus on a specific type of phylogenetic networks, namely *evolutionary phylogenetic networks*, which explicitly model reticulate evolutionary events. Rather than surveying tools and implementations, in this chapter we address issues that are central to accurate detection of reticulate evolution and reconstruction of phylogenetic networks. The rest of this chapter is organized as follows. In Section 2, we define evolutionary phylogenetic networks, discuss their relationships with trees, and outline the general approach for their reconstruction from gene trees. In Section 3, we discuss extensions of three popular optimization criteria, maximum parsimony (MP), maximum compatibility, and maximum likelihood (ML), to the domain of phylogenetic networks. In Section 4, we address various processes that result in patterns that resemble those resulting from reticulate evolutionary events and the need for a framework to distinguish among those processes as a prerequisite to accurate reconstruction of phylogenetic networks. In Section 5, we provide a set of exercises for the reader to gain more understanding of the issues surrounding phylogenetic networks. We conclude in Section 6 with a list of further reading materials that provide in-depth details about other aspects of phylogenetic networks.

2 Phylogenetic Networks and the Trees Within

In this work, we focus on *evolutionary* phylogenetic networks, i.e., networks that model reticulate evolutionary events explicitly. An important assumption underlying all results in this section as well as Section 3 is that the sole cause of gene tree incongruence is reticulate evolution and that a phylogenetic network reconciles gene trees by explicitly modeling reticulate evolutionary events while ignoring discord processes such as lineage sorting. We discuss the implications of incorporating lineage sorting into the framework in Section 4. While much of the literature is on unrooted, undirected networks (and trees), we focus exclusively in this chapter on rooted networks (and trees).

Definition 0.1. A *phylogenetic $\mathscr{X}$-network*, or $\mathscr{X}$-network for short, N is an ordered pair (G, f), where

- $G = (V, E)$ is a directed, acyclic graph (DAG) with $V = \{r\} \cup V_L \cup V_T \cup V_N$, where
  - $indeg(r) = 0$ (r is the *root* of N);
  - $\forall v \in V_L$, $indeg(v) = 1$ and $outdeg(v) = 0$ (V_L are the *leaves* of N);
  - $\forall v \in V_T$, $indeg(v) = 1$ and $outdeg(v) \geq 2$ (V_T are the *tree-nodes* of N); and,
  - $\forall v \in V_N$, $indeg(v) = 2$ and $outdeg(v) = 1$ (V_N are the *network-nodes* of N),

  and $E \subseteq V \times V$ are the network's edges (we distinguish between *network-edges*, edges whose heads are network-nodes, and *tree-edges*, edges whose heads are tree-nodes.
- $f : V_L \to \mathscr{X}$ is the *leaf-labeling* function, which is a bijection from V_L to $\mathscr{X}$.

Figure 1(a) shows an example of a phylogenetic $\mathscr{X}$-network. Clearly, Definition 0.1 generalizes that of a phylogenetic $\mathscr{X}$-tree; an $\mathscr{X}$-tree is a phylogenetic network with $V_N = \emptyset$.

The semantics of network-nodes are context dependent. For example, in phylogenetics, a network-node may represent a hybrid speciation event, whereas in evolutionary population genetics it may represent a recombination event. While Definition 0.1 requires a network-node to have two parents and a single child, this definition may be relaxed so as to allow for three or more (graph-theoretic) parents and two or more (graph-theoretic) children, which correspond to in- and out-polytomies, respectively; e.g., see the discussions in [70, 76].

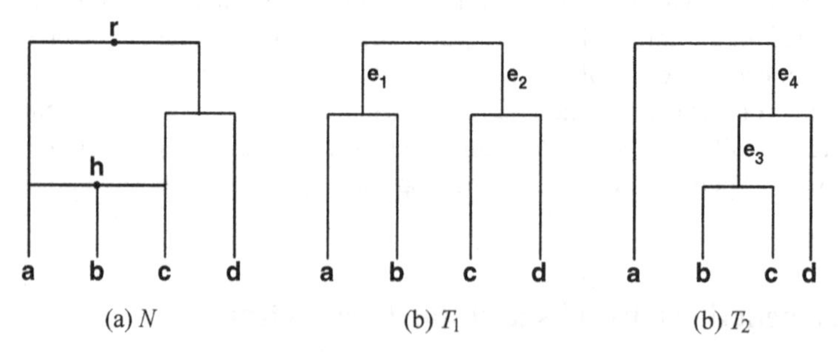

Fig. 1 (a) A phylogenetic $\mathscr{X}$-network, rooted at node r, with a single network-node, h, and with $\mathscr{X} = \{a, b, c, d\}$. The trees T_1 (b) and T_2 (c) are the elements of $\mathscr{T}(N)$.

A fundamental observation that underlies reticulate evolution is that the network modeling the evolutionary history of a set of organisms *contains*[1] a set of trees, each of which models the evolutionary histories of certain genomic regions of those organisms. At the lowest level of "atomicity," each nucleotide in the genomes of these organisms has evolved down exactly one of those trees[2]. *The descent of a single nucleotide in a set of organisms could not have followed two different evolutionary histories from the most recent common ancestor of those organisms.* Formally, we can define the set of $\mathscr{X}$-trees contained inside a phylogenetic $\mathscr{X}$-network. Procedure **Induce** in Figure 2 yields one of the trees contained inside a phylogenetic $\mathscr{X}$-network. All trees that can be obtained by applying this procedure to a given phylogenetic $\mathscr{X}$-network N is denoted by $\mathscr{T}(N)$. For the $\mathscr{X}$-network N in Figure 1(a), the set $\mathscr{T}(N) = \{T_1, T_2\}$, where T_1 and T_2 are the two trees shown in Figure 1(b) and 1(c), respectively. Notice that $|\mathscr{T}(N)| = O(b^\ell)$, where b is the maximum indegree of a node in N, and ℓ is the number of network-nodes in N. A tighter bound can be obtained as

Induce(N)
Input: Phylogenetic $\mathscr{X}$-network $N = (G, f)$, where $G = (V, E)$.
Output: Phylogenetic $\mathscr{X}$-tree $T = (G', f)$.

1. For each node $v \in V_N$, remove all but one of the edges incoming into v; let $T = (G', f)$, where $G' = (V', E')$, be the resulting tree.
2. While $\exists u \in V'$ such that $indeg(u) = outdeg(u) = 1$
   a. Let u be such a node with $\{(p,u),(u,c)\} \subseteq E'$;
   b. $V' = V' - \{u\}$; (* *remove a node of indegree and outdegree 1* *)
   c. $E' = E' - \{(p,u),(u,c)\}$; (* *remove its incident edges* *)
   d. $E' = E' \cup \{(p,c)\}$; (* *connect its parent to its child* *)
3. Return T;

Fig. 2 Procedure **Induce** for computing a tree in $\mathscr{T}(N)$ for a given phylogenetic $\mathscr{X}$-network N. Observe the random choice of an incoming edge to keep in Step 1. This procedure can be iterated in a deterministic fashion to produce all trees in $\mathscr{T}(N)$ or run non-deterministically a certain number of times to sample from the trees in $\mathscr{T}(N)$.

$$|\mathscr{T}(N)| \leq \prod_{u \in V_N} (indeg(u)). \tag{1}$$

Given an $\mathscr{X}$-network N and an $\mathscr{X}$-tree T, the problem of deciding whether $T \in \mathscr{T}(N)$ is NP-complete [53].

[1] In this context, the term *contain* has been used in the literature interchangeably with two other terms: *induce* and *display*.

[2] Some argue that a forest, rather than a tree, may be a more appropriate model at this atomic level, to allow for events such as insertions and deletions.

Notice that both the **Induce** procedure and the result on the cardinality of $\mathscr{T}(N)$ do not apply when events such as lineage sorting occur; we discuss this in more detail in Section 4.

While computing the set $\mathscr{T}(N)$ for a given $\mathscr{X}$-network is straightforward, computing an $\mathscr{X}$-network N from a set $\mathscr{T}$ of trees is not as straightforward. In fact, this problem is the holy grail of reticulate evolution. First, observe that for a given set $\mathscr{T}$ of $\mathscr{X}$-trees, there may not exist an $\mathscr{X}$-network N such that $\mathscr{T} = \mathscr{T}(N)$ (see Exercise 1); in this case, it is desirable to find an $\mathscr{X}$-network N such that $\mathscr{T} \subseteq \mathscr{T}(N)$. A trivial way to obtain such a network $N = (G, f)$, where $G = (V, E)$, is as follows:

1. $V = \{v_x : x(\neq \emptyset) \subseteq \mathscr{X}\}$. In other words, create one node for each non-empty subset of taxa.
2. $E = \{(v_x, v_y) : v_x, v_y \in V, \ y \subset x\}$.

Clearly, N is an $\mathscr{X}$-network[3] and $\mathscr{T} \subseteq \mathscr{T}(N)$. Baroni *et al.* proposed another "direct" method for constructing a phylogenetic network from a collection of trees [3]. However, while the networks obtained by the method of Baroni *et al.* are smaller in size than those obtained by the method described here, both methods result in a gross overestimation of the extent of reticulation in the evolutionary history.

These observations have been the basis for much work on phylogenetic networks, particularly those with explicit evolutionary implications. In the case of reconstructing *ancestral recombination graphs* (ARGs), the problem has been investigated from the perspective of reconciling the "evolutionary trees" that model the evolution of *single nucleotide polymorphisms*, or SNPs. For reconstructing reticulate evolutionary histories of species, single nucleotides clearly do not provide enough information, and the atomic unit used in this context is a gene. Hereafter, we refer to these units, such as SNPs, genes, haplotype blocks, etc., as *markers*, which are, in essence, the observed biological data from which the phylogenetic network is inferred.

Definition 0.2. The Phylogenetic Network Reconstruction (PNR) Problem

> **Input:** A set of *markers*, $\mathscr{M} = \{M_1, M_2, \ldots, M_k\}$, from a set $\mathscr{X}$ of organisms and a criterion Φ.
> **Output:** A phylogenetic $\mathscr{X}$-network N that models the evolution of $\mathscr{M}$ and that is optimal under criterion Φ.

For example, one version of the problem of inferring ancestral recombination graphs (ARGs) can be formulated as an instance of PNR if one takes $\mathscr{M}$ to be the set of SNPs, and Φ to be the criterion "N contains the minimum number of network-nodes and every SNP is compatible with at least one tree in $\mathscr{T}(N)$." As another example, one version of the problem of inferring species evolutionary networks can be formulated as an instance of PNR if, given a set $W = \{T_1, \ldots, T_k\}$ of trees with T_i being the *gene tree* of gene M_i, the criterion Φ is taken to be "N contains the minimum number of network-nodes and $T_i \in \mathscr{T}(N)$ for every $T_i \in W$."

[3] This construction does not ensure that the leaves have indegree of 1, which is one of the requirements in Definition 0.1, but the construction can be extended in a straightforward manner to take care of this.

2.1 Combining Trees Into a Network via SPR Operations

One of the most commonly pursued approaches for reconstructing phylogenetic networks is based on reconciling "gene trees," under the assumption that incongruities, or disagreements, among these trees are caused by only reticulate evolutionary events, such as horizontal gene transfer or hybrid speciations. In this case, several methods have been developed for inferring a lower bound on the number of reticulation events by identifying the minimum number of *subtree prune and regraft*, or SPR, operations required to transform one tree into the other. As the name indicates, an SPR operation applied to tree T cuts, or prunes, a subtree t of T, yielding a tree T', and attaches, or regrafts, it from its root to another branch in T' [1]; see Figure 3 for an illustration. The SPR distance between two trees is the minimum

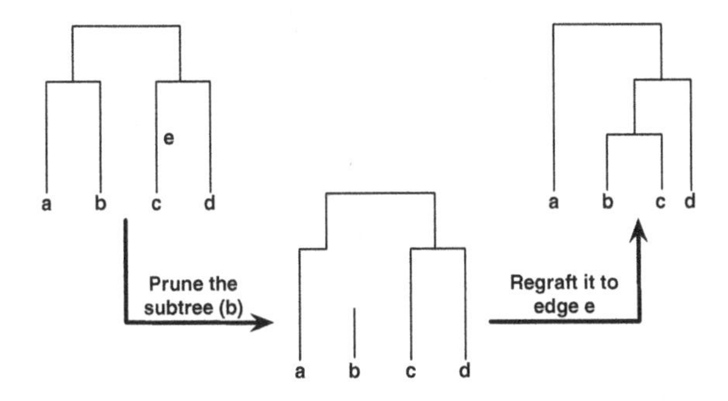

Fig. 3 An illustration of the *subtree prune and regraft*, or SPR, operation. The subtree that contains only the leaf b is pruned from the tree on the left, thus resulting in a forest of two trees, shown in the middle, and then the subtree is regrafted as a sibling of c, resulting in the tree on the right. Observe that the SPR distance between the two trees on the left and right is 1, and both trees can be reconciled in a phylogenetic network with one network-node, as can be seen in Figure 1.

number of SPR moves required to transform one tree into the other. For example, the SPR distance between the two trees in Figure 1 is 1, since a single SPR move is required, as illustrated in Figure 3.

The problem of computing the SPR distance between two rooted trees has been shown to be NP-hard as well as fixed-parameter tractable [7]. Examples of exact algorithms and heuristics for reconciling trees via SPR operations include the exact algorithm of Bordewich and Semple [7], the exact algorithm of Wu [105], HorizTrans [38], RIATA-HGT [78], EEEP [4], HorizStory [62], and the method of Goloboff as implemented in the TNT software package [33]. For the most part, these methods are aimed at finding the phylogenetic network N with the minimum number of network-nodes that contains the pair of input trees. For example, the network N in Figure 1(a) is the only phylogenetic network with a single network-node that contains both trees T_1 and T_2 in Figure 1.

There are several limitations with using the SPR distance as a proxy for the amount of reticulation, as well as with the methods listed above for estimating this distance. We discuss some of those here, and discuss the issue of *time-consistency* of SPR moves in the next section.

It is worth mentioning that methods that attempt to find minimal sets of SPR moves to reconcile a pair of trees are in fact attempts at approximating the true number of reticulation events in the evolutionary history. However, while the SPR distance provides a lower bound on this number, recent results have shown that the SPR distance can provide a value that is arbitrarily smaller than the true amount of reticulation [2, 46].

The tools listed above all assume $k = 2$ (i.e., they solve the problem for a pair of trees) and assume that each of the two trees has exactly $|\mathscr{X}|$ leaves, each labeled uniquely by one label from $\mathscr{X}$. In other words, these tools do not solve the problem, in terms of computing a minimal network, for more than two trees, nor do they allow for trees with different leaf-sets. Both of these present practical limitations to the use of the methods in practice, particularly the latter, since, in general, there is no guarantee that a 1-1 correspondence exists between the leaves of the (species) phylogenetic network and those of the gene trees.

A very important issue that tools for combining trees into a network must account for is the potential multiplicity of different, optimal (minimal, in this case) networks. Than *et al.* [100] showed that the number of minimal networks that reconcile a given pair of trees may be exponential in the minimum number of reticulation events required.

Last but not least, reconstructed gene trees are often non-binary (which mostly indicates soft polytomies[4]). The reconciliation problem becomes more complicated when non-binary trees are concerned. In this case, one objective is to *simultaneously* resolve the trees and infer the minimum number of reticulation events. The number of resolutions of non-binary tree is exponential in the degree of the nodes, and hence efficient techniques are required for solving this problem. Than and Nakhleh [99] provided a heuristic for solving several cases of this problem, which are implemented in the PhyloNet package [101] as an extension of the RIATA-HGT method [78].

2.2 Totally-ordered Trees and Time-consistent SPR Operations

In our discussion thus far of the SPR operation and its induced distance, we have considered only the topologies of a pair of trees. However, when times at the internal nodes of the species and gene trees are known (in the former case, those times would indicate the divergence time of the species from their common ancestors, and in the latter case those times would indicate the times of the coalescence events),

[4] In a rooted phylogenetic tree, a polytomy is a node with more than two children. There are two types of polytomies: a *hard polytomy* indicates the hypothesis that the speciation event gave rise to multiple lineages, whereas a *soft polytomy* indicates the lack of knowledge to resolve a multifurcating node into a sequence of bifurcating nodes.

the situation becomes more complicated. Rooted trees in which internal nodes are totally ordered are called *ordered tree* [90].

When ordered trees are considered, two crucial issues arise:

1. Topologically identical or similar trees may be very different when branch lengths are considered (S. Edwards recently labeled such phenomenon "branch length heterogeneity" [22], though in the different context of lineage sorting), and
2. certain SPR moves may not be *time consistent*.

We elaborate on these two issues in a few examples. Consider the two trees in Figure 4. Topologically, the two trees are identical. However, considering the trees on the left and right to be the species and gene trees, respectively, the species a and b diverged at time T_2 (similarly for species c and d), while their genes coalesced at time T_1, which is different from T_2. This is a scenario of branch length heterogeneity, and the trees, when viewed as ordered trees, are different.

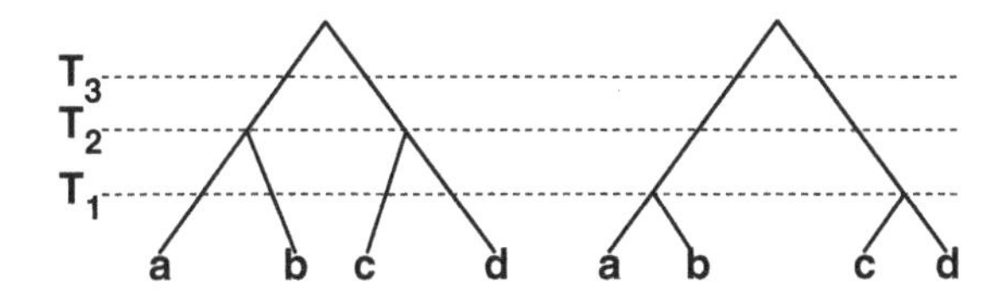

Fig. 4 Two phylogenetic trees that require no SPR moves to transform into each other, when only the topologies are considered. However, when times at internal nodes are considered, the two trees are different, and require a minimum number of two SPR moves, as shown in Figure 5. The horizontal dashed lines represent times.

In this case, the true SPR distance is not zero, but rather two, as illustrated in Figure 5.

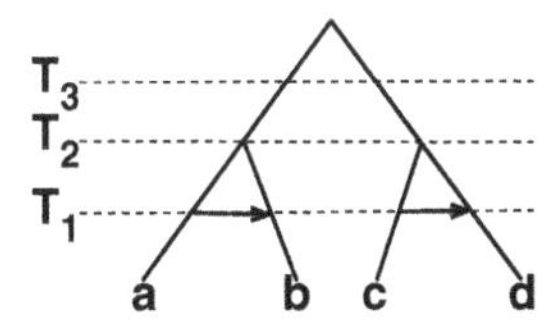

Fig. 5 Two SPR moves required to transform the tree on the left in Figure 4 into the one on the right, when times at internal nodes are taken into account. The horizontal dashed lines represent times.

For the second issue, consider the species and gene trees shown in Figure 6 (left and right, respectively). When their topologies are compared, a single SPR move suffices to transform the species tree into the gene tree, as shown in Figure 7. However, notice that in this scenario, the transfer of the genetic material took place between two organisms that do not co-exist in time. In other words, this SPR move is not time consistent.

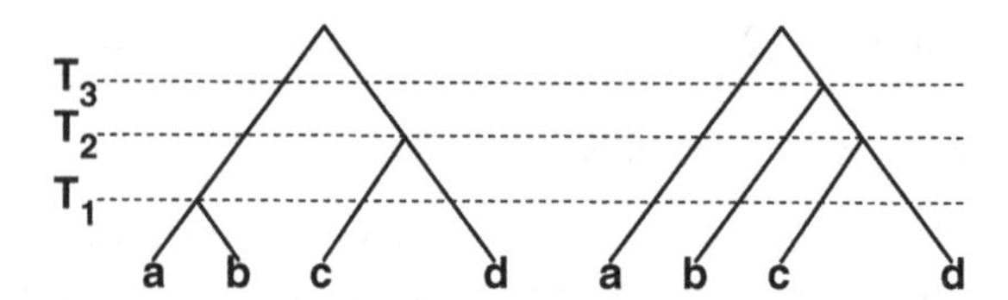

Fig. 6 Two trees that differ in the placement of b, thus requiring a single SPR move to transform the tree on the left into the one on the right, as shown in Figure 7(a), when only the topologies of the trees are considered. However, such a move is not time-consistent since the "donor" (tail of the HGT edge) and "recipient" (head of the HGT edge) do not co-exist in time. The horizontal dashed lines represent times.

An important question in this case is whether such an SPR move should be ruled out in a species/gene tree reconciliation scenario. While the scenario, as drawn in Figure 7(a), contains a time inconsistent SPR move, this inconsistency may be explained as an artifact of *incomplete taxon sampling*, as we now illustrate.

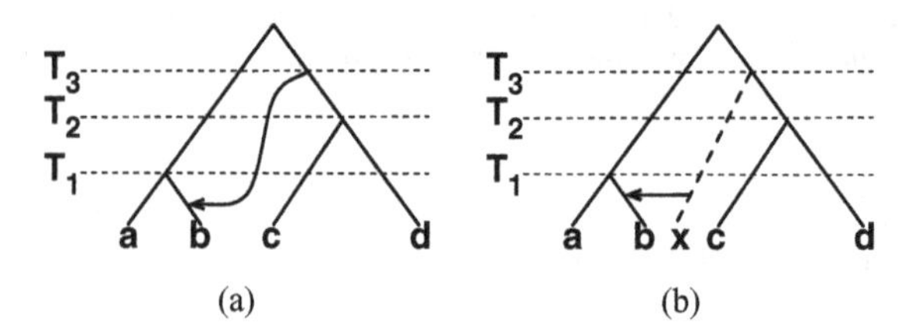

Fig. 7 (a) The SPR move required for the scenario in Figure 6. This SPR move is not time-consistent. (b) The SPR move can be viewed as time-consistent if incomplete taxon sampling occurs; in this case, the horizontal transfer occurred from a taxon x, which is contemporaneous with b yet was not sampled when the the species and gene trees were reconstructed. The horizontal dashed lines represent times.

Consider the case in which the horizontal transfer occurred from species x to species b, where x was a sibling of the clade (c,d) yet was not sampled (or became extinct after the horizontal transfer event) in the evolutionary analysis. This case is shown in Figure 7(b). In this scenario, while the SPR moves necessary to transform the species tree into the gene tree is seemingly time inconsistent, it is in fact a reflection of incomplete taxon sampling, or even a true biological hypothesis—that of the extinction of species x. Determining whether a time inconsistent SPR move is truly so or is merely a reflection of incomplete taxon sampling (or extinction) is a very challenging question.

It is important to note, though, that not all time inconsistent SPR moves can be justified with the incomplete taxon sampling scenario. Consider the species and gene trees in Figure 8 (left and right, respectively). In this case, a single SPR move, pruning the clade (b,c) and regrafting it as a sibling of d, would reconcile the two trees, as shown in Figure 9(a). Clearly, this SPR move is time inconsistent. Unlike the previous case, incomplete taxon sampling cannot explain the inconsistency in this scenario, since no matter how we augment the species tree with "phantoms" of missing taxa, the source and destination of the SPR move cannot be made contemporaneous. Instead, a scenario involving two time consistent SPR moves may be the correct one, as illustrated in Figure 9(b).

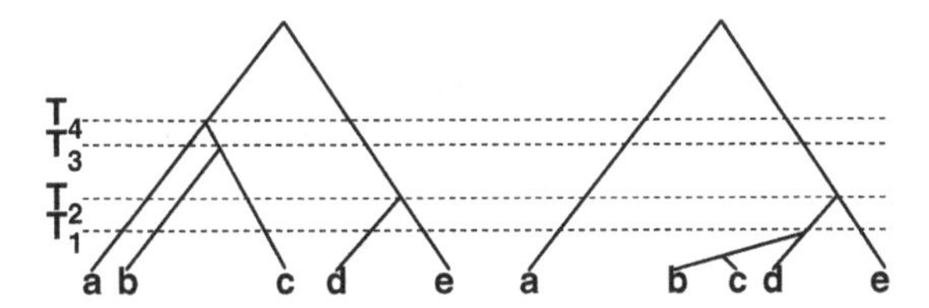

Fig. 8 Two trees that differ in the placement of clade (b,c), thus requiring a single SPR move to transform the tree on the left into the one on the right, as shown in Figure 9(a), when only the topologies of the trees are considered. However, such a move is not time-consistent since the donor and recipient do not co-exist in time. The horizontal dashed lines represent times.

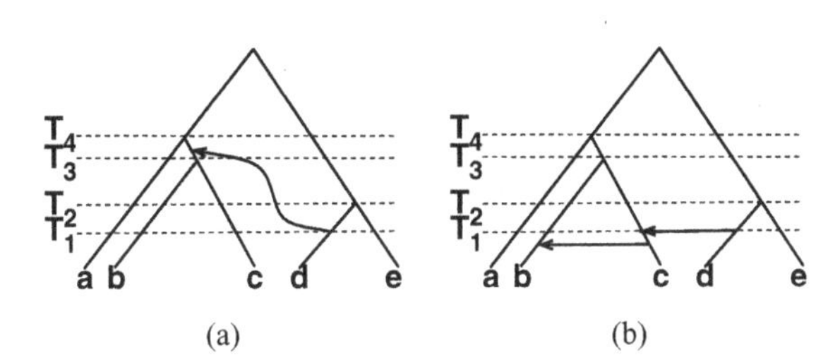

Fig. 9 (a) The SPR move required for the scenario in Figure 8. This SPR move is not time-consistent. (b) A solution of two time-consistent SPR moves that explains the scenario in Figure 8. The horizontal dashed lines represent times.

3 Optimization Criteria for Inferring and Evaluating Phylogenetic Networks

The relationship between a phylogenetic network N and its constituent trees, $\mathcal{T}(N)$, allows for extending sequence-based optimization criteria from phylogenetic trees to phylogenetic networks. Such extensions are based on the fact that, at the lowest level of atomicity in genetic inheritance, a nucleotide in the genomes of a set of species evolves down a single tree, even if the evolutionary history of the species is best modeled by a network[5]. This, in essence, is the concept of *positional homology*. In this section, we discuss extensions to the maximum parsimony, maximum compatibility, and maximum likelihood criteria.

Let T be an $\mathcal{X}$-tree with leaf-set $\mathcal{L}$, and let Σ be an alphabet (e.g., $\Sigma = \{A,C,T,G\}$ for DNA). A function $\lambda : \mathcal{L} \rightarrow \Sigma$ is called a *state assignment function* for tree T over alphabet Σ. The function $\hat{\lambda} : V(T) \rightarrow \Sigma$ is an extension of λ on T if it agrees with λ on the leaves of T (i.e., if $\hat{\lambda}(v) = \lambda(v)$ for every $v \in \mathcal{L}$). In a similar way, we define a function $\lambda^k : \mathcal{L} \rightarrow \Sigma^k$ and an extension $\hat{\lambda}^k : V(T) \rightarrow \Sigma^k$. The latter function is called a *labeling* of T, and it denotes the labeling of all nodes of a tree T with sequences of length k over alphabet Σ. Given a labeling $\hat{\lambda}^k$, we denote by $d_e(\hat{\lambda}^k)$ the Hamming distance (or any edit distance) between the two sequences labeling the two endpoints of edge $e \in E(T)$. We define the state assignment and labeling functions for an $\mathcal{X}$-network similarly. The difference between the labeling of a tree and that of a network lies in the interpretation of sequence evolution. Let (u,v) be an edge in a phylogenetic tree with $x = \hat{\lambda}^k(u)$ and $y = \hat{\lambda}^k(v)$. Then, the state at position i in sequence y is the result of zero or more mutations on the state at position i in sequence x. In a phylogenetic network, this interpretation is slightly more involved. Assume edge (u,v) in a phylogenetic network, with x and y defined as before. If $indeg(v) = 1$, then the relationship between the states at position i in sequences x and y is identical to that in trees. However, if $indeg(v) = m$, where $m > 1$, then the state at position i in sequence y is the result of zero or more mutations on the state at position i in exactly one of the sequences labeling the m parents of v. This labeling and interpretation serve as the basis for extending sequence-based optimization criteria from trees to networks.

3.1 *Maximum Parsimony of Phylogenetic Networks*

Roughly speaking, the maximum parsimony criterion is a reflection of Occam's razor; that is, the best solution is the simplest. In the context of phylogenetics, the maximum parsimony criterion seeks the tree on a given set of genomic sequences such that the tree minimizes the overall number of mutations along all edges of the tree. This is formalized as follows.

[5] The same comment in Footnote 2 applies here.

Definition 0.3. The parsimony length of a phylogenetic tree T with a labeling λ^k is $PS(T,\lambda^k) = \min_{\hat{\lambda}^k \in \hat{\Lambda}^k}[\sum_{e\in E(T)} d_e(\hat{\lambda}^k)]$, where $\hat{\Lambda}^k$ is the set of all possible extensions of λ^k.

We denote by $PS_i(T,\lambda^k)$ the parsimony length of tree T with respect to site i. Given a labeling λ^k of a set $\mathscr{X}$ of taxa, the maximum parsimony (MP) problem for phylogenetic trees amounts to solving

$$T^* = \text{argmin}_T PS(T,\lambda^k), \tag{2}$$

where T ranges over all $\mathscr{X}$-trees. There is a polynomial time algorithm for computing the parsimony length of a fixed $\mathscr{X}$-tree [29], while solving the MP problem in general is NP-hard [18, 30].

In the early 1990's, Jotun Hein introduced an extension of the maximum parsimony (MP) criterion to model the evolutionary history of a set of sequences in the presence of recombination [40, 41]. Recently, Nakhleh and colleagues gave a mathematical formulation of the MP criterion for phylogenetic networks and devised computationally efficient solutions aimed at reconstructing and evaluating the quality of phylogenetic networks under the MP criterion [49, 51, 52]. The parsimony length of a phylogenetic network with respect to a set of sequences is defined as follows.

Definition 0.4. The parsimony length of a phylogenetic network N with a labeling λ^k of the leaves of N is

$$PS(N,\lambda^k) = \sum_{1\leq i\leq k}\left[\min_{T\in\mathscr{T}(N)} PS_i(T,\lambda^k)\right].$$

Notice that this definition of the parsimony length allows for the rather biologically unrealistic scenario of switching back and forth between different trees for consecutive sites. For example, for $k = 10$, the definition may lead to the scenario in which sites 1, 3, 5, 7, and 9 are best fit by tree T' and sites 2, 4, 6, 8, and 10 are best fit by tree T'', for two different trees T' and T''. This was addressed in practice in the sequence of papers by Jin *et al.* by doing the computation on a block-by-block, rather than site-by-site, basis. Another way to address this issue is to introduce a penalty for switching among trees. As the parsimony criterion is based on the assumption of rare events (e.g., [26, 27]), a reticulation event may be best modeled as causing a penalty of one change (J. Felsenstein, personal communication).

Given a labeling λ^k of a set $\mathscr{X}$ of taxa, the maximum parsimony (MP) problem for phylogenetic networks amounts to solving

$$N^* = argmin_N PS(N,\lambda^k), \tag{3}$$

where N ranges over all $\mathscr{X}$-networks. Unlike the case of trees, the problem of computing the parsimony length of a fixed $\mathscr{X}$-network is NP-hard [49], and the problem of solving the MP problem for phylogenetic networks is NP-hard as well, as it contains the MP problem for trees as a special case.

Let N be an $\mathscr{X}$-network, and let N' be another $\mathscr{X}$-network obtained by adding a set H of edges to N, where each edge in H is posited between a pair of edges whose heads are tree-nodes in N. Then, we have

$$\mathscr{T}(N) \subseteq \mathscr{T}(N').$$

This result is illustrated in Figure 10.

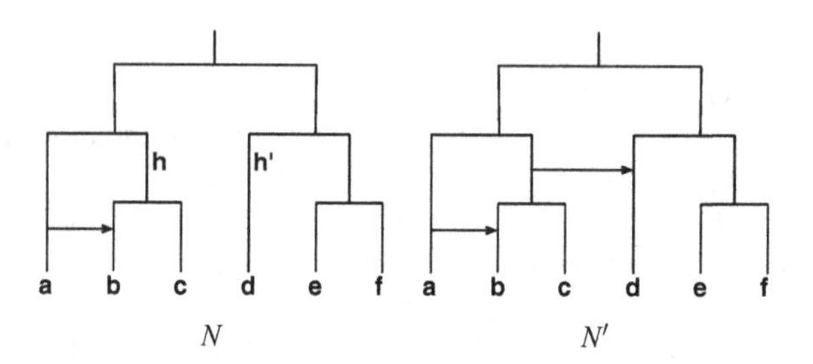

Fig. 10 Two $\mathscr{X}$-networks N and N' such that N' is obtained by adding an additional edge to N from edge h to edge h'. We have $\mathscr{T}(N) = \{T_1, T_2\}$ and $\mathscr{T}(N') = \{T_1, T_2, T_3, T_4\}$, where $T_1 = ((a,(b,c)),(d,(e,f)))$, $T_2 = (((a,b),c),(d,(e,f)))$, $T_3 = ((a,((b,c),d)),(e,f))$, and $T_4 = (((a,b),(c,d)),(e,f))$. Clearly, $\mathscr{T}(N) \subseteq \mathscr{T}(N')$.

From this fact it follows that, for a given labeling λ^k of a set $\mathscr{X}$ of taxa, we have

$$PS(N', \lambda^k) \leq PS(N, \lambda^k).$$

This simple observation has a significant implication on the use of the MP criterion for inferring networks, as defined above. It basically implies that adding more edges to a network "never hurts" under the MP criterion as defined above: the parsimony length either decreases or stays the same as more edges are added. This in turn implies that while making networks more "complex" improves their parsimony lengths, using the MP criterion in this fashion would inevitably result in a gross over-estimation of the amount of reticulation in the evolutionary history of a data set. This had led to refining the definition of the MP criterion so that adding edges to a network is accepted only if the parsimony length is improved beyond a given threshold [51]. Currently, such a threshold is dataset-specific and is determined by inspection of the trend of parsimony length decrease as the complexity of networks is increased. Such an approach has produced very promising results, on both synthetic and biological data sets [51, 98].

3.2 Character Compatibility of Phylogenetic Networks

Two models of sequence evolution that have been central in population genetics, and which have been assumed to underlie a special type of phylogenetic networks are the *infinite-allele model* and *infinite-site model*. The infinite-allele model, proposed by Kimura and Crow [56], assumes that each mutation at a site results in a state that is different from any preexisting state at that site in the population. The infinite-site model, proposed by Kimura [55], assumes that the sequences are very long and that the mutation rate per site is low so that each site mutates at most once. These two models can be formulated within the parsimony framework. If a site i evolves down a tree T under the infinite-allele model, and m distinct states are observed at site i in the leaves of T, then the parsimony length of T with respect to site i is $m-1$. If site i evolves under the infinite-site model, then the parsimony length of T with respect to site i is either 0 (no mutations occurred at site i) or 1 (exactly one mutation occurred). In the phylogenetics jargon, a site that evolves down a tree T under either infinite-allele or infinite site model is said to be *compatible* with the tree T. A tree T for which all sites in the sequences labeling its leaves are compatible is called a *perfect phylogeny*. Gusfield provided an $O(nm)$ algorithm for determining whether there exists a perfect phylogeny for a set of n binary sequences, each of length m, and reconstructing such a perfect phylogeny if it exists [35], thus improving on an earlier $O(nm^2)$ algorithm [24, 68].

Barring any (meiotic) *recombination* events, the evolutionary history of a sequence of sites under the infinite-site model is modeled by a tree. However, when recombination occurs, the evolutionary histories of sites to the left and right of a recombination breakpoint follow different paths in their ancestries, thus giving rise to a phylogenetic network model. The compatibility criterion can be extended to phylogenetic networks in a fashion similar to that of extending the MP criterion. We say that a site is compatible with a phylogenetic network N if it is compatible with at least one of the trees in $\mathscr{T}(N)$. Determining if a site is compatible with a phylogenetic network is NP-Complete [53]. An *ancestral recombination graph* [34], or ARG for short, is a phylogenetic network that models the evolution of a set of sequences under the infinite-site model, in which:

- each edge is labeled by a set of numbers denoting the sites that mutate along that edge,
- each node of indegree 2 is labeled by a number denoting the recombination breakpoint giving rise to that network-node, and
- each site in the sequences is compatible with the network.

Figure 11 shows an ARG modeling the evolutionary history of a set of four sequences under the infinite-site model. ARGs have also been referred to as *perfect phylogenetic networks* [102]. Much work has been done on reconstructing *minimal* ARGs, i.e., ARGs with the minimum number of nodes of indegree 2 to model the evolution of a set of binary sequences under the infinite-site model; e.g., see [36, 37, 91, 92, 93, 94]. Recently, Willson provided a new method for

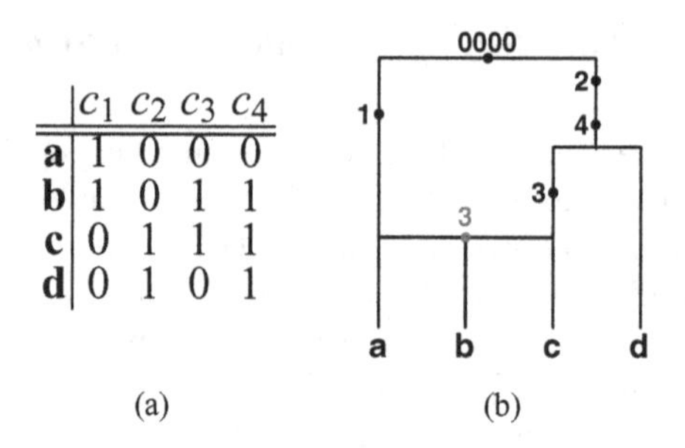

	c_1	c_2	c_3	c_4
a	1	0	0	0
b	1	0	1	1
c	0	1	1	1
d	0	1	0	1

Fig. 11 (a) A data set of four binary sequences **a**, **b**, **c**, and **d**. (b) An ARG showing the evolutionary history of the four sequences from the ancestral sequence 0000 under the infinite-site model. The solid circle at the node of indegree 2 indicates a recombination event, and the value 3 indicates that the states of sites 1 and 2 (which are 1 and 0, respectively) were inherited from the left parent, whereas the states of sites 3 and 4 (which are 1 and 1, respectively) were inherited from the right parent, thus forming the sequence 1011 at **b**. The other solid circles indicate mutations, where the numbers associated with them indicate the site at which each mutation occurred.

reconstructing certain phylogenetic networks from binary sequences when *back-mutations* are allowed to occur at network-nodes [104].

While we focused on binary characters in the preceding discussion, perfect phylogenetic networks can be defined on multi-state characters as well. Let $\lambda : \mathscr{L} \to \Sigma$ be a leaf-labeling of a tree T, with $\Sigma' \subseteq \Sigma$ being the character states that are observed at the leaves of T (not all character states in Σ may be observed at the leaves, and hence the need for Σ'). We say that λ is compatible on T if there exists an extension $\hat{\lambda}$ such that

$$\sum_{e \in E(T)} d_e(\hat{\lambda}) = |\Sigma'| - 1.$$

We say that λ is compatible with a phylogenetic network N if it is compatible with at least one of the trees in $\mathscr{T}(N)$. Character compatibility on a tree and on a network can be extended in a straightforward manner to sequences of characters (λ^k). Figure 12(a) shows a tree whose leaves are labeled by sequences of length 2 over the alphabet $\Sigma = \{1,2,3,4\}$. For the first character (site), we have $\Sigma_1' = \{1,2,3\}$ and for the second we have $\Sigma_2' = \{2,3,4\}$. The first character is compatible with the tree, whereas the second is not. When a single reticulation event is added to the tree, as shown in Figure 12(b), we obtain a perfect phylogenetic network for the sequences labeling the leaves; see Exercise 5.

Nakhleh *et al.* proposed multi-state perfect phylogenetic networks[6] to model the evolutionary histories of natural languages in the presence of borrowing [77]. The *Character Compatibility on Phylogenetic Networks* Problem is to decide whether a given phylogenetic network is a perfect phylogenetic network for a set C of characters (alternatively, a leaf-labeling λ^k). This problem has been shown to be NP-

[6] In [77], network-edges were allowed to be bi-directional.

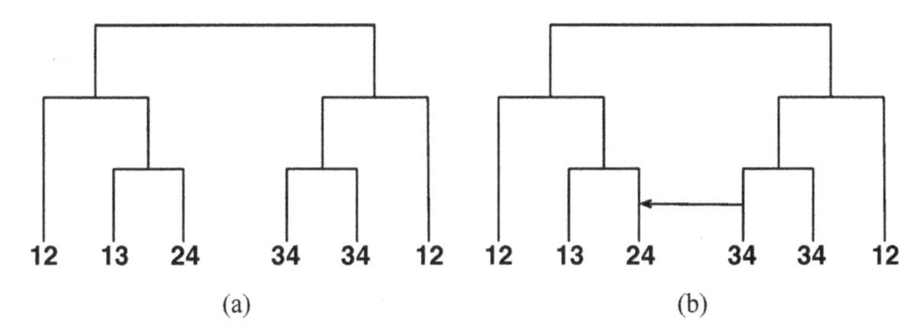

Fig. 12 (a) A phylogenetic tree leaf-labeled by sequences of length 2 over the alphabet $\Sigma = \{1,2,3,4\}$. The first character is compatible on the tree, whereas the second is not. (b) A perfect phylogenetic network obtained from the tree by adding a single reticulation event.

hard [75] even for binary characters. Kanj *et al.* provided an efficient parameterized algorithm for the binary case of this problem [54].

3.2.1 Binary Character Compatibility and Combining Trees into a Network

There is an elegant connection between the problem of combining a set of trees into a network and the problem of inferring a perfect phylogenetic network (with only uni-directional edges) for a set of binary sequences. Let $\mathscr{T} = \{T_1, T_2, \ldots, T_m\}$ be a set of (rooted) $\mathscr{X}$-trees. For each edge e in a tree $T_i \in \mathscr{T}$, define a binary site c_e with its states assigned as follows for each $x \in \mathscr{X}$:

$$c_e(x) = \begin{cases} 1, & x \text{ under } e; \\ 0, & \text{otherwise.} \end{cases} \tag{4}$$

The collection $C = \cup_{T_i \in \mathscr{T}} \{c_e : e \in E(T_i)\}$ is called the *character encoding* of the trees in $\mathscr{T}$. If the trees in $\mathscr{T}$ contain p distinct edges (two edges are distinct if they define different clusters of leaves), then C contains p distinct sites, and each taxon $x \in \mathscr{X}$ is associated with a binary sequence s_x of length p. The main result here is that if N is a network such that $\mathscr{T} \subseteq \mathscr{T}(N)$ then N is a perfect phylogenetic network for the set C, which is the character encoding of $\mathscr{T}$.

	e_1	e_2	e_3	e_4
a	1	0	0	0
b	1	0	1	1
c	0	1	1	1
d	0	1	0	1

Fig. 13 The character encoding of the two trees in Figure 1(b) and 1(c). The resulting matrix is identical to that of the binary sequences in Figure 11.

Figure 13 shows the character encoding of the two trees in Figure 1. Indeed, the sequences in Figure 13 are compatible with the phylogenetic network in Figure 11, which is identical (in terms of topology) to the phylogenetic network N in Figure 1(a) that contains the two trees.

3.3 Maximum Likelihood of Phylogenetic Networks

Extending the maximum likelihood (ML) criterion to phylogenetic networks is done in a similar fashion to that used in the MP criterion, with the additional details about the probabilistic setting in which to interpret the trees of a network and summarize the likelihood scores computed on these trees.

Assuming independence among sites, the overall likelihood of a set S of aligned sequences, given by the labeling function λ^k, given a tree topology ψ and a model M (branch lengths and model of sequence evolution), is the product of the probability of the labeling of every site i given ψ and M:

$$L(\lambda^k|\psi,M) = \prod_{i=1}^{k} L(\lambda^k[i]|\psi,M), \tag{5}$$

where k is the number of sites, and $L(\lambda^k[i]|\psi,M)$ can be defined in two ways:

- For *(average) likelihood* [95], L_{avg}, we have:

$$\sum_{\hat{\Lambda}^k}\left[\mathbf{P}(root)\cdot\prod_{e\in E(T)}\mathbf{P}_e(t_e)\right], \tag{6}$$

  where $\hat{\Lambda}^k$ is the set of all possible extensions of λ^k, and $\mathbf{P}_e(t_e)$ denotes the probability of observing the sequences at the two endpoints of edge e whose branch length is t_e.
- For *ancestral likelihood* [84], L_{anc}, we have:

$$\max_{\hat{\Lambda}^k}\left[\mathbf{P}(root)\cdot\prod_{e\in E(T)}\mathbf{P}_e(t_e)\right]. \tag{7}$$

Given a labeling λ^k of a set $\mathscr{X}$ of taxa, the maximum likelihood (ML) problem for phylogenetic trees amounts to solving

$$(\psi^*,M^*) = \text{argmax}_{\psi,M} L(\lambda^k|\psi,M), \tag{8}$$

where ψ ranges over all $\mathscr{X}$-tree topologies, and M ranges over all combinations of branch lengths and models of sequence evolution. When all elements of this combination are specified, scoring the likelihood can be done in polynomial time using

Felsenstein's "pruning" algorithm [28]. Solving the ML problem in general is NP-hard [16].

Lathrop defined a maximum likelihood criterion for phylogenetic inference of populations when some of those populations are hybridized (in this context, hybridization corresponds to *admixture*) [59]. Strimmer and Moulton defined the maximum likelihood criterion for *splits networks*, once their edges are oriented so as to produce a rooted, directed, acyclic, graph [96]. Jin *et al.* defined ML criteria for evolutionary phylogenetic networks [50], which we review here.

Let N be an $\mathscr{X}$-network in which network-nodes have indegree 2 (the results can be generalized in a straightforward way to networks with nodes whose indegree is higher than 2), and let $\mathscr{R} = \{p_i = (e_l^i, e_r^i) : e_l^i, e_r^i \in E(N),\ e_l^i = (x,v),\ e_r^i = (y,v),$ and $x \neq y\}$, with $r = |\mathscr{R}|$. In other words, $\mathscr{R}$ is the set of pairs of edges where each pair is incident into the same network node. Further, we associate with each pair $p_i \in \mathscr{R}$ parameter $\gamma_i \in [0,1]$ which denotes the probability of choosing the "left" edge e_l^i (the probability of choosing the "right" edge e_r^i is $(1-\gamma_i)$). These probabilities are to be estimated from the sequence data, and can be interpreted as the proportion of sites (of the sequence at a network-node) inherited from one of the parents [96]. When multiple loci are involved in the analysis, these probabilities can denote the proportion of the genome arising from a particular parent [69]; see Section 4.3. In the case of admixture, these probabilities correspond to the proportion of the population derived from a particular ancestral population [59]. For example, consider the phylogenetic network N in Figure 14. For this network, we have $\mathscr{R} = \{p_1 = ((u,x),(v,x)), p_2 = ((w,y),(z,y))\}$, parameter γ_1 associated with p_1 (which denotes the probability of taking edge (u,x) for certain sites in the sequence at node x), and parameter γ_2 associated with p_2 (which denotes the probability of taking edge (w,y) for certain sites in the sequence at node y).

Let $T \in \mathscr{T}(N)$. A *characteristic set* of tree T is a set φ_T of size r that contains exactly one edge from every pair in $\mathscr{R}$ such that when all network-edges except for those in φ_T are removed from network N in Step 1 of procedure **Induce** in Figure 2, the procedure yields tree T. For the network N and its induced trees shown in Figure 14, we have $\varphi_{T_1} = \{(v,x),(z,y)\}$, $\varphi_{T_2} = \{(u,x),(z,y)\}$, $\varphi_{T_3} = \{(v,x),(w,y)\}$, and $\varphi_{T_4} = \{(u,x),(w,y)\}$.

Notice that multiple characteristic sets may exist for the same tree T; in this case, we denote the set of all characteristic sets by Φ_T. Then, the probability of a tree T, given network N and leaf-labeling λ^k is

$$\mathbf{P}(T|N,\lambda^k) = \sum_{\varphi_T \in \Phi_T} \left[\prod_{e_l^i \in \varphi_T} \gamma_i \prod_{e_r^j \in \varphi_T} (1-\gamma_j) \right]. \tag{9}$$

In other words, the probability of inducing a tree T by network N is the product of the probabilities of all the network-edges used to induce T. The summation in the formula is to account for cases when there exist multiple ways to induce the tree T. The probabilities of the four trees in Figure 14 are given in the caption of the figure.

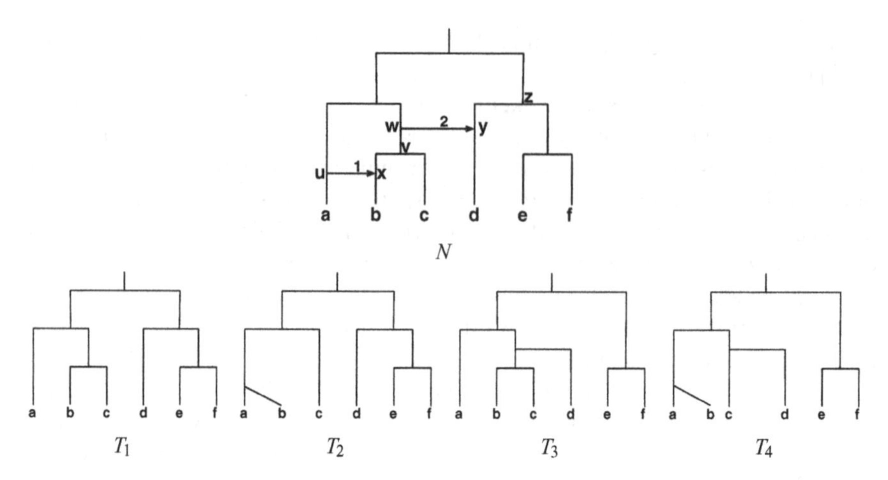

Fig. 14 A phylogenetic network (top) and the four trees it induces (bottom). Using Formula (9), we have $\mathbf{P}(T_1|N) = (1-\gamma_1)(1-\gamma_2)$, $\mathbf{P}(T_2|N) = \gamma_1(1-\gamma_2)$, $\mathbf{P}(T_3|N) = (1-\gamma_1)\gamma_2$, and $\mathbf{P}(T_4|N) = \gamma_1\gamma_2$.

We are now in position to define likelihood criteria for phylogenetic networks. The likelihood of a phylogenetic network with respect to a set of sequences is defined as follows.

Definition 0.5. The likelihood of a phylogenetic network N with a labeling λ^k of the leaves of N is

$$L(\lambda^k|N,M) = \sum_{T\in\mathscr{T}(N)} \left[\mathbf{P}(T|N,\lambda^k)\cdot L(\lambda^k|T,M_T)\right], \tag{10}$$

where M is the model (branch lengths, probabilities γ_i, and model of sequence evolution), M_T is the "restriction" of M to tree T, and $L(\lambda^k|N,M)$ can be either the average or ancestral likelihood functions.

Given a labeling λ^k of a set $\mathscr{X}$ of taxa, the maximum likelihood problem for phylogenetic networks can be defined so as to solve

$$(N^*,M^*) = \text{argmax}_{N,M} L(\lambda^k|N,M), \tag{11}$$

where N ranges over all $\mathscr{X}$-network topologies, and M ranges over all combinations of branch lengths, probabilities γ_i, and models of sequence evolution.

Notice that, while the likelihood of a network, as given by Definition 0.5, is an average of the likelihood of all trees within the networks, we can modify this definition so that the likelihood of a network is the best over all trees, which is analogous to the way we defined the parsimony length of a network above. In this case, we have

$$L(\lambda^k|N,M) = \max_{T\in\mathscr{T}(N)} \left[\mathbf{P}(T|N,\lambda^k)\cdot L(\lambda^k|T,M_T)\right].$$

This definition would be more appropriate for inferring ancestral states on a phylogenetic network.

Finally, the type of input data further refines the versions of the ML problems, as outlined in [50]. This results in several formulations of ML criteria for phylogenetic networks, where these formulations amount to the combinations of tree likelihood type (ancestral vs. average), tree selection criterion (average vs. maximum), and input data.

Problem 0.1. (The Tiny ML Problem)

> **Input:** The full model M of an $\mathscr{X}$-network N, and a labeling λ^k of the leaves.
> **Output:** The labeling $\hat{\lambda}^k$ that maximizes the likelihood of the network.

Problem 0.2. (The Small ML Problem)

> **Input:** The topology of a phylogenetic network N and a labeling λ^k of the leaves.
> **Output:** The branch lengths, edge probabilities, and labeling $\hat{\lambda}^k$ that maximize the likelihood of the network.

Problem 0.3. (The Big ML Problem)

> **Input:** The labeling λ^k of a set $\mathscr{X}$ of taxa.
> **Output:** A full model M of an $\mathscr{X}$-network N that maximizes $L(\lambda^k|N,M)$.

4 To Network, or Not to Network, That Is the Question

In our discussion thus far, we have made an important assumption: incongruities and incompatibilities in the data are due to reticulate evolutionary events and therefore should be reconciled by using a phylogenetic network. We assumed that gene trees disagree due to the occurrence of events such as horizontal gene transfer, and sought a network that reconciles them. In the case of ancestral recombination graphs and perfect phylogenetic networks, we assumed that if a perfect phylogenetic tree does not exist for a set of sequences, then that is an indication of the occurrence of intralocus recombination [45], and hence a network, rather than a tree, is sought as a model of the evolutionary history. However, this assumption must be inspected carefully and thoroughly before phylogenetic network reconstruction is attempted. Several ways exist for explaining the evolution of a data set without invoking reticulate evolutionary events:

- In the analysis of biological data, gene trees are unknown and reconstructed from sequence data. These reconstructions of the trees may have errors in them, in the form of wrong edges. When compared to a species tree, these wrong edges masquerade as true incongruities, triggering the false inference of reticulate evolutionary events, and sometimes they may in fact hide true incongruities, thus resulting in an underestimation of the amount of reticulation in the data; e.g., see [100].

- As Figure 15(a) shows, a gene tree may disagree with a species tree due to a combination of duplication and loss events that took place during the evolution of the gene. In this case, and notwithstanding the incongruities among gene trees, these trees need be reconciled into a tree, not a network.

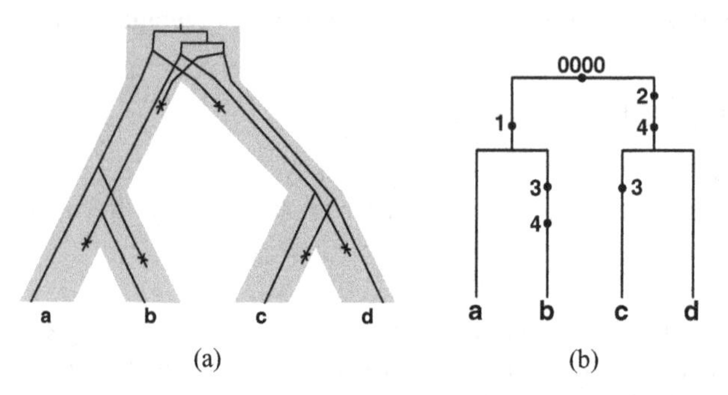

Fig. 15 (a) A gene tree (solid lines) evolving within the branches of the species tree, where the gene tree topology is identical to that of T_2 in Figure 1(b). The gene tree differs from the species tree due to multiple gene duplication and loss events. (b) A phylogenetic tree that models the evolutionary history of the sequences in Figure 11(a) from the ancestral sequence 0000, while violating the infinite-site model assumptions. In this scenario, sites c_3 and c_4 mutated twice, yet no recombination events were invoked.

- As Figure 15(b) shows, the evolution of a set of sequences may be explained by multiple mutations at a site, rather than inferring putative recombination events. In this case, the evolutionary history is still a tree, albeit relaxing the infinite-site model to allow recurrent mutations.
- As Figure 16 shows, a gene tree may differ from the species tree due to *lineage sorting* . Informally, lineage sorting happens when two alleles of a gene from two species fail to coalesce, or "merge" at a common ancestral gene, at the divergence time of the two species, and instead they coalesce deeper. We elaborate on this process further below.

Notice that when gene trees disagree with each other, or with the species tree, it is crucial to determine the cause, or causes, of incongruence first, and then use the appropriate reconciliation method. What is needed in practice is a unified, probabilistic framework that, given a set of gene trees, determines the causes of incongruence. It has been argued that a combination of techniques from population genetics and phylogenetics is needed to achieve this goal, particularly to distinguish between reticulate evolutionary events and lineage sorting as probable causes of incongruence [60]. A natural choice for approaching this issue has been to augment the standard *coalescent* theory so as to allow for computing the probabilities of gene trees assuming the presence of events such as horizontal gene transfer.

In a seminal paper, Maddison proposed a framework for inferring the species tree such that both mutations at the nucleotide level and incongruence among gene trees are taken into account [63]. The likelihood of a given species tree, according to [63],

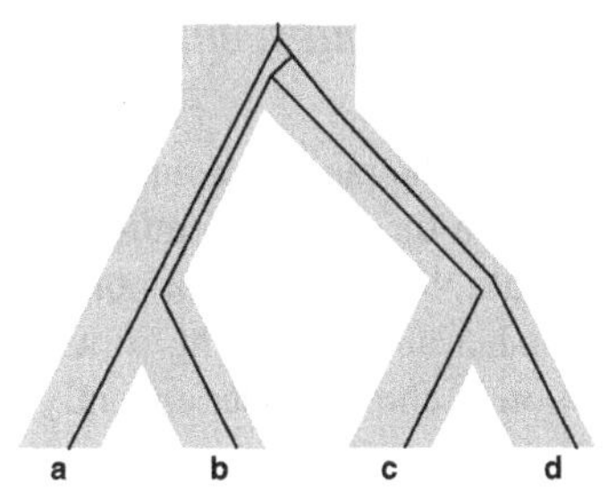

Fig. 16 A gene tree (solid lines) evolving within the branches of the species tree, where the gene tree topology is identical to that of T_2 in Figure 1(b). The gene tree differs from the species tree due to (incomplete) lineage sorting.

is the product, over all loci, of the probability of obtaining the observed sequences at the locus:

$$\prod_{\text{loci}} \sum_{\text{possible gene trees}} [\mathbf{P}(\text{sequences}|\text{gene tree}) \cdot \mathbf{P}(\text{gene tree}|\text{species tree})]. \qquad (12)$$

The probability $\mathbf{P}(\text{gene tree}|\text{species tree})$, when deep coalescence is allowed, can be calculated using coalescence theory, as we briefly review in Section 4.1. However, in the most general setting, the species phylogeny may not be a tree. Therefore, an extension to Maddison's framework is necessary to account for reticulate evolutionary events. The ML formulation given in Section 3.3 is similar to Maddison's proposal, but it explicitly models reticulate evolution and ignores lineage sorting. What is needed is an extension to the coalescent to allow for calculating the probability of a gene tree given a species phylogeny assuming any combination of the three discord processes (lineage sorting, reticulate evolution, and gene duplication/loss) could be involved. Preliminary work that simultaneously accounts for lineage sorting and horizontal gene transfer events has been proposed in [100] and another that simultaneously accounts for lineage sorting and hybrid speciation has been proposed in [69]; we review these two in Sections 4.2 and 4.3, respectively.

It is worth mentioning that other approaches for distinguishing reticulate evolution from lineage sorting without explicit modeling of the coalescent process have been introduced. For example, Sang and Zhong proposed a test statistic for distinguishing between lineage sorting and hybridization based on the divergence time of the two parents of a hybrid [89]. However, Holder *et al.* showed later that this statistic fails to reliably distinguish between the two processes [42]. More recently, Holland *et al.* proposed to use *supernetworks* for this task [43].

4.1 Lineage Sorting and the Coalescent

Lineage sorting occurs because of the random contribution of genetic material from each individual in a population to the next generation. Some fail to have offspring while some happen to have multiple offspring. In population genetics, this process was first modeled by R. A. Fisher and S. Wright, in which each gene of the population at a particular generation is chosen independently from the gene pool of the previous generation, regardless of whether the genes are in the same individual or in different individuals. Under the Wright-Fisher model, "the coalescent" considers the process backward in time [44, 57, 97]. That is, the ancestral lineages of genes of interest are traced from offspring to parents. A coalescent event occurs when two (or sometimes more) genes "merge" at the same parent, which is called the most recent common ancestor (MRCA) of the two genes.

The basic process can be treated as follows. Consider a pair of genes at time τ_1 in a randomly mating haploid population. The population size at time τ is denoted by $N(\tau)$. The probability that both genes are from the same parental gene at the previous generation (time τ_1+1) is $1/N(\tau_1+1)$. Therefore, starting at τ_1, the probability that the coalescence between the pair occurs at τ_2 is given by

$$Prob(\tau_2) = \frac{1}{N(\tau_2)} \prod_{\tau=\tau_1+1}^{\tau_2-1} \left(1 - \frac{1}{N(\tau)}\right). \tag{13}$$

When $N(\tau)$ is constant, the probability density distribution (pdf) of the coalescent time (i.e., $t = \tau_2 - \tau_1$) is given by a geometric distribution and can be approximated by an exponential distribution for large N:

$$Prob(t) = \frac{1}{N} e^{-t/N}. \tag{14}$$

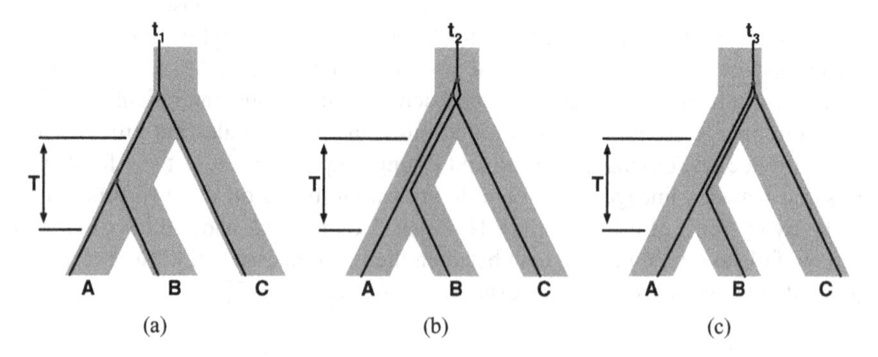

Fig. 17 A species tree on three species A, B, and C. Shown within the branches of the species tree are the three possible gene tree topologies that may result due to different coalescence histories.

Under the three-species model (Figure 17), there are three possible types of gene tree: *(AB)C*, *(AC)B* and *A(BC)*. Let *Prob*[(AB)C], *Prob*[(AC)B] and *Prob*[A(BC)] be the probabilities of the three types of gene tree. These three probabilities are simply expressed with a continuous time approximation when all populations have equal and constant population sizes, N, where N is large:

$$\mathbf{P}(t_1) = 1 - \frac{2}{3}e^{-T/N}, \tag{15}$$

and

$$\mathbf{P}(t_2) = \mathbf{P}(t_3) = \frac{1}{3}e^{-T/N}. \tag{16}$$

Recently, Rosenberg and colleagues showed that the most likely gene tree may be different from the species tree, when the number of leaves is four or more [19, 88]. It is worth mentioning, however, that when the number of leaves is three, the result does not apply, since the expression in (15) is greater than the expression in (16) for all strictly positive, finite values of T and N.

Observe that in the presence of lineage sorting (in addition to reticulate evolutionary events), the number of gene trees given a (species) phylogenetic network is no longer bounded, as given above by Inequality (1). Rather, the number of possible gene trees now equals the number of possible rooted trees (with the same number of leaves as that of the network). For example, let us consider how the tree $(((a,b),c),d)$ could be one of the gene trees inside the phylogenetic network in Figure 1(a). To obtain this tree, consider the scenario under which b inherits its gene from the a lineage, the genes of c and d fail to coalesce before they reach the root r; instead, c first coalesces with the ancestral gene of a and b, and then the ancestral copy of all three coalesces with that of d. This scenario is illustrated in Figure 18.

4.2 Augmenting the Coalescent with Horizontal Gene Transfer

We now review the model of [100] for extending the coalescent to allow HGT as a cause of incongruence. Suppose that each haploid individual in a population with size N has a lifespan that follows an exponential distribution with mean l. When an individual dies, another individual randomly chosen from the population replaces it to keep the population size constant. In other words, one of the $N-1$ alive lineages is duplicated to replace the dead one. Under the Moran model, the ancestral lineages of individuals of interest can be traced backward in time, and the coalescent time between a pair of individuals follows an exponential distribution with mean $lN/2$ [25, 87]. While phylogeny-based detection of HGT is usually based on quantifying incongruence between a species and a gene tree, the situation becomes more complicated when lineage sorting may be a cause of the incongruence as well.

Consider a model with three species, A, B, and C, in which an HGT event occurs from species B to C, as illustrated in Figure 19. Suppose the MRCA of all three species has a single copy of a gene x. Let a, b and c be the orthologous genes in

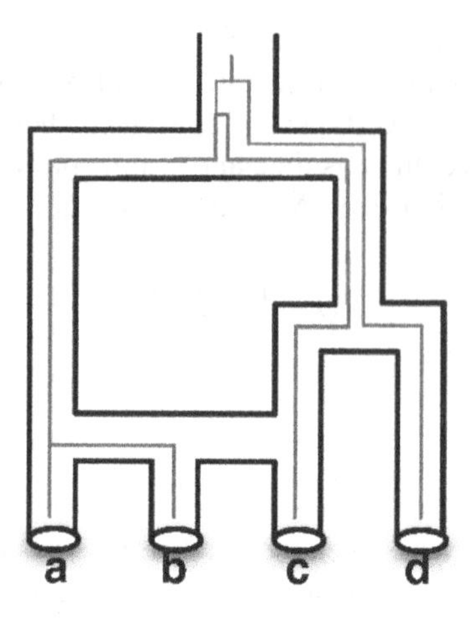

Fig. 18 Illustration of the combined effect of reticulate evolution and lineage sorting. The tubes represent a phylogenetic network in which b is a hybrid taxon (the same as the one in Figure 1(a)), and shown within the tubes is gene tree $(((a,b),c),d)$. Notice that this gene tree cannot be obtained using the **Induce** procedure described in Figure 2, and it is not one of the two trees shown in Figure 1.

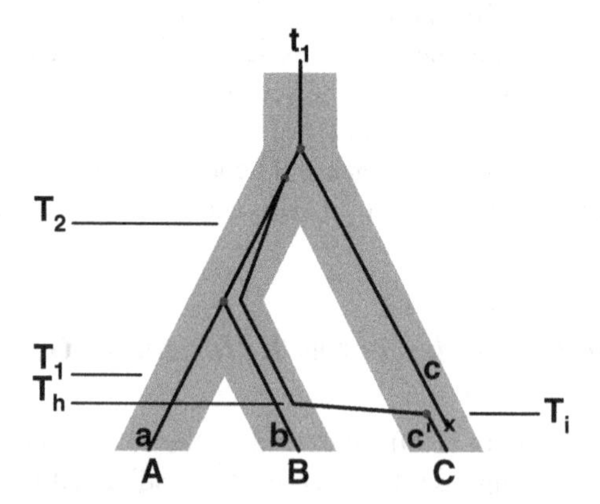

Fig. 19 A three bacterial species model with an HGT event. A demonstration that a congruent tree could be observed even with HGT.

the three species, respectively, whose ancestral gene at the MRCA is x. At time T_h, a gene was transferred from species B and was inserted in a genome in species C at T_i, which is denoted by c'. Since HGT is assumed to be instantaneous at the scale of evolution, in reality, it is always the case that $T_i = T_h$. However, since these times are estimated in practice, it may be the case that $T_h < T_i$. For example, if a gene duplication occurs in lineage b in Figure 19, and one of the two in-paralogs is transferred to c, then the estimated time T_h would be the duplication time, which is earlier than the actual time of the HGT events, T_i.

Following the HGT event, c was physically deleted from the genome, so that each of the three species currently has a single copy of the focal gene. If there is no lineage sorting, the gene tree should be $a(bc')$. Since this tree is incongruent with the species tree, $(AB)C$, we could consider it as an evidence for HGT. However, lineage sorting could also produce the incongruence between the gene tree and species tree without HGT. It is also important to note that lineage sorting, coupled with HGT, could produce a congruent gene tree, as illustrated in Figure 19. Although b and c' have a higher chance to coalesce first, the probability that the first coalescence occurs between a and b or between a and c' may not be negligible especially when $T_1 - T_h$ is short. The probabilities of the three types of gene tree can be formulated under this tri-species model with HGT as illustrated in Figure 19. Here, T_h could exceed T_1; in such a case it can be considered that HGT occurred before the speciation between A and B. Assuming that all populations have equal (constant) population sizes, N, the three probabilities can be obtained modifying (15) and (16):

$$\mathbf{P}[\mathrm{(AB)C}] = \begin{cases} \frac{1}{3}e^{-(T_1-T_h)/N}, & \text{if } T_h \leq T_1 \\ 1-\frac{2}{3}e^{-(T_h-T_1)/N}, & \text{if } T_h > T_1 \end{cases}, \tag{17}$$

$$\mathbf{P}[\mathrm{(AC)B}] = \begin{cases} \frac{1}{3}e^{-(T_1-T_h)/N}, & \text{if } T_h \leq T_1 \\ \frac{1}{3}e^{-(T_h-T_1)/N}, & \text{if } T_h > T_1 \end{cases}, \tag{18}$$

and

$$\mathbf{P}[\mathrm{A(BC)}] = \begin{cases} 1-\frac{2}{3}e^{-(T_1-T_h)/N}, & \text{if } T_h \leq T_1 \\ \frac{1}{3}e^{-(T_h-T_1)/N}, & \text{if } T_h > T_1 \end{cases}. \tag{19}$$

4.3 Augmenting the Coalescent with Hybrid Speciation

We now review the model of [69] for extending the coalescent to allow hybrid speciation as a cause of incongruence, using the scenario depicted in Figure 20 as an example. The issue at hand is, given a collection of genes whose trees may be incongruent, whether their incongruence due to hybrid speciation or lineage sorting. In the former case, their reconciliation would result in the phylogenetic network depicted by the wide bands in Figure 20. However, as the time T between the MRCA of any two of the species and the MRCA of all three becomes smaller, the probability of gene tree disagreement due to lineage sorting increases.

Let a, b, and c be three orthologous genes randomly sampled from the three species A, B, and C, respectively, where B is a hybrid of A and C. The model of Meng and Kubatko assumes that when a gene b is arbitrarily selected from species B, then its most recent common ancestor occurs with species A with probability γ and with species C with probability $1-\gamma$. These two possible trees are t_1 and t_2, respectively, discussed in the caption of Figure 20. Once one of these two trees is selected, the model treats the tree as a species tree and allows the coalescent process

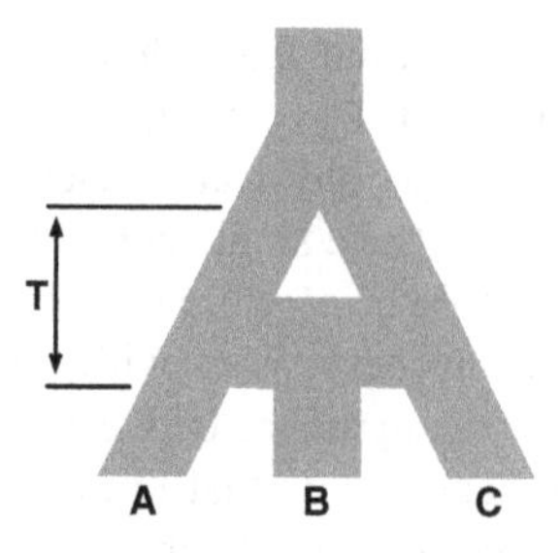

Fig. 20 A phylogenetic network representing a hybrid speciation event involving species A and C, and producing species B. The two possible (alternative) species trees are $t_1 = ((A,B),C)$ and $t_2 = (A,(B,C))$.

to operate for that gene.[7] Using Equations (15) and (16) above for calculating the probabilities of gene trees given a species tree, and assuming t_1 as a species tree, we have

$$\mathbf{P}[((\mathrm{A},\mathrm{B}),\mathrm{C})] = 1 - \tfrac{2}{3}e^{-T/N},$$
$$\mathbf{P}[((\mathrm{A},\mathrm{C}),\mathrm{B})] = \mathbf{P}[(\mathrm{A},(\mathrm{B},\mathrm{C}))] = \tfrac{1}{3}e^{-T/N}.$$

Assuming t_2 as the species tree, we have

$$\mathbf{P}[(\mathrm{A},(\mathrm{B},\mathrm{C}))] = 1 - \tfrac{2}{3}e^{-T/N},$$
$$\mathbf{P}[((\mathrm{A},\mathrm{C}),\mathrm{B})] = \mathbf{P}[((\mathrm{A},\mathrm{B}),\mathrm{C})] = \tfrac{1}{3}e^{-T/N}.$$

The question is to determine, given a collection of genes sampled from the genomes of the three species, whether the evolutionary history of the three species is the phylogenetic network in Figure 20, the species tree t_1, or the species tree t_2. One way of answering this question is to estimate the probability γ. If $\gamma = 1$, then the evolutionary history of the three species is the species tree t_1. If $\gamma = 0$, then the evolutionary history of the three species is the species tree t_2. If $0 < \gamma < 1$, then the evolutionary history is the phylogenetic network shown in Figure 20, with the proportions of the genome of B inherited from A and C are γ and $(1-\gamma)$, respectively.

Meng and Kubatko provided a maximum likelihood estimation of the parameters γ and T, as well as a Bayesian estimation technique [69]. We briefly review the main points of the maximum likelihood estimation. Let $\mathscr{G} = \{gt_1, gt_2, \ldots, gt_k\}$ be an *i.i.d.* sample of gene trees, where gt_i is the tree of gene i, sampled so that their topologies are independent and follow the hybridization model described in Figure 20. The likelihood function for a given phylogenetic network with a specified location for the hybrid speciation event (as shown in Figure 20) is given by:

$$L(\gamma, T|\mathscr{G}) = \prod_{i=1}^{k} \mathbf{P}(gt_i|\gamma, T) = \prod_{i=1}^{k} [\gamma \mathbf{P}(gt_i|t_1, T) + (1-\gamma)\mathbf{P}(gt_i|t_2, T)] \tag{20}$$

[7] Notice the similarity between this and the probability of a tree as given by Equation (9).

Notice that this formula is a special case of Formula (10) (when taken for multiple genes) given in Definition 0.5. Formula (10) is defined for networks with any number of hybrid speciation events, and the parameter M_T in the formula is a generalization of the pairs (t_1, T) and (t_2, T) in Formula (20), since M_T is the model, which includes the tree topology, its branch lengths, and the model of evolution.

The question now becomes one of estimating the parameters γ and T that maximize the likelihood function and determining, based on these (particularly γ), whether the phylogenetic network or tree is the evolutionary history of the species, and, if the latter, which of the two (t_1 or t_2).

5 Exercises

Here we give a set of exercises for the reader to gain a better understanding of evolutionary phylogenetic networks and issues related to their reconstruction and evaluation.

1. Show an example of a set $\mathscr{T}$ of trees, with $|\mathscr{T}| = 2$, and a minimal network N that reconciles both trees in $\mathscr{T}$ such that $\mathscr{T} \neq \mathscr{T}(N)$.
2. Figure 11 shows one minimal ARG for the given sequence data set. Draw all other minimal ARGs.
3. a. Show a phylogenetic network N with $|\mathscr{T}(N)| = 2^k$, where k is the number of network-nodes in N.
   b. Show a phylogenetic network N with $|\mathscr{T}(N)| < 2^k$, where k is the number of network-nodes in N.
4. Show two trees, each with nine leaves, whose SPR distance is 3, and for which the number of minimal phylogenetic networks that reconcile the two trees is 27. (Hint: Consider trees with three clades, each clade with three leaves, and each clade requires a single SPR move.)
5. For each of the two characters labeling the leaves of the network N in Figure 12(b), show a tree in $\mathscr{T}(N)$ on which the character is compatible, by also showing the labeling of internal nodes of the tree.
6. Using the illustration in Figure 18, describe one coalescence scenario for each of the possible gene trees that are induced by the phylogenetic network in Figure 1 assuming lineage sorting could occur.

6 Further Reading

An excellent resource on phylogenetic networks is *Who is Who in Phylogenetic Networks* [31], which, as of the date of writing this manuscript, catalogs 264 publications and 34 software tools dedicated to phylogenetic networks. There have been several recent detailed surveys of phylogenetic reconstruction methods [32, 47, 48, 61, 64, 72], some of which identify their similarities and differences.

Recently, several results have appeared on measures for comparing phylogenetic network topologies and quantifying their dissimilarities; we refer the reader to [3, 8, 9, 10, 11, 14, 15, 70, 76, 79]. Further, some proposals have been made on representing phylogenetic networks for I/O operations using an *extended Newick*, or eNewick, format; e.g., see [12, 13, 71, 101].

Acknowledgements The author would like to thank the editors for the invitation to contribute this chapter, Joe Felsenstein, Lenny Heath, and Mike Steel for extensive and insightful comments that helped improve the manuscript significantly, and Guohua Jin, Laura S. Kubatko, Kevin Liu, Richard Manning, Francesc Rosselló, and Yun Yu for helpful comments on the text. The illustration in Figure 18 and its accompanying discussion are the result of a discussion with Laura S. Kubatko and James Degnan. This work was supported in part by DOE grant DE-FG02-06ER25734, NSF grant CCF-0622037, and grant R01LM009494 from the National Library of Medicine. The contents are solely the responsibility of the authors and do not necessarily represent the official views of the DOE, NSF, National Library of Medicine or the National Institutes of Health.

References

1. Allen, B., Steel, M.: Subtree transfer operations and their induced metrics on evolutionary trees. Annals of Combinatorics **5**, 1–13 (2001)
2. Baroni, M., Grunewald, S., Moulton, V., Semple, C.: Bounding the number of hybridisation events for a consistent evolutionary history. J. Math. Biol. **51**, 171–182 (2005)
3. Baroni, M., Semple, C., Steel, M.: A framework for representing reticulate evolution. Annals of Combinatorics **8**(4), 391–408 (2004)
4. Beiko, R., Hamilton, N.: Phylogenetic identification of lateral genetic transfer events. BMC Evolutionary Biology **6** (2006)
5. Bergthorsson, U., Adams, K., Thomason, B., Palmer, J.: Widespread horizontal transfer of mitochondrial genes in flowering plants. Nature **424**, 197–201 (2003)
6. Bergthorsson, U., Richardson, A., Young, G., Goertzen, L., Palmer, J.: Massive horizontal transfer of mitochondrial genes from diverse land plant donors to basal angiosperm Amborella. Proc. Nat'l Acad. Sci., USA **101**, 17,747–17,752 (2004)
7. Bordewich, M., Semple, C.: On the computational complexity of the rooted subtree prune and regraft distance. Annals of Combinatorics **8**, 409–423 (2004)
8. Cardona, G., Llabrés, M., Rosselló, F., Valiente, G.: A distance metric for a class of tree-sibling phylogenetic networks. Bioinformatics **24**(13), 1481–1488 (2008)
9. Cardona, G., Llabrés, M., Rosselló, F., Valiente, G.: Metrics for phylogenetic networks I: Generalizations of the robinson-foulds metric. IEEE/ACM Transactions on Computational Biology and Bioinformatics **6**(1), 1–16 (2009)
10. Cardona, G., Llabrés, M., Rosselló, F., Valiente, G.: Metrics for phylogenetic networks II: Nodal and triplets metrics. IEEE/ACM Transactions on Computational Biology and Bioinformatics (2009)
11. Cardona, G., Llabrés, M., Rosselló, F., Valiente, G.: On Nakhleh's latest metric for phylogenetic networks. IEEE/ACM Transactions on Computational Biology and Bioinformatics (2009). To appear
12. Cardona, G., Rosselló, F., Valiente, G.: Extended Newick: It is time for a standard representation of phylogenetic networks. BMC Bioinformatics **9**, 532 (2008)
13. Cardona, G., Rossello, F., Valiente, G.: A Perl package and an alignment tool for phylogenetic networks. BMC Bioinformatics **9**(1), 175 (2008)
14. Cardona, G., Rosselló, F., Valiente, G.: Tripartitions do not always discriminate phylogenetic networks. Mathematical Biosciences **211**(2), 356–370 (2008)

15. Cardona, G., Rosselló, F., Valiente, G.: Comparison of tree-child phylogenetic networks. IEEE/ACM Transactions on Computational Biology and Bioinformatics (2009). To appear
16. Chor, B., Tuller, T.: Maximum likelihood of evolutionary trees is hard. Proc. 9th Ann. Int'l Conf. Comput. Mol. Biol. (RECOMB05) pp. 296–310 (2005)
17. Darwin, C.: On the origin of species by means of natural selection. J. Murray, London (1859)
18. Day, W.: Computationally difficult parsimony problems in phylogenetic systematics. Journal of Theoretical Biology **103**, 429–438 (1983)
19. Degnan, J., Rosenberg, N.: Discordance of species trees with their most likely gene trees. PLoS Genetics **2**, 762–768 (2006)
20. Doolittle, W.: Lateral genomics. Trends in Biochemical Sciences **24**(12), M5–M8 (1999)
21. Doolittle, W.: Phylogenetic classification and the universal tree. Science **284**, 2124–2129 (1999)
22. Edwards, S.: Is a new and general theory of molecular systematics emerging? Evolution **63**(1), 1–19 (2009)
23. Ellstrand, N., Whitkus, R., Rieseberg, L.: Distribution of spontaneous plant hybrids. Proc. Nat'l Acad. Sci., USA **93**(10), 5090–5093 (1996)
24. Estabrook, G., McMorris, F.: When are two qualitative taxonomic characters compatible? J. Math. Biosci. **4**, 195–200 (1977)
25. Ewens, W.: Mathematical Population Genetics. Springer-Verlag, Berlin (1979)
26. Felsenstein, J.: Cases in which parsimony or compatibility methods will be positively misleading. Systematic Zoology **27**, 401–410 (1978)
27. Felsenstein, J.: Alternative methods of phylogenetic inference and their interrelationship. Systematic Zoology **28**, 49–62 (1979)
28. Felsenstein, J.: Evolutionary trees from DNA sequences: A maximum likelihood approach. J. Mol. Evol. **17**, 368–376 (1981)
29. Fitch, W.: Toward defining the course of evolution: Minimum change for a specified tree topology. Syst. Zool. **20**, 406–416 (1971)
30. Foulds, L., Graham, R.: The Steiner problem in phylogeny is NP-complete. Adv. Appl. Math. **3**, 43–49 (1982)
31. Gambette, P.: Who is who in phylogenetic networks: Articles, authors and programs. http://www.lirmm.fr/∼gambette/PhylogeneticNetworks/
32. Gemeinholzer, B.: Phylogenetic networks. In: B.H. Junker, F. Schreiber (eds.) Analysis of Biological Networks, pp. 255–282. John Wiley and Sons Ltd (2008)
33. Goloboff, P.: Calculating SPR distances between trees. Cladistics **24**, 591–597 (2007)
34. Griffiths, R., Marjoram, P.: An ancestral recombination graph. In: P. Donnelly, S. Tavare (eds.) Progress in Population Genetics and Human Evolution, *IMA Volumes in Mathematics and its Applications*, vol. 87, pp. 257–270. Springer-Verlag, Berlin (1997)
35. Gusfield, D.: Efficient algorithms for inferring evolutionary trees. Networks **21**, 19–28 (1991)
36. Gusfield, D., Bansal, V., Bafna, V., Song, Y.: A decomposition theory for phylogenetic networks and incompatible characters. Journal of Computational Biology **14**, 1247–1272 (2007)
37. Gusfield, D., Eddhu, S., Langley, C.: Efficient reconstruction of phylogenetic networks with constrained recombination. In: Proceedings of Computational Systems Bioinformatics (CSB 03) (2003)
38. Hallett, M., Lagergren, J.: Efficient algorithms for lateral gene transfer problems. In: Proc. 5th Ann. Int'l Conf. Comput. Mol. Biol. (RECOMB01), pp. 149–156. ACM Press, New York (2001)
39. Hao, W., Golding, G.: Patterns of bacterial gene movement. Mol. Biol. Evol. **21**(7), 1294–1307 (2004)
40. Hein, J.: Reconstructing evolution of sequences subject to recombination using parsimony. Math. Biosciences **98**, 185–200 (1990)
41. Hein, J.: A heuristic method to reconstruct the history of sequences subject to recombination. J. Mol. Evol. **36**, 396–405 (1993)
42. Holder, M., Anderson, J., Holloway, A.: Difficulties in detecting hybridization. Systematic Biology **50**(6), 978982 (2001)

43. Holland, B., Benthin, S., Lockhart, P., Moulton, V., Huber, K.: Using supernetworks to distinguish hybridization from lineage-sorting. BMC Evolutionary Biology **8**, 202 (2008)
44. Hudson, R.: Properties of the neutral allele model with intergenic recombination. Theor. Popul. Biol. **23**, 183–201 (1983)
45. Hudson, R., Kaplan, N.: Statistical properties of the number of recombination events in the history of a sample of DNA sequences. Genetics **111**, 147–164 (1985)
46. Humphries, P., Semple, C.: Note on the hybridization number and subtree distance in phylogenetics. Applied Mathematics Letters (2009). In press
47. Huson, D.H.: Split networks and reticulate networks. In: O. Gascuel, M. Steel (eds.) Reconstructing Evolution, New Mathematical and Computational Advances, pp. 247–276. Oxford University Press (2007)
48. Huson, D.H., Bryant, D.: Application of phylogenetic networks in evolutionary studies. Molecular Biology and Evolution **23**(2), 254–267 (2006)
49. Jin, G., Nakhleh, L., Snir, S., Tuller, T.: Efficient parsimony-based methods for phylogenetic network reconstruction. Bioinformatics **23**, e123–e128 (2006). Proceedings of the European Conference on Computational Biology (ECCB 06)
50. Jin, G., Nakhleh, L., Snir, S., Tuller, T.: Maximum likelihood of phylogenetic networks. Bioinformatics **22**(21), 2604–2611 (2006)
51. Jin, G., Nakhleh, L., Snir, S., Tuller, T.: Inferring phylogenetic networks by the maximum parsimony criterion: A case study. Molecular Biology and Evolution **24**(1), 324–337 (2007)
52. Jin, G., Nakhleh, L., Snir, S., Tuller, T.: A new linear-time heuristic algorithm for computing the parsimony score of phylogenetic networks: Theoretical bounds and empirical performance. In: I. Mandoiu, A. Zelikovsky (eds.) Proceedings of the International Symposium on Bioinformatics Research and Applications, *Lecture Notes in Bioinformatics*, vol. 4463, pp. 61–72 (2007)
53. Kanj, I., Nakhleh, L., Than, C., Xia, G.: Seeing the trees and their branches in the network is hard. Theoretical Computer Science **401**, 153–164 (2008)
54. Kanj, I., Nakhleh, L., Xia, G.: The compatibility of binary characters on phylogenetic networks: Complexity and parameterized algorithms. Algorithmica **51**, 99–128 (2008)
55. Kimura, M.: The number of heterozygous nucleotide sites maintained in a finite population due to steady flux of mutations. Genetics **61**, 893–903 (1969)
56. Kimura, M., Crow, J.: The number of alleles that can be maintained in a finite population. Genetics **49**, 725–738 (1964)
57. Kingman, J.F.C.: The coalescent. Stochast. Proc. Appl. **13**, 235–248 (1982)
58. Kurland, C., Canback, B., Berg, O.: Horizontal gene transfer: A critical view. Proc. Nat'l Acad. Sci., USA **100**(17), 9658–9662 (2003)
59. Lathrop, G.: Evolutionary trees and admixture: Phylogenetic inference when some populations are hybridized. Ann. Hum. Genet. **46**, 245–255 (1982)
60. Linder, C., Rieseberg, L.: Reconstructing patterns of reticulate evolution in plants. American Journal of Botany **91**, 1700–1708 (2004)
61. Linder, C.R., Moret, B.M.E., Nakhleh, L., Warnow, T.: Network (reticulate) evolution: Biology, models, and algorithms. In: The Pacific Symposium on Biocomputing (2004)
62. MacLeod, D., Charlebois, R., Doolittle, F., Bapteste, E.: Deduction of probable events of lateral gene transfer through comparison of phylogenetic trees by recursive consolidation and rearrangement. BMC Evolutionary Biology **5** (2005)
63. Maddison, W.: Gene trees in species trees. Systematic Biology **46**(3), 523–536 (1997)
64. Makarenkov, V., Kevorkov, D., Legendre, P.: Phylogenetic network construction approaches. In: Applied Mycology and Biotechnology, pp. 61–97 (2006)
65. Mallet, J.: Hybridization as an invasion of the genome. TREE **20**(5), 229–237 (2005)
66. Mallet, J.: Hybrid speciation. Nature **446**, 279–283 (2007)
67. McClilland, M., Sanderson, K., Clifton, S., Latreille, P., Porwollik, S., Sabo, A., Meyer, R., Bieri, T., Ozersky, P., McLellan, M., Harkins, C., Wang, C., Nguyen, C., Berghoff, A., Elliott, G., Kohlberg, S., Strong, C., Du, F., Carter, J., Kremizki, C., Layman, D., Leonard, S., Sun, H., Fulton, L., Nash, W., Miner, T., Minx, P., Delehaunty, K., Fronick, C., Magrini, V., Nhan,

M., Warren, W., Florea, L., Spieth, J., Wilson, R.: Comparison of genome degradation in Paratyphi A and Typhi, human-restricted serovars of *salmonella enterica* that cause typhoid. Nature Genetics **36**(12), 1268–1274 (2004)
68. Meacham, C.: Theoretical and computational considerations of the compatibility of qualitative taxonomic characters. NATO ASI Series **G1 on Numerical Taxonomy** (1983)
69. Meng, C., Kubatko, L.: Detecting hybrid speciation in the presence of incomplete lineage sorting using gene tree incongruence: A model. Theoretical Population Biology **75**(1), 35–45 (2009)
70. Moret, B., Nakhleh, L., Warnow, T., Linder, C., Tholse, A., Padolina, A., Sun, J., Timme, R.: Phylogenetic networks: Modeling, reconstructibility, and accuracy. IEEE/ACM Transactions on Computational Biology and Bioinformatics **1**(1), 13–23 (2004)
71. Morin, M., Moret, B.: NetGen: Generating phylogenetic networks with diploid hybrids. Bioinformatics **22**(15), 1921–1923 (2006)
72. Morrison, D.A.: Networks in phylogenetic analysis: new tools for population biology. International Journal of Parasitology **35**, 567–582 (2005)
73. Mower, J., Stefanovic, S., Young, G., Palmer, J.: Gene transfer from parasitic to host plants. Nature **432**, 165–166 (2004)
74. Nakamura, Y., Itoh, T., Matsuda, H., Gojobori, T.: Biased biological functions of horizontally transferred genes in prokaryotic genomes. Nature Genetics **36**(7), 760–766 (2004)
75. Nakhleh, L.: Phylogenetic networks. Ph.D. thesis, The University of Texas at Austin (2004)
76. Nakhleh, L.: A metric on the space of reduced phylogenetic networks. IEEE/ACM Transactions on Computational Biology and Bioinformatics (2009). To appear
77. Nakhleh, L., Ringe, D., Warnow, T.: Perfect phylogenetic networks: A new methodology for reconstructing the evolutionary history of natural languages. LANGUAGE, Journal of the Linguistic Society of America **81**(2), 382–420 (2005)
78. Nakhleh, L., Ruths, D., Wang, L.: RIATA-HGT: A fast and accurate heuristic for reconstructing horizontal gene transfer. In: L. Wang (ed.) Proceedings of the Eleventh International Computing and Combinatorics Conference (COCOON 05), pp. 84–93 (2005). LNCS #3595
79. Nakhleh, L., Sun, J., Warnow, T., Linder, R., Moret, B., Tholse, A.: Towards the development of computational tools for evaluating phylogenetic network reconstruction methods. In: Proceedings of the 8th Pacific Symposium on Biocomputing, pp. 315–326. World Scientific Pub. (2003)
80. Noor, M., Feder, J.: Speciation genetics: Evolving approaches. Nature Review Genetics **7**, 851–861 (2006)
81. Ochman, H., Lawrence, J., Groisman, E.: Lateral gene transfer and the nature of bacterial innovation. Nature **405**(6784), 299–304 (2000)
82. Posada, D., Crandall, K.: The effect of recombination on the accuracy of phylogeny estimation. J. Mol. Evol. **54**(3), 396–402 (2002)
83. Posada, D., Crandall, K., Holmes, E.: Recombination in evolutionary genomics. Annu. Rev. Genet. **36**, 75–97 (2002)
84. Pupko, T., Pe'er, I., Shamir, R., Graur, D.: A fast algorithm for joint reconstruction of ancestral amino-acid sequences. Mol. Biol. Evol. **17**(6), 890–896 (2000)
85. Rieseberg, L., Baird, S., Gardner, K.: Hybridization, introgression, and linkage evolution. Plant Molecular Biology **42**(1), 205–224 (2000)
86. Rieseberg, L., Carney, S.: Plant hybridization. New Phytologist **140**(4), 599–624 (1998)
87. Rosenberg, N.: Gene genealogies. In: C. Fox, J.B. Wolf (eds.) Evolutionary Genetics: Concepts and Case Studies, chap. 15. Oxford Univ. Press University Press (2005)
88. Rosenberg, N., Tao, R.: Discordance of species trees with their most likely gene trees: The case of five taxa. Systematic Biology **57**, 131–140 (2008)
89. Sang, T., Zhong, Y.: Testing hybridization hypotheses based on incongruent gene trees. Systematic Biology **49**(3), 422434 (2000)
90. Song, Y.: Properties of subtree-prune-and-regraft operations on totally-ordered phylogenetic trees. Annals of Combinatorics **10**, 129–146 (2006)

91. Song, Y., Ding, Z., Gusfield, D., Langley, C., Wu, Y.: Algorithms to distinguish the role of gene-conversion from single-crossover recombination in the derivation of SNP sequences in populations. Journal of Computational Biology **14**, 1273–1286 (2007)
92. Song, Y., Hein, J.: Parsimonious reconstruction of sequence evolution and haplotype blocks: Finding the minimum number of recombination events. In: Proc. 3rd Int'l Workshop Algorithms in Bioinformatics (WABI03), vol. 2812, pp. 287–302. Springer-Verlag (2003)
93. Song, Y., Hein, J.: On the minimum number of recombination events in the evolutionary history of DNA sequences. Journal of Mathematical Biology **48**, 160–186 (2004)
94. Song, Y., Hein, J.: Constructing minimal ancestral recombination graphs. Journal of Computational Biology **12**, 147–169 (2005)
95. Steel, M., Penny, D.: Parsimony, likelihood, and the roles of models in molecular phylogenetics. Mol. Biol. Evol. **17**, 839–850 (2000)
96. Strimmer, K., Moulton, V.: Likelihood analysis of phylogenetic networks using directed graphical models. Mol. Biol. Evol. **17**, 875–881 (2000)
97. Tajima, F.: Evolutionary relationship of DNA sequences in finite populations. Genetics **105**, 437–460 (1983)
98. Than, C., Jin, G., Nakhleh, L.: Integrating sequence and topology for efficient and accurate detection of horizontal gene transfer. In: Proceedings of the Sixth RECOMB Comparative Genomics Satellite Workshop, *Lecture Notes in Bioinformatics*, vol. 5267, pp. 113–127 (2008)
99. Than, C., Nakhleh, L.: SPR-based tree reconciliation: Non-binary trees and multiple solutions. In: Proceedings of the Sixth Asia Pacific Bioinformatics Conference (APBC), pp. 251–260 (2008)
100. Than, C., Ruths, D., Innan, H., Nakhleh, L.: Confounding factors in HGT detection: Statistical error, coalescent effects, and multiple solutions. Journal of Computational Biology **14**(4), 517–535 (2007)
101. Than, C., Ruths, D., Nakhleh, L.: PhyloNet: A software package for analyzing and reconstructing reticulate evolutionary relationships. BMC Bioinformatics **9**, 322 (2008)
102. Wang, L., Zhang, K., Zhang, L.: Perfect phylogenetic networks with recombination. Journal of Computational Biology **8**(1), 69–78 (2001)
103. Welch, R., Burland, V., Plunkett, G., Redford, P., Roesch, P., Rasko, D., Buckles, E., Liou, S., Boutin, A., Hackett, J., Stroud, D., Mayhew, G., Rose, D., Zhou, S., Schwartz, D., Perna, N., Mobley, H., Donnenberg, M., Blattner, F.: Extensive mosaic structure revealed by the complete genome sequence of uropathogenic *escherichia coli*. Proc. Nat'l Acad. Sci., USA **99**(26), 17,020–17,024 (2002)
104. Willson, S.: Reconstruction of certain phylogenetic networks from the genomes at their leaves. Journal of Theoretical Biology **252**, 338–349 (2008)
105. Wu, Y.: A practical method for exact computation of subtree prune and regraft distance. Bioinformatics **25**(2), 190–196 (2009)

Genome Wide Association Studies

Paola Sebastiani and Nadia Solovieff

Abstract The availability of high throughput technology for parallel genotyping has opened the field of genetics to genome-wide association studies (GWAS). These studies generate massive amount of genetic data that challenge investigators with issues related to data management, statistical analysis of large data sets, visualization, and annotation of results. We will review the common approach to analysis of GWAS data and then discuss options to learn more from these data.

1 An overview of population genetics

One of the surprising findings of the Human Genome Project was the discovery that humans share the same DNA with the exception of about 0.1% of nucleotide bases [18]. These variations are called *single nucleotide polymorphisms* (SNPs) and occur when a single nucleotide (A, T, C, or G) in the genome sequence differs between individuals. Figure 1 provides an example. Some SNPs in genes are known to cause "monogenic disease" in which one of the SNP alleles determines a change in the protein produced by the gene that results in the disease. Sickle cell anemia is a well known example of a monogenic disease that is caused by a single mutation of the β-globin gene. The mutation determines a variant of the hemoglobin protein [35], and, although the disease is severe and affected subjects have a short life expectancy, the mutation has been maintained in the population because it is associated with resistance to malaria. Sickle cell anemia was the first monogenic disease ever described and led to Pauling's theory of molecular disease that opened a new chapter in the history of medicine [24].

Paola Sebastiani
Boston University, Boston MA, e-mail: sebas@bu.edu

Nadia Solovieff
Boston University, Boston MA, e-mail: ntimofee@bu.edu

L.S. Heath and N. Ramakrishnan (eds.), *Problem Solving Handbook in Computational Biology and Bioinformatics*, DOI 10.1007/978-0-387-09760-2_8,

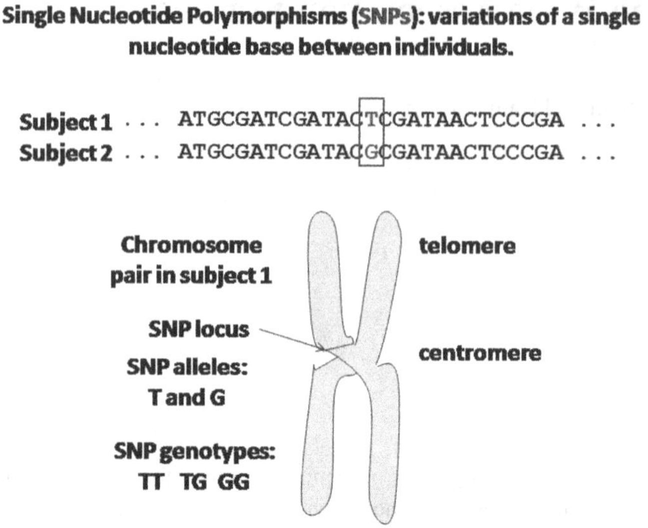

Fig. 1 *Example of a single nucleotide polymorphism (SNP) in which the nucleotide T is replaced by G in some subjects. The variants T and G are also called the SNP alleles, and the alleles in each chromosome pairs are called the genotypes.*

Our cells contain two copies of each chromosome, and monogenic diseases are classified as "dominant" or "recessive" based on the number of copies of the mutated alleles that are necessary for the disease to manifest. Disease with a genetic basis is also classified as autosomal or X-linked according to whether the mutation is on one of the autosomal chromosomes (1–22) or on the chromosome X. An autosomal dominant disease needs only one mutated allele in one of the chromosome pairs to manifest, while a recessive disease needs two mutated variants to manifest, and one mutated variant makes a subject simply a carrier of the disease. A carrier can transmit the mutated variant to the offspring but not the disease, and a recessive disease can only be transmitted when both parents are carriers. Sickle cell anemia is an example of a recessive autosomal disease because the mutated gene is on chromosome 11, and only subjects who carry two copies of the mutation and are therefore homozygous for the mutation are affected. This classification changes when the SNP is on chromosome X, so that one single variant can make an individual a carrier or affected based on gender. Haemophilia and the Duchenne and Becker forms of muscular dystrophy are known examples of X-linked recessive diseases in which one mutation on chromosome X in males causes the disease [23, 40].

Over the past decade, about 1,200 disease-causing genes have been identified by studying well characterized phenotypes — the physical manifestation of the genotype or, equivalently, the combination of variants in the chromosome pair — and using gene mapping techniques [4, 19]. Monogenic diseases are usually rare, while there are common diseases that have a genetic component shown by familial aggregation but do not follow the inheritance rules of dominant or recessive disorders. Examples include many common age-related diseases, such as diabetes [15, 34], cardiovascular disease [9], and dementia [38], that are presumed to be determined by the interaction of several genes (epistasis), and their interaction with environmental factors (gene $\times$ environment interaction). These common, complex traits are a large public health burden and the discovery of genetic profiles that can be used for both disease risk prediction as well as for the development of treatments and cures is one of the current priorities. However, the search for the genetic bases of complex traits faces two major difficulties. The first difficulty is the challenge of discovering a potentially very large number of genetic variants that are associated with the disease, and their modifications due to exposures to environmental conditions. The second difficulty is the definition of the correct phenotype to be used in the design of the study and the analysis of the data. In this chapter, we will focus on the first challenge. Modeling complex phenotypes is very challenging and there are only a few successful examples [21, 31, 33].

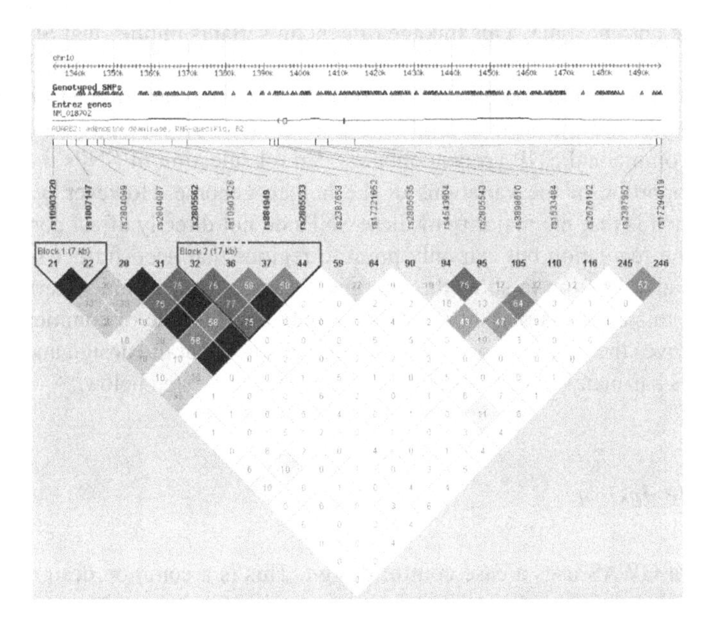

Fig. 2 *Linkage disequilibrium (LD) map of a region of the gene ADARB2. The map was generated using the program HaploView and genotype data of 18 SNPs from the 30 trios of the HapMap CEPH and includes annotation of the physical location of the gene of the human genome (chromosome 10, between position 1,340,000 and 1,490,000 bases). The white bar shows the location of the SNPs. Each square represents the correlation of two SNPs, measured by the correlation coefficient r^2 between the alleles of the two SNPs, and the shades of grey indicate the strength of the correlation ranging from no correlation (white) to strong correlation (black). The correlation analysis identifies two blocks of LD highlighted by the black outlines. The blocks are identified using an algorithm designed by S Gabriel [13].*

2 Genome-Wide Association Studies

In the past, genotyping costs limited the search for disease associated genes to candidate genes that were selected based on prior biological knowledge of pathways related to the phenotype. Genome-wide association studies (GWAS) were recently made possible by the technology of SNP arrays [14] and provide an unbiased approach in which hundreds of thousands or even millions of SNPs across the genome are tested for association with a phenotype. Commercially available SNP arrays allow the simultaneous genotyping of hundreds of thousands of SNPs that comprise a small proportion of the estimated 10 million existing SNPs but provide sufficient coverage of the variations. For example, the "Human660W-Quad BeadChip " (Illumina, San Diego, CA) includes more than 660,000 SNPs, providing comprehensive genomic coverage across multiple populations. This coverage is made possible by the block structure of the human genome that is due to linkage disequilibrium (LD) and was described by Gabriel in [13]. LD is the effect of non-random association of SNPs that results in the inheritance of blocks or "haplotypes" of nearby SNPs in the same chromosome. This linkage of genetic variants implies that SNPs in the same haplotypes are mutually informative so that a subset of them is sufficient to tag blocks of the human genome with high precision [8, 30], and to identify genomic regions that are associated with disease. Figure 2 shows an example. Based on this property, commercial SNP arrays contain a careful selection of SNPs that capture a large proportion of the variations of the human genome. However, a caveat of this approach is that the majority of these SNPs do not directly affect gene expression or gene regulation but can only point to regions that need further search and experimental validation to show the biological relevance. A technical term used to emphasize this aspect is that these SNPs are markers. The use of commercial SNP arrays removes the task of choosing SNPs for the study, but the design and analysis of a GWAS has many challenges that we describe step by step below.

2.1 Study design

Typically, a GWAS uses a case control design. This is a common design in many epidemiology studies with a dichotomous outcome, for example the presence or absence of a disease. Subjects are labeled as cases if they have the disease and controls if they do not [5, 16]. The definition of controls is not always obvious. A control subject should be free of disease but must also be free of other traits that are not shared by cases, to avoid confounding. For example, if the phenotype of interest is known to manifest within a certain age, it is tempting to choose controls that are much older than the cases to limit the chance that they never develop the phenotype. This choice for controls would introduce confounding by age, because cases and controls are different not only by presence/absence of disease but also by age, and the consequence of this confounding is that the genetic differences between cases and controls may be related to aging rather than the disease. A common approach to avoid confounding is to match controls on variables that are not of interest in the study, such as exposure to some environmental conditions.

Some investigators have used referent cohort subjects used in other genetic studies. For example, the first large GWAS published by the Wellcome Trust Case-Control Consortium used the same pool of 3,000 controls chosen from the British population to search for genetic modifiers of seven common traits [9]. This strategy is becoming more and more feasible with the increasing availability of GWAS data from dbGaP, the database of genotype-phenotype associations (http://www.ncbi.nlm.nih.gov/gap), and the Illumina control database. However, this approach can introduce confounding due to population stratification, which occurs when allele frequencies differ between cases and controls because of ancestral differences [25]. This topic will be described in detail in section 2.6. When the trait of interest is a quantitative measure, such as blood pressure or fetal hemoglobin concentration, it is recommended that subjects included in the study represent sufficient variability of the trait.

2.2 *Quality control*

To avoid erroneous findings, one must carefully clean the data prior to analysis. SNPs and samples with low quality can be identified with a low SNP call rate, defined as the proportion of non-missing genotypes per SNP, and a low individual call rate, defined as the proportion of non-missing genotypes per subject, respectively [42]. SNPs and samples with low call rates should be removed from all subsequent analyses. Processing large numbers of samples can result in sample swaps and errors in sample tracking in the lab. There are 2 tests that can be performed to identify some of these types of errors. The first involves comparing the gender assignment in the study database to the gender predicted by the heterozygosity of the SNPs on chromosome X. Clusters of misclassified subjects can identify batches of samples that may have flipped or batches of samples with poor DNA quality. The second test will be discussed in detail in Section 2.3 and involves comparing known familial relations from the study database to the relations inferred from the genetic data. Both of these analyses can be easily performed by a publicly available software package PLINK [29], which will be discussed in Section 3.

Investigators often remove SNPs with low minor allele frequencies to avoid false positive findings and SNPs not meeting Hardy Weinberg Equilibrium (HWE) in the controls. HWE states that both allele and genotype distributions should remain stable in a population when there are no external perturbations such as emigration or large number of deaths due, for example, to epidemics [5]. If the allele M of a SNP has frequency $p(M) > 0.5$, and the other allele m has frequency $1 - p(M)$, HWE states that the genotypes MM, Mm and mm should have frequencies $p(M)^2$, $2p(M)(1 - p(M))$ and $(1 - p(M))^2$. Statistical tests can be used to check that a SNP genotype distribution follows HWE. However, we advise analyzing all SNPs regardless of whether or not they are in HWE and then carefully examining the

	IBD Probabilities			
Relative Pair	0	1	2	P(IBD)
MZ Twins	0	0	1	1
Full Sibs	0.25	0.5	0.25	0.5
Parent-Offspring	0	1	0	0.5
Grandparent-Grandchild	0.5	0.5	0	0.25
Half-Sibs	0.5	0.5	0	0.25
Avuncular	0.5	0.5	0	0.25
First Cousin	0.75	0.25	0	0.125
Unrelated	1	0	0	0

Fig. 3 *Relation between probabilities of genome-wide alleles shared IBD and relatedness. Column 1 describes the type of relation, columns 2–4 report the genome-wide proportion of alleles shared by IBD that can be 0 (column 2), 1 (column 3) and 2 (column 4). The last column indicates the expected probability of alleles shared by IBD for various relations. For example, monozygotic twin (row one) should share the same DNA and therefore the probability of any two alleles shared IBD is 1, while siblings will share 0 alleles IBD with probability* $1/4$, *1 allele IBD with probability* $1/2$ *and 2 alleles IBD with probability* $1/4$. *Unrelated samples sharing 2 alleles with a probability of 1 can point to sample swaps.*

validity of these SNPs not in HWE if they are found significantly associated with the phenotype.

2.3 Discovering hidden relations

Familial relations between pairs of subjects can be quantified by the number of alleles shared between subjects identically by descent (IBD). An allele is IBD when it is the same allele from the same ancestor while an allele is shared identically by state (IBS) when it is the same allele but not from the same recent ancestor [20]. Using Mendel's law, we can estimate the probability that two family members share 0,1 or 2 alleles IBD, and Figure 3 provides some examples. These probabilities can be extended to any degree of relatives and can be estimated using a genome-wide panel of SNPs. IBD estimation, implemented in PLINK [29], is based on the number of alleles shared genome wide IBS and is very computationally expensive when computed on all pairs of subjects but is a valuable quality control measure. Comparing known familial relations to the estimated probabilities of IBD can help identify errors in sample tracking or swapped samples. For example, we expect that a parent-offspring pair will have IBD probabilities approximately equal to $1/2, 1/2$ and 0 for 0,1 and 2 alleles shared IBD. The IBD analysis can also identify unknown related individuals. One must remove related individuals prior to the analysis or appropriately account for the family structure in a family based association analysis to avoid inflating the false positive rate. Additionally, duplicate samples can easily be identified as sharing 2 alleles IBD with a probability close to 1. Duplicates are not uncommon when comparing individuals in different studies, for example a discovery and replication study, of a rare trait since individuals participating in a study are likely to enroll in multiple studies.

2.4 Single SNP analysis

The common approach to statistical analysis of GWAS data is a single SNP analysis: a statistical test is conducted to verify, for each SNP, the null hypothesis that there is no association between the SNP and the phenotype against the alternative hypothesis that there is an association. The frequentist approach weighs the evidence against the null hypothesis by the p-value that is defined as the probability of observing a stronger association than that estimated from the data under the null hypothesis of no association. If the null hypothesis is true, the p-value should be large because estimating a strong association by chance is unlikely to happen. Therefore, a small p-value is taken as evidence against the null hypothesis, and it is the rationale for the decision rule to reject the null hypothesis when the p-value is smaller than a fixed significance level [7].

This approach is not error free, because we could observe by chance an extreme association even when the null hypothesis is true. Rejecting the null hypothesis when it is true is known as a Type I error, while accepting the null hypothesis when the alternative hypothesis is true is a Type II error. The power of the test is defined as the probability of accepting the alternative hypothesis of association when the alternative hypothesis is true [7].

$$\text{Type I error: } R\, H_0 | H_0 \text{ True}$$

$$p \text{ value: } P(R\, H_0 | H_0 \text{ True })$$

$$\text{Power: } P(R\, H_0 | H_0 \text{ False })$$

This approach requires a method to estimate the association and a significance level α to be used as the threshold for the p-value. We consider two cases, based on whether the phenotype is a continuous or a categorical trait.

The genetic effect of a SNP X on a continuous trait Y can be modeled using a linear regression model:

$$E(Y|X_1, X_2) = \beta_0 + \beta_1 X_1 + \beta_2 X_2$$

where X_1 and X_2 are variables that take values $X_1 = 1$ when the SNP genotype is Mm and 0 otherwise, and $X_2 = 1$ when the SNP genotype is mm and 0 otherwise. The regression coefficient β_0 represents the average value of the phenotype when the SNP genotype is MM, while the combinations of coefficients $\beta_0 + \beta_1$ and $\beta_0 + \beta_2$ represent the average values of the phenotype when the SNP genotypes are Mm and mm respectively. Therefore, β_1 is the average change of phenotype between genotypes Mm and MM, and β_2 is the average change of phenotype between genotypes mm relative to MM. The two parameters represent the genetic effect. As long as one of the regression coefficients is statistically different from 0, there is a *genotypic association* between the SNP and the phenotype, and the significance of the association can be tested using standard least squares methods when the phenotype follows a normal distribution [2].

A popular alternative is to represent the three genotypes by the variable X taking values $0 = MM$, $1 = Mm$, and $2 = mm$ and use this variable in linear regression. This parameterization is known as the additive genetic model [20], and the regression coefficient of the variable X represents the average change in the trait for each extra copy of the allele m. The additive model is easy to interpret and therefore most commonly used. Other parsimonious parameterizations include dominant or recessive models in which genotypes are aggregated in two groups. The dominant model for the allele M tests the associations of genotypes grouped as MM and Mm versus mm, while the recessive model for the allele M uses the grouping MM versus Mm and mm. The former can be implemented by using a simple regression model with $X = 1$ if the SNP genotype is mm and 0 otherwise, while the latter parameterization can use $X = 1$ when the SNP genotype is either Mm or mm and 0 otherwise [6, 20]. All these parameterizations are more parsimonious but less general than the genotypic association and failure to detect association with the additive, or dominant, or recessive model does not imply lack of any association between the SNP and the trait [33]. When the trait correlates with other covariates, such as gender, the genetic association can be adjusted for the covariates by adding them to the regression equation, or by modeling the residuals from the regression model that includes only the covariates. Adjustment should be done only for covariates that are significantly associated with the trait to avoid unnecessary loss of power.

When the subjects are grouped as cases and controls, general genotype association can be tested using the traditional χ^2 test of independence for a 2x3 contingency table [7]. More parsimonious procedures include the Armitage trend test [16], in which genotypes are recoded to model a linear increase in the odds of the disease on the logarithmic scale for each different genotype , or associations of dominant or recessive models in which genotypes are aggregated in two groups as described above. Allelic association can be tested by recoding the data from genotypes into alleles [20].

Logistic regression can also be used to model any of these associations [16]. The idea is to model the odds for the disease in the logarithmic scale as the linear regression function:

$$log\left\{\frac{p(Disease|X_1,X_2)}{1-p(Disease|X_1,X_2))}\right\} = \beta_0 + \beta_1 X_1 + \beta_2 X_2$$

where the variables code for different genotypes as described earlier. The regression coefficients have direct interpretation as log-odds ratios for disease. For example,

$$\beta_1 = log\left\{\frac{p(Disease|X_1=1,X_2)}{1-p(Disease|X_1=1,X_2))}\right\} - log\left\{\frac{p(Disease|X_1=0,X_2)}{1-p(Disease|X_1=0,X_2))}\right\}$$

represents the log-odds ratio for disease in subjects with the genotype *Mm* relative to *MM*. Similarly, β_2 represents the log-odds ratio for disease in subjects with the genotype *mm* relative to *MM*. Besides the genetic effects, the regression equation can include covariates, and it can be extended to include multiple interacting SNPs as well as gene-environment interactions. The significance of the association is tested using the likelihood ratio test with large sample approximations [22]. When the frequency of some genotypes is < 5, the large sample approximation may fail, and permutation methods should be used.

Bayesian methods are also becoming popular in genetic epidemiology [36]. The Bayesian approach to hypothesis testing uses the data to update the prior probabilities of the two hypotheses of no association and association into their posterior probabilities. The decision to reject the null hypothesis is based on an "ad hoc" threshold on the odds of the posterior probabilities that trades off sensitivity and specificity. The review in [33] discusses some examples in details. It is important to emphasize that, in the frequentist approach, the decision to reject the null hypothesis of no association uses a threshold on the p-value to minimize the probability of the type I error. This procedure does not assess per se whether the null hypothesis is true or false. In the Bayesian approach, the decision to reject the null hypothesis is based directly on the probability that the null hypothesis is false, given the data and only the Bayesian inference allows for a direct assessment of the likelihood of parameters and hypotheses [2].

The results of the analysis are usually visualized using a Manhattan plot that displays the $\log_{10}(p\text{-value})$ of the tested association. Figure 4 shows an example. The plot highlights the regions of the genome that are associated with the trait and informs about the robustness of the associations when clusters of nearby SNP are simultaneously associated.

2.5 *Power and multiple comparisons*

Because the number of SNPs analyzed in a GWAS can be as large as several hundred thousands, if each SNP is tested with a significance level $\alpha = 0.05$ the probability of rejecting one of more null hypotheses by chance is very large. This number is the "family-wise error rate" and can be calculated with the formula

$$\text{probability}(\text{number of Type I error } > 0) = (1-\alpha)^N,$$

where N is the number of hypotheses that are tested [39]. The family-wise error rate approaches 1 very quickly when $\alpha = 0.05$ and N is relatively large. An equivalent way of assessing the magnitude of the problem is to consider the fact that the number of false positive associations that are expected by chance when testing N null hypotheses — assuming all are true — is $\alpha \times N$. For example this number is 25,000 when $\alpha = 0.05$ and $N = 500,000$. The Bonferroni correction is a conservative solution that limits the number of false positive associations by dividing the

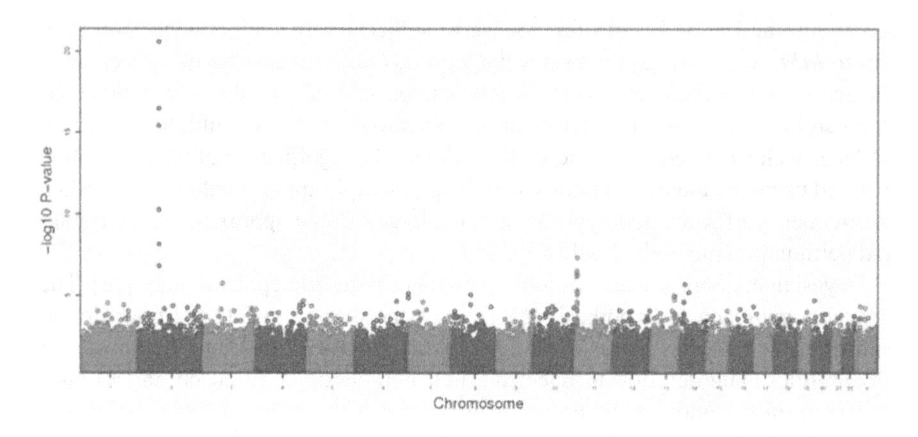

Fig. 4 *Manhattan plot that summarizes the results of a GWAS of fetal hemoglobin in sickle cell anemia patients. The x-axis reports the SNPs in each chromosome ordered by physical position on the reference human genome sequence, and the y-axis reports the* $\log_{10}$*(p-value) of each SNP association that was tested using the additive model. Different chromosomes are visualized by alternating color bands. (See online version for color figure). High levels of fetal hemoglobin is associated with lower rate of vaso-occlusive complications in patients with the disease and the discovery of genetic modulators of fetal hemoglobin is important for prognostic and therapeutic reasons. We conducted a GWAS of fetal hemoglobin in more than 1,000 patients with sickle cell anemia and the Manhattan plot shows a large spike on chromosome 2 that corresponds to a region containing the gene BCL11A. The cluster of SNPs with high statistical significance suggests the there may be some functional variant in the region marked by these SNPs that affects changes in fetal hemoglobin expression.*

significance level used in each individual test by the number of tests N, so that α is replaced by α/N. This solution requires significance levels of the order 10^{-6} or smaller to meet genome-wide significance and therefore very large sample sizes are necessary to have the statistical power to detect small genetic effects. For example, Wang and colleagues showed that to estimate an allelic odds ratio of 1.5 with 80% power when the disease allele has a frequency between 0.4 and 0.5, we need a sample size of 1,000 cases and 1,000 controls. The sample size necessary to detect the same genetic effect when the disease allele frequency is 10% is 2000 cases and 2000 controls and increases almost exponentially with smaller disease allele frequencies [37].

The Bonferroni correction is too conservative and controlling the false discovery rate rather than the overall false positive rate has been proposed as a less conservative method. The false discovery rate is the proportion of false positive associations among the associations detected as significant and can be controlled by using a simple algorithm [3]. Neither the Bonferroni correction nor the false discovery rate changes the rank of the p-values, but each simply provides additional guidance as to which associations are most significant across the entire study. The bias of the frequentist approaches to control the false positive rate (or the false discovery rate) means that the power of a GWAS can only be increased by increasing the sample size. The Bayesian approach offers a different solution and often a gain of power by using a more sophisticate decision theoretic approach to hypothesis testing. The review in [33] provides several examples.

Replication of the results from GWAS in at least one independent study is a common procedure to remove false positive associations [26]. However, the choice of replication set is very important, and the subjects should be chosen from an independent study population, with the same genetic background as the primary study population, the same definition of the phenotype. Furthermore, the replication should confirm the association of the same SNPs with the same genetic model and show the same genetic effect. Replication of findings in a population with different genetic backgrounds can strengthen the evidence of true associations and identify variants that are robust to different genetic background and environmental exposures. However, failure to reproduce an association in a genetically different population should not be taken as evidence of a false positive. Another emerging approach to replication of GWAS is the use of meta-analysis, which combines the results of different studies using formal statistical procedures [12].

2.6 Population stratification

Population stratification is a major confounder in GWAS and occurs when the allele frequencies differ between cases and controls because of ancestral differences rather than differences due to the phenotype [25]. If the frequency of an allele for a SNP differs across ethnic groups, and the phenotype also differs with respect to ethnicity, then a false positive association will be found between the phenotype and the SNP if one does not appropriately account for ethnicity. This is a classic example of confounding where ethnicity is associated with both the predictor and the outcome and thus confounds the relationship between the two variables.

Across the genome, there are many SNPs that differ across ethnic groups, and thus many of the test statistics from a GWAS will be inflated as can be seen in the QQ plot of Figure 5. Note that this will only occur if the population substructure differs with respect to the phenotype. The amount of inflation in the test statistics is often summarized by the genomic control inflation factor λ that is defined as the median test statistic divided by the median test statistic assuming no association for any SNPs [10]. Since one expects only a small percentage, certainly less than 50%, of the SNPs to be associated with the phenotype the median test statistic in the analysis should be approximately equal to the median test statistic under the null and thus λ should be approximate equal to 1 and values greater than 1 suggest inflation.

There are a number of methods to detect and account for population substructure including genomic control, structured association and principal component analysis. The method of genomic control advocates dividing each test statistic by the genomic control inflation factor to reduce the inflation. However, some SNPs exhibit large differences in allele frequencies across ethnic groups, while other SNPs exhibit no

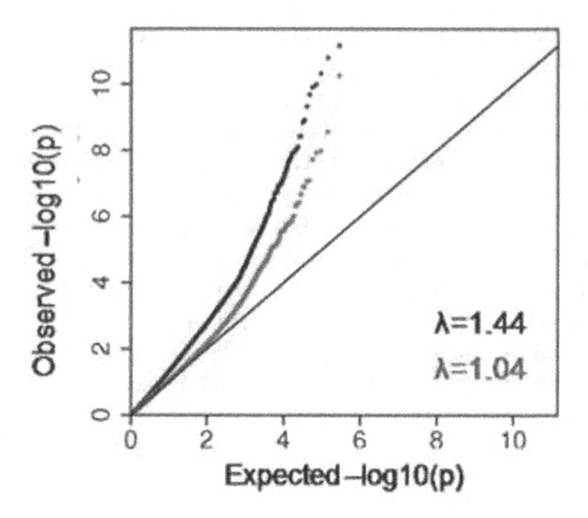

Fig. 5 *The figure displays what is known as a QQ-plot. The x-axis displays the ordered* $-\log_{10}(p-$ *values) that are expected when testing N null hypothesis and all are true. The y-axis displays the ordered* $-\log_{10}(p-$ *values) that are calculated from the GWAS. Typically, the number of true significant associations is small, and so we expected that the plot of the observed versus expected p-values aligns with the solid black line. The dotted, darker line displays the results of a GWAS in the presence of inflation due to population stratification as seen by the values pulling away from the solid black line and a large genomic control of* $\lambda = 1.44$. *The paler line (red in the online color figure) displays the results of a GWAS with little to no inflation with a much smaller genomic control value and with values deviating from the solid black line only in the tail consistent with real findings.*

differences, and thus by dividing all test statistics by the same amount results in a loss of power.

Structured association is a model-based clustering technique that groups individuals into clusters based on a subset of SNPs. It is advantageous to use SNPs that are known to be ancestrally informative for the population of interest. The analysis, implemented in the program STRUCTURE [28], assigns a probability of being in a particular cluster to each individual.

Principal components analysis (PCA) is one of the most widely used methods in GWAS since it uses information from a genome wide set of SNPs and because of its convenient implementation in the software EIGENSOFT [27]. PCA summarizes the variability of a genome wide set of SNPs by creating principal components which are linear combinations of all of the SNPs. The first principal component (PC) will capture the largest amount of variability in the data and each consecutive PC will capture less and less variability. The top principal components generally capture the population substructure due to ethnicity in GWA data. Figure 6 plots the top 2 PCs from a PCA of a case control study of exceptional longevity in which all subjects are Caucasian. The pattern observed is typical for a cohort of Caucasians of European descent and has been observed in a number of different studies [25]. The cluster in the top left corner contains Ashkenazi Jewish subjects and the line of subjects spanning from the top right to the bottom middle of the plot represents a cline ranging from northwestern to southeastern Europe. Higher order PCs capture even finer gradients of population substructure. Based on the PC plot, it is evident that there are many more controls from southeastern Europe than cases and this imbalance causes severe inflation in the test statistics with GC value of 1.44 (see QQ plot). To control for population stratification, investigators often adjust for the top PCs in a regression model between the phenotype and the SNP and the adjustment generally reduces the inflation substantially. Alternatively, one can also create clusters from the top PCs and then match cases and controls within each cluster to balance the proportions of cases and controls and reduce population stratification bias.

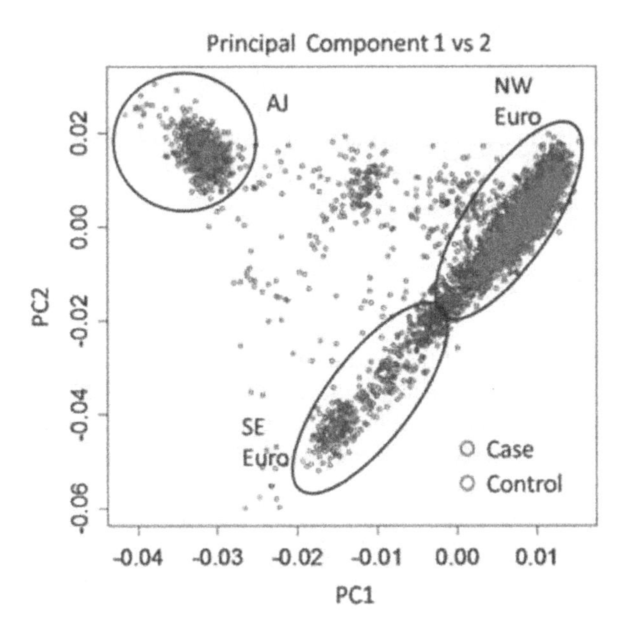

Fig. 6 *Plot of PC1 vs PC2 of a case control study of exceptional longevity. Cases are depicted as darker dots (blue in the online color figure) and cases are depicted as paler dots (red in the online color figure). The top 2 PCs separate the Ashkenazi Jewish subjects in the top left corner of the plot. The line formed by subjects from the top right to the bottom middle of the plot represents a northwestern to southeastern cline across Europe. Although the structure of the cases and controls is similar there are many more controls from southeastern Europe than cases, which will cause inflation in the test statistics.*

3 Resources

There are a number of publicly available programs to aid the analysis and interpretation of GWAS and genetic studies in general. The statistical package R has several tools for analysis of GWASs, and some tools for visualization and annotation of results are available through Bioconductor. PLINK is a free, open-source whole genome association analysis toolset, which allows the user to perform a wide range of data management and quality control tasks and a number of basic analyses including IBD estimation, standard association analysis, LD estimation, and haplotype analysis in a very timely and efficient fashion [29]. The software package EIGENSOFT includes a module to perform principal component analysis, described in Section 2.6, and can conveniently accept a number of different file formats including the format required by PLINK [27]. IMPUTE is a program for genome wide imputation that is very accurate and useful to synchronize genotype data collected with different SNP arrays [41].

Issues of patient confidentiality make it difficult to share genetic data without appropriate certification and an institutional review board approved protocol. Many data sets from GWASs are available from dbGAP, the database of genotype and phenotype. Access to the data has to be approved to qualified applicants in order to guarantee patients protection. Annotated results from GWASs are maintained at the Genome Research Institute (http://www.genome.gov/gwastudies/) and a graphical display of the results mapped on the human genome is available from the HapMap website (http://www.hapmap.org/karyogram/gwas.html).

4 Advanced Topics

GWASs have rapidly changed the field of genetics and in the last few years have produced massive amounts of data. The yield of discovery from GWASs has been, however, modest and several conjectures have been proposed to justify the lack of results. Possible explanations include the possibility that commercial arrays are not sufficiently comprehensive and many more variants remain to be discovered using next generation sequencing approaches [1], and the fact that many GWASs were not sufficiently powered to identify genetic variants with modest effects [11]. The stringent thresholds imposed to control the false positive rate reduce the statistical power of a GWAS even further, and family based studies may be a better alternative for dissecting the genetic basis of common diseases. We did not discuss statistical approaches to family based studies in this chapter, and it is important to realize that, when related people are included in a study, the statistical analysis needs to take into account the correlation between their genetic backgrounds. A comprehensive review to this topic is in reference [17].

Another important fact is that the bulk of data produced by GWASs remains largely unexplored, because of the challenge of mining and modeling massive data sets. Genetic data can be used for risk prediction modeling, and integration of genetic data with other gene products can speed up the discovery of disease mechanism and identify targets for further studies. This powerful use of genetic data is however challenging traditional statistical methods. Models that are suitable to describe complex systems can be used to describe complex genetic diseases. Of all possible choices including classification and regression trees, random forests and other machine learning algorithms, we find Bayesian networks a more flexible method of analysis for complex genetic disease, and we used this model formalism to build a risk prediction model of stroke in patients with sickle cell anemia [32]. Adapting structure learning of Bayesian Networks to GWAS data is a non trivial problem and substantial work is needed in this area.

5 Exercise

1. Genotype data were collected for a SNP on 1,000 diseased patients and 1,000 healthy patients. The genotype frequencies are displayed below.

	AA	*AG*	*GG*	*Total*
Diseased	525	300	175	1000
Not Diseased	640	245	115	1000
	1165	545	290	2000

   a. What is the frequency of the minor allele in the general population? What is the frequency of the minor allele in diseased subjects?
   b. We are interested in testing the association between disease status and the SNP using a logistic regression model. Code the SNP genotypes for the additive, dominant, recessive and genotypic models. Assume that m in this chapter refers to the minor allele.
   c. Compute the odds ratio for the dominant and recessive models. What can you conclude based on the odds ratios?

2. The allele frequencies for a SNP in the population are $p(G) = 0.85$ and $p(T) = 0.15$. What are the expected genotype counts in a random sample of 1000 individuals, assuming Hardy Weinberg Equilibrium?
3. If the estimated IBD proportions between 2 people are $0.20, 0.55$, and 0.25 for sharing $0, 1$ and 2 alleles IBD, what is their relation? What if the IBD proportions are $0.44, 0.52,\ 0.04$ for sharing $0, 1$ and 2 alleles IBD?
4. If we test $350{,}000$ SNPs and implement 0.10 significant level, what is the family-wise error rate? What is the significance cutoff using a Bonferroni correction?

Acknowledgements Research supported by NIH/NHLBI R01 HL87681 and R01 HL068970.

References

1. Altshuler, D., Daly, M.J., Lander, E.S.: Genetic mapping in human disease. Science **322**(5903), 881–888 (2008)
2. Balding, D.J.: A tutorial on statistical methods for population association studies. Nat Rev Genet **7**(10), 781–791 (2006)
3. Benjamini, Y., Drai, D., Elmer, G., Kafkafi, N., Golani, I.: Controlling the false discovery rate in behavior genetics research. Behav Brain Res **125**(1-2), 279–284 (2001)
4. Botstein, D., Risch, N.: Discovering genotypes underlying human phenotypes: Past successes for Mendelian disease, future approaches for complex disease. Nat. Genet. **33**, 228–237 (2003). Suppl.
5. Cardon, L.R., Bell, J.I.: Association study designs for complex diseases. Nat Rev Genet. **2**, 91–99 (2001)
6. Carey, G.: Human Genetics for the Social Sciences. Sage Publications (2003)
7. Casella, G., Berger, R.L.: Statistical Inference. Duxbury Press, Belmont, Ca (1990)
8. Consortium, I.H.: A haplotype map of the human genome. Nature **429**, 1300–1320 (2005)
9. Consortium, W.T.C.C.: Genome-wide association study of 14,000 cases of seven common diseases and 3,000 shared controls. Nature **447**(7145), 661–678 (2007)
10. Devlin, B., Roeder, K.: Genomic control for association studies. Biometrics **55**(4), 997–1004 (1999)
11. Donnelly, P.: Progress and challenges in genome-wide association studies in humans. Nature **456**(7223), 728–731 (2008)
12. Egger, M., Smith, G.D.: Meta-analysis. potentials and promise. BMJ **315**(7119), 1371–1374 (1997)
13. Gabriel, S., Schaffner, S., Nguyen, H., Moore, J., Roy, J., Blumenstiel, B., Higgins, J., DeFelice, M., Lochner, A., Faggart, M., Liu-Cordero, S., Rotimi, C., Adeyemo, A., Cooper, R., Ward, R., Lander, E., Daly, M., Altshuler, D.: The structure of haplotype blocks in the human genome. Science **296**(5576), 2225–2229 (2002)
14. Gunderson, K.L., Steemers, F.J., Lee, G., Mendoza, L.G., Chee, M.S.: A genome-wide scalable snp genotyping assay using microarray technology. Nat Genet **37**(5), 549–554 (2005)
15. Hakonarson, H., Grant, S.F.A., Bradfield, J.P., Marchand, L., Kim, C.E., Glessner, J.T., Grabs, R., Casalunovo, T., Taback, S.P., Frackelton, E.C., Lawson, M.L., Robinson, L.J., Skraban, R., Lu, Y., Chiavacci, R.M., Stanley, C.A., Kirsch, S.E., Rappaport, E.F., Orange, J.S., Monos, D.S., Devoto, M., Qu, H.Q., Polychronakos, C.: A genome-wide association study identifies KIAA0350 as a type 1 diabetes gene. Nature **448**(7153), 591–594 (2007)
16. Jewell, N.P.: Statistics for Epidemiology. CRC/Chapman and Hall, Boca Raton (2003)
17. Laird, N.M., Lange, C.: Family-based designs in the age of large-scale gene-association studies. Nat Rev Genet **7**(5), 385–394 (2006)
18. Lander, E.S., Consortium, I.H.G.S.: Initial sequencing and analysis of the human genome. Nature **409**(6822), 860–921 (2001)
19. Lander, E.S., Schork, N.J.: Genetic dissection of complex traits. Science **265**, 2037–2048 (1994)
20. Lewis, C.M.: Genetic association studies: Design, analysis and interpretation. Brief Bioinform **3**(2), 146–153 (2002)
21. Loscalzo, J., Kohane, I., Barabasi, A.L.: Human disease classification in the postgenomic era: A complex systems approach to human pathobiology. Mol Syst Biol **3**, 124 (2007)
22. McCullagh, P., Nelder, J.: Generalized Linear Models, 2nd edn. Chapman and Hall, London and New York (1989)
23. Monaco, A.P., Neve, R.L., Colletti-Feener, C., Bertelson, C.J., Kurnit, D.M., Kunkel, L.M.: Isolation of candidate cDNAs for portions of the Duchenne muscular dystrophy gene. Nature **323**(6089), 646–650 (1986)
24. Nagel, R.L.: Pleiotropic and epistatic effects in sickle cell anemia. Curr. Opin. Hematol. **8**, 105–110 (2001)

25. Patterson, N., Price, A.L., Reich, D.: Population structure and eigenanalysis. PLoS Genet **2**(12), e190 (2006)
26. Pearson, T.A., Manolio, T.A.: How to interpret a genome-wide association study. JAMA **299**(11), 1335–1344 (2008)
27. Price, A.L., Patterson, N.J., Plenge, R.M., Weinblatt, M.E., Shadick, N.A., Reich, D.: Principal components analysis corrects for stratification in genome-wide association studies. Nat Genet **38**(8), 904–909 (2006)
28. Pritchard, J.K., Stephens, M., Donnelly, P.: Inference of population structure using multilocus genotype data. Genetics **155**(2), 945–959 (2000)
29. Purcell, S., Neale, B., Todd-Brown, K., Thomas, L., Ferreira, M.A.R., Bender, D., Maller, J., Sklar, P., de Bakker, P.I.W., Daly, M.J., Sham, P.C.: PLINK: A tool set for whole-genome association and population-based linkage analyses. Am J Hum Genet **81**(3), 559–575 (2007)
30. Sebastiani, P., Lazarus, R., Weiss, S.T., Kunkel, L.M., Kohane, I.S., Ramoni, M.F.: Minimal haplotype tagging. Proc Natl Acad Sci U S A **100**(17), 9900–9905 (2003)
31. Sebastiani, P., Nolan, V.G., Baldwin, C.T., Abad-Grau, M.M., Wang, L., Adewoye, A.H., McMahon, L.C., Farrer, L.A., Taylor, J.G., Kato, G.J., Gladwin, M.T., Steinberg, M.H.: A network model to predict the risk of death in sickle cell disease. Blood **110**(7), 2727–2735 (2007)
32. Sebastiani, P., Ramoni, M.F., Nolan, V., Baldwin, C.T., Steinberg, M.H.: Genetic dissection and prognostic modeling of overt stroke in sickle cell anemia. Nat Genet **37**(4), 435–440 (2005)
33. Sebastiani, P., Timofeev, N., Dworkis, D.A., Perls, T.T., Steinberg, M.H.: Genome-wide association studies and the genetic dissection of complex traits. Am J Hematol **84**(8), 504–515 (2009)
34. Sladek, R., Rocheleau, G., Rung, J., Dina, C., Shen, L., Serre, D., Boutin, P., Vincent, D., Belisle, A., Hadjadj, S., Balkau, B., Heude, B., Charpentier, G., Hudson, T.J., Montpetit, A., Pshezhetsky, A.V., Prentki, M., Posner, B.I., Balding, D.J., Meyre, D., Polychronakos, C., Froguel, P.: A genome-wide association study identifies novel risk loci for type 2 diabetes. Nature **445**(7130), 881–885 (2007)
35. Steinberg, M.H.: Predicting clinical severity in sickle cell anaemia. Br J Haematol. **129**, 465–481 (2005)
36. Stephens, M., Balding, D.J.: Bayesian statistical methods for genetic association studies. Nat Rev Genet **10**(10), 681–690 (2009)
37. Wang, W.Y.S., Barratt, B.J., Clayton, D.G., Todd, J.A.: Genome-wide association studies: Theoretical and practical concerns. Nat Rev Genet **6**(2), 109–118 (2005)
38. Waring, S.C., Rosenberg, R.N.: Genome-wide association studies in Alzheimer disease. Arch Neurol **65**(3), 329–334 (2008)
39. Yang, Q., Cui, J., Chazaro, I., Cupples, L.A., Demissie, S.: Power and type I error rate of false discovery rate approaches in genome-wide association studies. BMC Genet **6 Suppl 1**, S134 (2005)
40. Youssoufian, H., Kazazian, H.H., Phillips, D.G., Aronis, S., Tsiftis, G., Brown, V.A., Antonarakis, S.E.: Recurrent mutations in haemophilia A give evidence for CpG mutation hotspots. Nature **324**(6095), 380–382 (1986)
41. Zhao, Z., Timofeev, N., Hartley, S.W., Chui, D.H., Fucharoen, S., Perls, T.T., Steinberg, M.H., Baldwin, C.T., Sebastiani, P.: Imputation of missing genotypes: an empirical evaluation of impute. BMC Genet **9**, 85 (2008)
42. Zondervan, K.T., Cardon, L.R.: Designing candidate gene and genome-wide case-control association studies. Nat Protoc **2**(10), 2492–2501 (2007)

Part III
Proteins: Structure, Function, and Biochemistry

This third part of the book collects algorithms and techniques for understanding proteins, from the individual level to networks of proteins.

Novel Perspectives on Protein Structure Prediction

Bonnie Berger, Jérôme Waldispühl

Abstract Our understanding of the protein structure prediction problem is evolving. Recent experimental insights into the protein folding mechanism suggest that many polypeptides may adopt multiple conformations. Consequently, modeling and prediction of an *ensemble* of configurations is more relevant than the classical approach that aims to compute a single structure for a given sequence. In this chapter, we review recent algorithmic advances which enable the application of statistical mechanics techniques to predicting these structural *ensembles*. These techniques overcome the limitations of costly folding simulations and allow a rigorous model of the conformational landscape. To illustrate the strength and versatility of this approach, we present applications of these algorithms to various typical protein structure problems ranging from predicting residue contacts to experimental X-ray crystallography measures.

1 Introduction

The prediction of a protein's tertiary structure from its primary structure is one of the most important problems in computational biology and biochemistry [24, 53] yet also one of the most difficult [7]. Classical approaches to predicting protein structure follow the traditional schema, which aims to associate a single structure to each sequence. While this view of the problem seems supported by the the way that data have been accumulated over years in databases, the reality of the phenomena as described by experimentalists can be significantly more complex [4]. For instance,

Bonnie Berger
Department of Mathematics & Computer Science and AI Lab, MIT, Cambridge, MA, USA, e-mail: bab@mit.edu

Jérôme Waldispühl
School of Computer Science, McGill University, Montreal, QC, Canada, e-mail: jeromew@cs.mcgill.ca

L.S. Heath and N. Ramakrishnan (eds.), *Problem Solving Handbook in Computational Biology and Bioinformatics*, DOI 10.1007/978-0-387-09760-2_9,

some proteins are intrinsically unstructured and characterized by lack of stable tertiary structure [27]. Other proteins such as prions have multiple stable, distinct, and functionally-related conformations [26, 44, 65]. There is also evidence that some proteins fold in multiple step processes using intermediate meta-stable structures in the folding landscape [61, 63]. Thus it is not unlikely for proteins to have alternate folds.

Beyond these examples, considering the protein structure prediction problem in a broader context can also radically change our perspective. Indeed, a cell contains many duplicates of the same protein sequence, which are all folding independently, potentially into similar but not necessarily identical structures. A molecule is never frozen forever in a rigid structure. In vivo, a polypeptide is perpetually adapting its structure, jumping from one stable conformation to another.

All these observations suggest that the protein structure prediction problem needs to be revisited. Computing a single conformation cannot reflect the diversity of the folds that a protein may adopt *in-vivo*. A complete view of the phenomena requires an embodiment of all these varying aspects of the same molecule in the same comprehensive model. We illustrate the differences between classical and modern approaches in Fig. 1. While the classical approach aims to assign a single structure to a given protein sequence (Fig. 1(a)), modern techniques aim to compute the *ensemble* of conformations that a polypeptide can adopt (Fig. 1(b)).

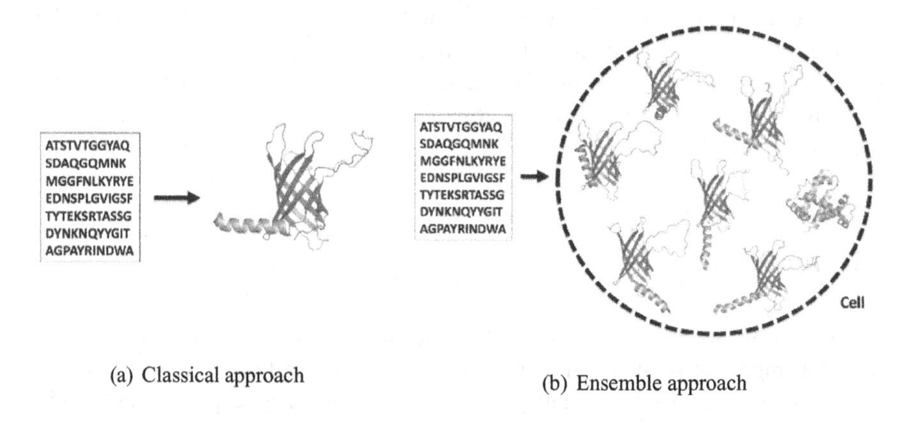

(a) Classical approach

(b) Ensemble approach

Fig. 1 Folding approaches. From a sequence, the classical method (a) aims to predict a single native structure, while the *ensemble* approach (b) aims to compute a picture of the complete ensemble of possible structures.

There has been substantial work on characterizing globular protein folding landscapes for lattice and non-lattice models – see Levitt and co-workers [55, 28, 29], Shakhnovich and co-workers [53, 40] and Dill and co-workers [16, 23, 52, 3, 37, 72]. In particular, it has been shown that the native state may be quite different from the predicted minimum energy conformation; indeed, Zhang and Skolnick [83] have shown that the native state is often closer to the centroid of the largest cluster of low energy conformations obtained by Monte Carlo sampling. Unfortunately, these studies could not complete a full description of the conformational landscape and were mostly restricted to computationally expensive folding simulations of single polypeptides.

In this chapter, we describe recent algorithmic advances that enable efficient computation of the complete *ensemble* of structures of a given polypeptide and prediction of stable conformations. These techniques aim to provide a realistic representation of the conformational landscape of a protein that could potentially be useful to study the folding dynamics of large polypeptides [69].

Seen at the cell level, the dynamic aspect of the system is indiscernible. The motion of individual molecules cannot be observed but the multiple conformational states remains visible. According to statistical mechanics principles, at equilibrium, the molecules achieving a particular fold are perpetually changing but the number of molecules in a specific state remains constant. Originally conceived for modeling the behavior of gas [13], the theory has been applied to other areas of computational biology, including the prediction of RNA secondary structure [50] and the study of transcription factor binding sites [6, 30, 54, 74].

We describe how statistical mechanical principles can be applied to modeling and predicting protein structures. As the theory is still progressing and a general discussion would be too long to conduct in this chapter, we will focus our discussion on the description of the first application of these techniques to a difficult but important class of proteins, namely the transmembrane β-barrel proteins [77, 75]. The techniques detailed in this chapter have potential to be extended to transmembrane α-helix bundles [79], certain β-sheet architectures [1, 11, 51, 21] and other structures that can be modeled using tree structures [16].

Transmembrane β-barrels (TMBs) constitute an important class of proteins typically found in the outer membrane of gram-negative bacteria, mitochondria and chloroplasts. These proteins display a wide variety of functions and are relevant to various aspects of cell metabolism. In particular, outer-membrane proteins (omps) are used in active ion transport, passive nutrient intake, membrane anchors, membrane-bound enzymes, and defense against membrane-attack proteins.

Since omps were discovered relatively recently and are difficult to crystallize, there are currently only about one hundred TMBs in the Protein Data Bank, and only 20 after the removal of homologous sequences. Some *in vitro* and *in vivo* mutation studies of omps [81, 46] have been performed, but, compared with the overwhelming amount of data on globular proteins, outer membrane proteins remain a biologically important but technically difficult area of research.

In this chapter, the ensemble of TMB structures of a given polypeptide is characterized by the Boltzmann partition function of the system. This achievement requires us (i) to provide a model of the structures to which we can apply dynamic programing principles for exploring the full conformational space, and (ii) to design an energy model which allow us to evaluate the stability of each conformation. From this quantity we show how to compute the Boltzmann pair probabilities $P(i,j)$ that residues i,j form an inter-β-strand contact, and rigorously sample conformations from the Boltzmann low energy ensemble. Additionally, we show how this partition function value can be used to estimate statistical mechanical parameters such as ensemble free energy, average internal energy, and heat capacity. Rigorously defined *stochastic contact maps*, sampling, and thermodynamic parameters give us insight into the folding landscape of outer membrane proteins — an insight that cannot be gained by methods solely dedicated to the prediction of native state conformations. This approach also provides a unified framework that allows us to simultaneously tackle a wide variety of structural prediction problems that were previously addressed by independent algorithms. This unified approach achieves a clear gain in accuracy, circumventing the problem of contradictory predictions encountered when interpreting the results of multiple, independent algorithms.

This chapter is organized as follows. In Section 2, we describe the combinatorial model used to represent the TMBs. Then, in Section 3, we introduce the energy model used to weight the structures, which consists of an extension of the state-of-the-art BETAWRAP energy model [11, 21] specialized for TMBs [75, 77].

In Section 4, we detail the algorithms used to compute the complete folding landscape. These algorithms run in polynomial time and space. These results have been obtained by taking advantage of the planarity imposed on a TMB by the cell membrane to derive a model that allows the computation of the partition function to be performed in polynomial time. A related approach was suggested by S. Istrail who proved that the partition function of an Ising model can be computed in polynomial time given a 2D lattice [42].

In Section 5, we illustrate the insight provided by these techniques by demonstrating its effectiveness on a variety of difficult protein prediction problems: (i) how to perform reliable residue contact predictions; (ii) how to provide a simple and intuitive representation of the folding landscape of a given polypeptide using *stochastic contact maps*; (iii) how X-ray crystal per-residue B-factors can be predicted with an accuracy rivaling that of leading specific B-factor prediction algorithms; and (iv) how Boltzmann-distributed structure sampling can be used to improve the accuracy of whole structure prediction over classical minimum folding energy approaches. In addressing this set of challenging structural prediction problems, we wish to underscore the strength and potential of this approach.

To conclude this chapter, we complete our review of recent protein structure ensemble analysis tools by addressing a related problem. Once a stable conformation has been identified, the question of how rigid or flexible the structure is remains. Thus, we present in Section 6 recent methods enabling efficient sampling of the local neighborhood of a given conformation.

2 Modeling transmembrane β-barrel structure

This section provides a simple and unambiguous representation of transmembrane protein structure that enables the design of dynamic programing equations for recursively enumerating all possible TMB structures. Originally, this modeling employed multi-tape context-free grammars [79, 75]; however, in this chapter we provide a more classical description using a graphical representation.

Transmembrane β-barrel (TMB) proteins are embedded in the outer membrane of Gram-negative bacteria, mitochondria and chloroplasts. The envelop of Gram-negative bacteria is built with two membranes (inner and outer) separated by a region called the periplasm. The composition of the bacterial outer membrane differs from that of the inner membrane by, among other things, the structure of its outer leaflet which include a complex lipopolysaccharide.

To accurately represent TMBs (in agreement with Schulz's summary [64]) three fundamental features of these structures are modeled: (i) the overall shape of the barrel (the number of TM β-strands and their relative arrangement); (ii) an exact description of the anti-parallel β-strand pairs which explicitly lists all residue contacts and their orientation (side-chains exposed toward the membrane or toward the lumen), as well as possible strand extensions; and (iii) the inclination of TM β-strands through the membrane plane. This decomposition of the structure into elementary units is illustrated in Figure 2.

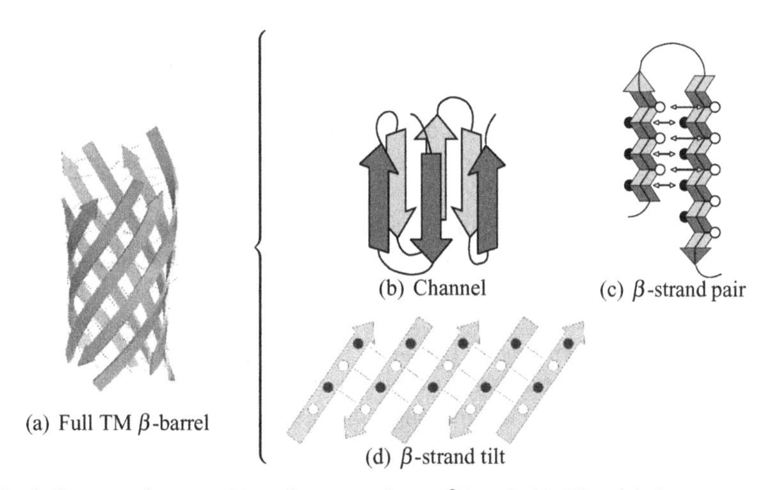

Fig. 2 Structure decomposition of transmembrane β-barrel. (a): The global structure of a transmembrane β-barrel. (b): Overall shape of the channel. (c): Anti-parallel β-strands. (d): Inclination of TM β-strands across the membrane plane.

The principle behind this modeling lies in a decomposition of the β-structure into individual blocks of β-strand pairs. In the case of TMBs, all these pairs are anti-parallel with the exception of the closing one in case the barrel has an odd number of strands. (Thus far only one TMB with an odd number of strands has been found via crystallization [5].) Consequently, the complete structure can be described as a sequence of individual β-strand pairs that will be used in Section 4 to design a dynamic programming algorithm for enumerating all the structures of the conformational landscape.

In TMBs, each strand is paired with two others. Graphically, our decomposition can be seen as follows: Instead of pairing each strand twice, we duplicate all strands and isolate each β-strand pair (see Fig. 3). Then, the barrel can be described as a sum of all its strand pairs.

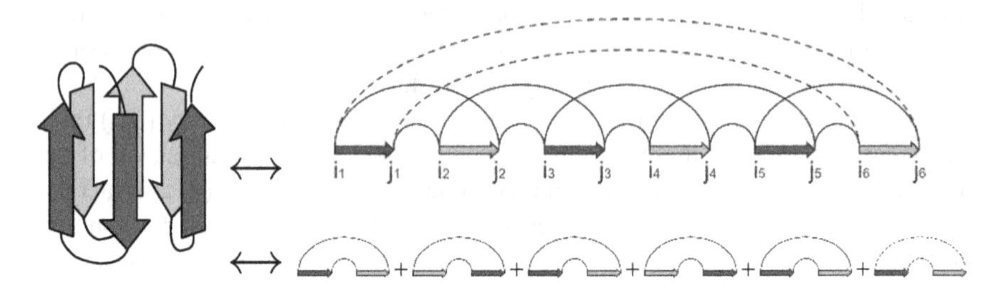

Fig. 3 Graphical decomposition of a transmembrane β-barrel. The strands are duplicated and each strand pair is isolated. The closing β-strand pair (with dashed lines) can be extracted and represented exactly as the others.

For TMBs, each strand is coupled with its two sequential neighbors (previous and next) and all pairings are anti-parallel with the exception of the closing strand pair that can be parallel if the barrel has an odd number of strands. With no restriction on generality, we will assume in this chapter that the TMBs have an even number of strands.

Formally, we define a β-strand pair (i.e. a block as seen in Fig. 3) with a 4-tuple $\binom{i_1,j_1}{i_2,j_2}$, where i_1 and j_1 (s.t. $i_1 < j_1$) are the indices of the left strand and i_2 and j_2 (s.t. $i_2 < j_2$) those of the right one (see Fig. 3). The left strand corresponds to the subsequence $[i_1, j_1]$, the right strand to $[i_2, j_2]$, and the loop connecting them corresponds to the subsequence $[j_1+1, i_2-1]$.

The length of the TM β-strands may vary. The number of residues in contact is $L_c = \min(j_1 - i_1 + 1, j_2 - i_2 + 1)$ and the length of the strand extension is $L_e = \big|(j_1 - i_1) - (j_2 - i_2)\big|$. To avoid invalid configurations, only one strand from each pair can be extended. In addition, for simplicity of description, we assume that the rightmost amino acid at index j_1 of the left strand is paired with the leftmost residue at index i_2 of the right strand. An example of a model freed from this constraint can be found in [78]. When an extension is done on the left strand, the right strand becomes shorter and the extension is called a *reduction* (Fig. 4(a)); when an extension occurs on the right strand, the latter is elongated and the operation, an *extension* (Fig. 4(b)).

The set $\mathscr{C}$ of residue-residue contacts involved in strand pairing can be defined as follows: $\mathscr{C} = \{(j_1 - k, i_2 + k) \mid 0 \le k < L_c\}$. The side-chain orientation alternates strictly around the strand backbone and can be labeled: *outwards*, that is facing toward the membrane, or *inwards*, that is facing toward the inside of the barrel, or channel (which can vary from entirely aqueous to mostly filled). Thus, we distinguish the subsets of residue contacts exposed to the same environment by $\mathscr{C}_0 = \{(j_1 - 2\cdot k, i_2 + 2\cdot k) \mid 0 \le k < \lfloor \frac{L_c}{2} \rfloor\}$ and $\mathscr{C}_1 = \{(j_1 - 1 - 2\cdot k, i_2 + 1 + 2\cdot k) \mid 0 \le k$

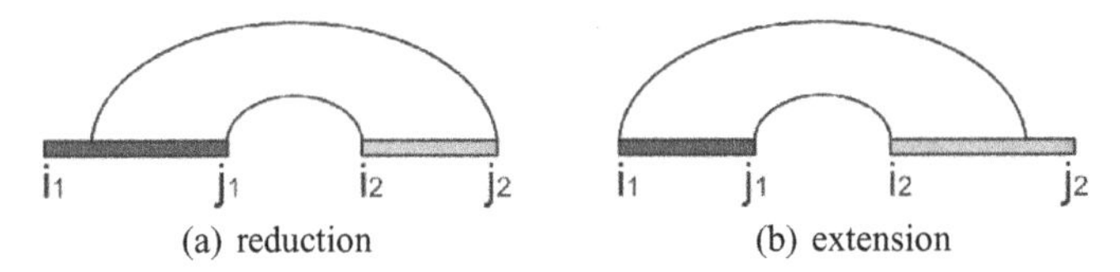

Fig. 4 (4(b)) Strand reduction: the left strand is elongated. (4(a)) Strand extension: the right strand is elongated.

$< \lfloor \frac{L_c}{2} \rfloor \}$. Assuming the location of the closest contact is known, we can also assign the nature of the milieu (i.e. membrane or channel).

Thus, we integrate these features in each block $\binom{i_1, j_1}{i_2, j_2}$ by annotating each residue appropriately. In practice, since residue labels strictly alternate, only the side-chain orientation of the first residue contact needs to be recorded. Figure 5 illustrates this modeling, although these details will be omitted when they are not crucial to the discussion.

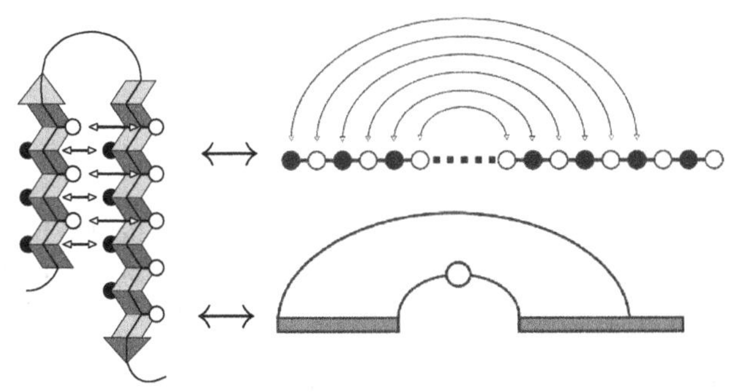

Fig. 5 Representation of a TM β-strand pair with extension on right strand. Residues are annotated by the side-chain orientation. Since the residue labels strictly alternate, only the first side-chain orientation needs to be indicated in a simplified representation (bottom).

The inclination of strands through the membrane is modeled using the *shear number*. This number represents the shift in the sequence of inter-strand residue contacts between consecutive β-strands, imposed by the inclination of these strands (cf. Fig. 6). This feature is implemented with the help of strand extension. Indeed, strictly alternating *reductions* and *extensions* to consecutive strand pairs allows us to obtain the desired configuration. Without loss of generality, and in conjunction with experimental observations [64], we assume that (i) the N-terminus is located on the periplasmic side and that (ii) the shear number is positive. It follows that the first loop (between the first and second TM strand) is on the extra-cellular side. Then we restrict *reductions* to occur around periplasmic loops and *extensions* around extra-cellular loops. Figure 6 illustrates how to proceed.

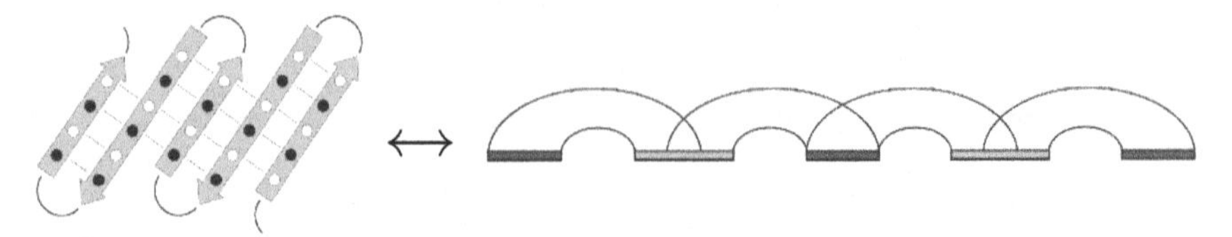

Fig. 6 Representation of strand inclination using shear number. *Reductions* and *extensions* alternate around periplasmic loops and extra-cellular loops in order to preserve the coherence of the orientation. The N-terminus of the protein sequence in the left diagram is at the right extremity.

It is noteworthy that, in principle, a similar representation could be used to include other classes of β-sheet protein domains as long as their structures follow similar topological rules. TMBs are well suited to this methodology since the cell membrane restricts the number of possible structural conformations that can arise, reducing the complexity of the representation. However, soluble β-barrel proteins can allow more flexibility in the barrel forming β-sheet and would thus require more sophisticated rules (such as consecutive strands that are out of sequence order) resulting in an increase in the computational complexity of the method.

3 Energy model

In this section, we describe a simple pseudo-energy model inherited from the state-of-the-art BETAWRAP energy model [11, 21] and specialized for TMBs [75]. In practice more refined versions have been designed [77]; however, all these rely on similar principles.

In the previous section, we defined a TMB as a sum of anti-parallel β-strand pairs, which are themselves defined as sequences of inter-strand residue contacts. These long-range interactions stabilize the β-strand pairs and thus the entire β-barrel. It follows that a reliable scoring function must explicitly integrate the stabilizing effect of these residue contacts.

We describe a simple model where the energy of the whole structure is the sum of the energies of each inter-strand residue contact found in the barrel. The challenge is thus to estimate reliable potentials for any pair of residues in contact. Unlike for RNAs, experimental measures to allow us to directly estimate the binding energies are not available for TMBs. Nevertheless, it is possible to estimate these potentials by computing residue contact statistics from known protein structures.

The statistical potentials for all possible amino acid pairs can be obtained by computing the probability of observing amino acid contacts in solved β-sheet structures with characteristics closely matching those found in TMBs. The classical approach used in [11, 21, 75, 77] takes a 50% non-redundant set of protein structures (`PDB50`) from the `PDB` [8], and uses `STRIDE` [31] to identify secondary structure features, solvent accessibility, and hydrogen bonds. Naturally, all solved structures of TMBs have to be removed as to not corrupt the testing.

In order to obtain the best possible estimate of these potentials, Berger and coworkers [11, 21] introduced a major conceptual advance with the BETAWRAP program. Instead of simply counting the occurrences of β-sheet amino acid pairings in all known proteins, the search is restricted to better match the environment associated with a given contact. Here, the barrel fold of TMBs is thought to consist of antiparallel, amphipathic β-sheets[1], with a hydrophobic environment in the outer membrane side of the barrel and a hydrophilic environment commonly existing within. Therefore, the anti-parallel bonded β-strands that exhibit an amphipathic pattern mimic relatively well the features of TM β-strand pairs, and thus can be used to count the frequency of pairs of residues. Alternating buried/exposed residues define amphipathicity. Usually, a buried residue is required to have less than 4% of the solvent accessible area as when that residue is in an extended G-X-G tripeptide [17], and an exposed residue is required to have an area greater than 15%.

The amino acid pair frequency counts are then used to estimate the probability $P(X,Y)$ of observing the amino acid pair (X,Y). Finer granularity information such as side-chain rotomers or atomic coordinates are not included in this model, but may be integrated into a more sophisticated model. Of note, Waldispühl *et al.* [77], introduce a variant of this model incorporating the notion of *stacking pairs* of adjacent pairs of residue contacts, which results in a significant gain of accuracy.

Once this amino acid pairs count is calculated, these frequency counts can be changed into statistical potentials. Let x, y be the indices of two amino acid that are in contact, and let $M \in \{0,1\}$ be a variable which represents the type of environment in which such a contact occurs (which side of the amphipathic sheet). Specifically, $M=0$ ($M=1$) when the side-chain orientation is toward the channel interior (membrane). Let $E(x,y,M)$ denote the energy of the contact between amino acids X and Y at positions x and y, with the environment M.

Pairwise frequencies are transformed into energy potentials using the standard procedure (taking the negative logarithm — see pp. 223–228 of [18] and [68] for details). Specifically, if $p_M(X,Y)$ is Boltzmann distributed, then $E(x,y,M) = -RT\log(p_M(X,Y)) - RT\log(Z_c)$, where $\log(Z_c)$ is a statistical re-centering constant that is chosen as a parameter. Further, although RT has no effect when computing the minimum folding energy structure [75], this is not the case when computing the partition function for β-barrel structures. For this reason, the current implementation in the program *partiFold* (`http://partiFold.csail.mit.edu`) [75, 76, 77] allows the user to stipulate an arbitrary Boltzmann constant. The folding pseudo-energy of the structure is the sum of all contact potentials. Formally, we have:

$$E = \sum_{(x,y)\in\mathscr{C}_0} E(x,y,0) + \sum_{(x,y)\in\mathscr{C}_1} E(x,y,1) \quad (1)$$

This model does not contain any energy contribution for periplasmic or extracellular loops, although such features can easily be computed and integrated with similar techniques.

[1] The amphipaticity defines a molecule which contains both polar (hydrophilic) and non-polar (hydrophobic) domains.

4 Algorithms

4.1 Computing the partition function

Since a TMB structure can be represented as a sequence of anti-parallel TM β-strand pairs, given any four indices i_1, j_1, i_2, j_2 and the environment M of the closing TM β-strand pair contact (i.e. "membrane" or "channel"), we can compute the energy $E(i_1, j_1, i_2, j_2, M)$ for the anti-parallel β-strand pairing of sequences $[i_1, j_1]$ with $[i_2, j_2]$. For all possible values of i_1, j_1, i_2, j_2 and M, we store the Boltzmann values $\exp\left(-E(i_1, j_1, i_2, j_2, M)/RT\right)$ in the array Q_{ap}. Since the length of TM strands, as well as those of strand extensions are bounded, the array can be filled in time $\mathcal{O}(n^2)$, where n is the sequence length.

$$Q_{ap}(i_1, j_1, i_2, j_2, M) = \prod_{k=0}^{L_c-1} \exp\left[-\frac{E(i_2-k, j_1+k, M+k \bmod 2)}{RT}\right] \tag{2}$$

Since the energy function is additive, we can decompose the energy of a TMB as the sum of the energies associated with each distinct anti-parallel TM β-strand pair. Let N be the number of TM β-strands of the TMB s and let i_k (resp. j_k) denote the index of the leftmost (resp. rightmost) residue of the k-th strand. In order to simplify the algorithms description, in the following we will omit the parameter M used to indicate the environment of the first contact of an anti-parallel TM β-strand pair. Therefore, the energy $E(s)$ of a given TMB structure s can be written as:

$$E(s) = E(i_N, j_N, i_1, j_1) + \sum_{k=1}^{N-1} E(i_k, j_k, i_{k+1}, j_{k+1}) \tag{3}$$

The Boltzmann partition function is defined as the sum $\sum_s e^{-\frac{E(s)}{RT}}$ taken over all the TMB structures s. To compute the partition function, we first introduce a dynamic table Q_{sheet} to store the partition function values for β-sheets built from concatenating anti-parallel TM β-strand pairs, i.e. TMB without closure. This table can be dynamically filled using the following recursion:

$$Q_{sheet}\begin{pmatrix} i_1, j_1 \\ i_k, j_k \end{pmatrix} = \sum_{(i_{k-1}, j_{k-1})} Q_{sheet}(i_1, j_1, i_{k-1}, j_{k-1}) \cdot Q_{ap}(i_{k-1}, j_{k-1}, i_k, j_k) \tag{4}$$

Once filled, we use this array to compute the partition function Q_{tmb} over all TMBs. Note that the index k can be used to control the number of strands in the barrels. This operation consists of adding the contributions of the anti-parallel β-strand pairs which close the extremities of the β-sheet. For this, we could use the values stored in the Boltzmann value array Q_{ap}; however, in practice, we use a special array which is better suited to the special rules for this last β-strand pair.[2]

[2] The rules for the closing pair, explicitly described in [75], mainly consist of relaxing some constraints, and allowing extensions on both sides of the strand.

$$Q_{tmb} = \sum_{(i_1,j_1)} \sum_{(i_N,j_N)} Q_{sheet}(i_1,j_1,i_N,j_N) \cdot Q_{ap}(i_N,j_N,i_1,j_1) \quad (5)$$

Note that in order to respect the pairwise orientation as well as strand inclination, the indices i_1, j_1 and i_N, j_N are swapped. Finally, it should be mentioned that in computing the partition function, the dynamic programming must ensure an exhaustive and non-overlapping count of all structures; in particular, the cases treated must be mutually exclusive, as is clearly the case in our algorithm.

We illustrate these equations and overview the complete procedure in Fig. 7. First, we initialize the dynamic arrays by computing all possible β-strand pairs (Fig. 7(a) and Equation 2). Then, we build the β-sheets by concatenating β-strand pairs (Fig. 7(b) and Equation 4). Finally, we close the TMBs by pairing the first and last β-strands (Fig. 7(c) and Equation 5)

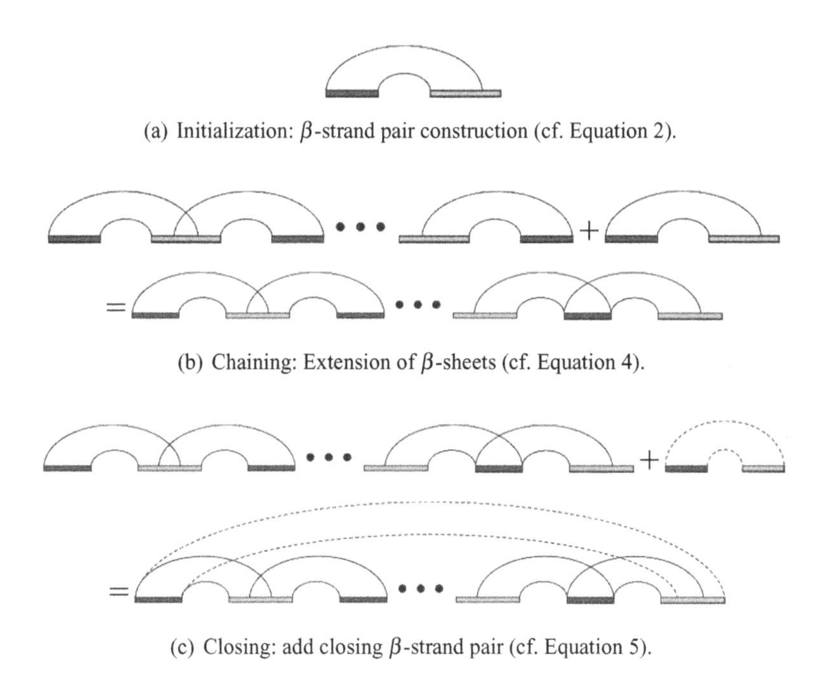

(a) Initialization: β-strand pair construction (cf. Equation 2).

(b) Chaining: Extension of β-sheets (cf. Equation 4).

(c) Closing: add closing β-strand pair (cf. Equation 5).

Fig. 7 Graphical representations of the recursive rules used to enumerate the conformational landscape.

Using formulas from classical statistical mechanics, a number of important thermodynamic parameters can be computed immediately from the partition function. These parameters, including ensemble free energy, heat capacity, average internal energy, etc. (see [22]), lead to a better understanding of the folding landscape. For example, as shown in [19], the average internal energy of the structures $\langle E(s) \rangle$ can be computed by

$$\langle E(s) \rangle = RT^2 \cdot \frac{\partial}{\partial T} \log Q(s), \quad (6)$$

while the standard deviation can be computed with a similar formula. Such thermodynamic parameters provide information on the stability of folds for a given sequence.

4.2 Computing the residue contact probability

In this section, we address the problem of computing the Boltzmann pair probabilities from the dynamic tables filled when computing the partition function value Q_{tmb}. First, we need to characterize the anti-parallel β-strand pairs that contain a given contact.

Proposition 0.1. *Let x and y ($x < y$) be two residues of two distinct consecutive anti-parallel β-strands, and j_1 and i_2 (s.t. $i_1 \leq x \leq j_1 < i_2 \leq y \leq j_2$) the two residues at the extremities of the connecting loop. Then, residues (x,y) are brought into contact if and only if $i_2 + j_1 = x + y$.*

It follows from this proposition that (x,y) is a valid contact if and only if the anti-parallel β-strands $\binom{i_1,j_1}{i_2,j_2}$ verify $x+y = j_1 + i_2$ and $i_1 \leq x \leq j_1 < i_2 \leq y \leq j_2$.

To evaluate the residue pair probability $p(x,y)$, we must compute the partition function value over all TMBs $Q(x,y)$ which contain this contact. Such TMBs can be decomposed into two, three, or four parts, depending on the strand pair where the contact occurs (i.e. in the the closing strand pair, the first and last pair of the sheet or in an intermediate one). All these cases are illustrated in Figure 8.

Let $\binom{i,j}{i',j'}$ be an index of a block modeling an anti-parallel TM β-strand pair. Then, we define $Q^{close}\binom{i,j}{i',j'}$, $Q^{first}\binom{i,j}{i',j'}$, $Q^{last}\binom{i,j}{i',j'}$ and $Q^{inter}\binom{i,j}{i',j'}$ to be the partition functions over all TMB structures which contain this anti-parallel TM β-strand pair as, respectively, the pair closing the barrel (Figure 8(a)), the first pair of the TM β-sheet (Figure 8(b)), the last pair of the TM β-sheet (Figure 8(c)) or any other intermediate pair (Figure 8(d)). Formally:

$$Q^{close}\begin{pmatrix} i_1,j_1 \\ i_N,j_N \end{pmatrix} = Q_{\text{sheet}}\begin{pmatrix} i_1,j_1 \\ i_N,j_N \end{pmatrix} \cdot Q_{\text{ap}}\begin{pmatrix} i_N,j_N \\ i_1,j_1 \end{pmatrix} \tag{7}$$

$$Q^{first}\begin{pmatrix} i_1,j_1 \\ i_2,j_2 \end{pmatrix} = \sum_{(i_{N-1},j_{N-1})} Q_{\text{ap}}\begin{pmatrix} i_1,j_1 \\ i_2,j_2 \end{pmatrix} \cdot Q_{\text{sheet}}\begin{pmatrix} i_2,j_2 \\ i_N,j_N \end{pmatrix} \cdot Q_{\text{ap}}\begin{pmatrix} i_N,j_N \\ i_1,j_1 \end{pmatrix} \tag{8}$$

$$Q^{last}\begin{pmatrix} i_{N-1},j_{N-1} \\ i_N,j_N \end{pmatrix} = \sum_{(i_1,j_i)} Q_{\text{sheet}}\begin{pmatrix} i_1,j_1 \\ i_{N-1},j_{N-1} \end{pmatrix} \cdot Q_{\text{ap}}\begin{pmatrix} i_{N-1},j_{N-1} \\ i_N,j_N \end{pmatrix} \cdot Q_{\text{ap}}\begin{pmatrix} i_N,j_N \\ i_1,j_1 \end{pmatrix} \tag{9}$$

$$Q^{inter}\begin{pmatrix} i_k,j_k \\ i_{k+1},j_{k+1} \end{pmatrix} = \sum_{\substack{(i_1,j_1) \\ (i_N,j_N)}} Q_{\text{sheet}}\begin{pmatrix} i_1,j_1 \\ i_k,j_k \end{pmatrix} \cdot Q_{\text{ap}}\begin{pmatrix} i_k,j_k \\ i_{k+1},j_{k+1} \end{pmatrix} \cdot Q_{\text{sheet}}\begin{pmatrix} i_{k+1},j_{k+1} \\ i_N,j_N \end{pmatrix} \cdot Q_{\text{ap}}\begin{pmatrix} i_N,j_N \\ i_1,j_1 \end{pmatrix} \tag{10}$$

Finally, using these functions, the partition function $Q(x,y) = \sum_S e^{-\frac{E(S)}{RT}}$, where the sum is over all TMBs that contain the residue contact (x,y), is computed as follows:

$$Q(x,y) = \sum_{\substack{(i,j)\\(i',j')}}^{x+y=j+i'} \left(Q^{close}\binom{i,j}{i',j'} + Q^{first}\binom{i,j}{i',j'} + Q^{last}\binom{i,j}{i',j'} + Q^{inter}\binom{i,j}{i',j'} \right) \tag{11}$$

Finally, the Boltzmann probability $p(x,y)$ of a contact between the residues at indices x and y can be obtained by computing the value $p(x,y) = \frac{Q(x,y)}{Q_{tmb}}$. However, we note that an extra field counting the number of strands in Q^{sheet} is required to ensure that the minimal number of strands in a TMB is not violated.

Assuming the length of TM β-strands and loops, as well as the shear number values, are bounded, the time complexity is $\mathcal{O}(n^3)$, where n is the length of the input sequence. When the maximal length of a loop is in $\mathcal{O}(n)$, this complexity should approach $\mathcal{O}(n^4)$. Similarly, the space complexity can be bounded by $\mathcal{O}(n^2)$.

4.3 Improved computation of the contact probabilities

The formidable time requirement for a brute force algorithm to compute Equation 10 prevents any immediate efficient application. Indeed, naively applying this equation to the $\mathcal{O}(n^2)$ possible residue pairs results in an overall time complexity of $\mathcal{O}(n^5)$. In this section, we show how a simple strategy using additional dynamic tables, has been used to reduce the time complexity by a factor of $\mathcal{O}(n^2)$.

Two basic observations lead to a natural improvement over a brute force algorithm. First, when the TM β-strand pair that contains the residue contact is not involved, the product of the partition function of two sub-structures is realized over all possible configurations (i.e. $Q_u\binom{i,j}{i',j'} \cdot Q_v\binom{i',j'}{i'',j''}$ is computed over all possible pairs of indices (i',j')). In equation 10, the pairs of indices (i_k,j_k) and (i_{k+1},j_{k+1}) are used for different residue contacts since the pair (i_N,j_N) varies. Thus we can precompute the values of $Q_{\text{sheet}}\binom{i_{k+1},j_{k+1}}{i_N,j_N} \cdot Q_{\text{ap}}\binom{i_N,j_N}{i_1,j_1}$ over all possible (i_N,j_N) and store them in a dynamic table for later retrieval. Given (i_1,j_1) and (i_{k+1},j_{k+1}), let Q_{tail} be the array storing the values $\sum_{(i_N,j_N)} Q_{\text{sheet}}\binom{i_{k+1},j_{k+1}}{i_N,j_N} \cdot Q_{\text{ap}}\binom{i_N,j_N}{i_1,j_1}$. This table can be filled in time $\mathcal{O}(n^3)$. Then, in place of equation 10, we now have equation 12.

$$Q^{inter}\binom{i_k,j_k}{i_{k+1},j_{k+1}} = \sum_{(i_1,i_2)} Q_{\text{sheet}}\binom{i_1,j_1}{i_k,j_k} \cdot Q_{\text{ap}}\binom{i_k,j_k}{i_{k+1},j_{k+1}} \cdot Q_{\text{tail}}\binom{i_{k+1},j_{k+1}}{i_1,j_1} \tag{12}$$

This improvement cannot be applied to Equations 8 and 9, since there is no redundancy in those cases. The time complexity for computing all possible contact probabilities $p(i,j)$ is now $\mathcal{O}(n^4)$. However, further observation allows us to save an additional factor of $\mathcal{O}(n)$ in the time complexity: when a TMB structure is considered in one of the equations 7, 8, 9 or 10, the TM β-strand pair which contains the contact (x,y) also involves many other contacts. Hence, instead of using these equations to compute the values $Q(x,y)$ and $p(x,y)$ separately, we consider each possible β-strand pair and immediately add its contribution to the partition function. From these improvements, we now have an algorithm to compute all the contact probabilities of a TMB, which runs in time $\mathcal{O}(n^3)$.

Although not explicitly mentioned thus far, we should emphasize that we can also compute the contact probability $p_M(x,y)$ for a specific environment M — i.e. membrane or channel (see Section 3 for an explanation of environment). To do so, we simply need to duplicate the dynamic tables in order to take into account the side-chain orientation for extremal TM β-strand pairs.

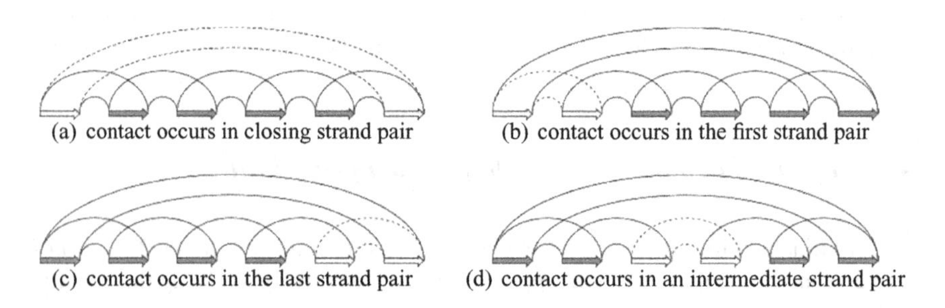

(a) contact occurs in closing strand pair
(b) contact occurs in the first strand pair
(c) contact occurs in the last strand pair
(d) contact occurs in an intermediate strand pair

Fig. 8 Decompositions of a transmembrane β-barrel, which allows us to isolate the antiparallel TM β-strand pair that contains the residue contact. The block that corresponds to this strand pair is indicated with white β-strands connected with dashed lines. The blocks in gray represent TM β-sheets (i.e. a sequence of anti-parallel TM β-strands).

4.4 Rigorous sampling of transmembrane β-barrels

In this section, we describe a rigorous sampling algorithm for TMBs. Given an amino acid sequence ω, it randomly generates, according to the distribution of structures in the Boltzmann ensemble, low energy TMB structures for ω. By sampling, we expect to be able to efficiently estimate non-trivial features concerning the ensemble of potential TMB folds, with the long-term goal of potentially contributing to drug design engineering.

The sampling algorithm uses the dynamic table filled during the computation of the partition function. It essentially proceeds in two steps illustrated in Figure 9. First, the “closing” anti-parallel strand pair is sampled according to the weight of all TMBs that contain it over all possible TMBs. Then, we sample each anti-parallel strand pair of the TM β-sheet from left to right (or alternatively from right to left) until the last one, according to the weight of that structure over all possible TM β-sheets. The full procedure is depicted in Figure 9. The correctness of the algorithm is ensured by construction of the dynamic table in Equations 4 and 5.

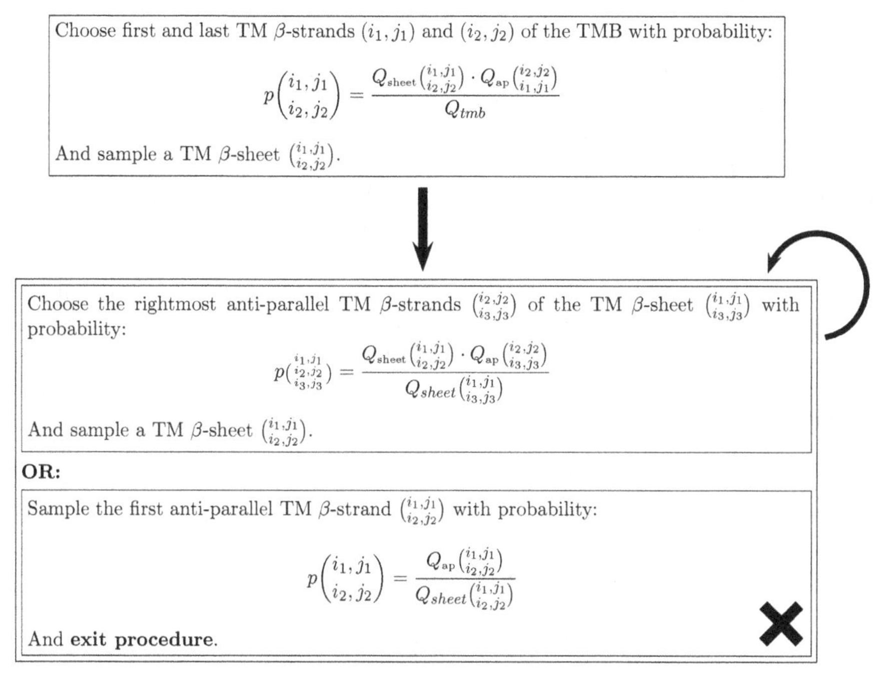

Choose first and last TM β-strands (i_1, j_1) and (i_2, j_2) of the TMB with probability:
$p\binom{i_1, j_1}{i_2, j_2} = \frac{Q_{\text{sheet}}\binom{i_1,j_1}{i_2,j_2} \cdot Q_{\text{ap}}\binom{i_2,j_2}{i_1,j_1}}{Q_{tmb}}$
And sample a TM β-sheet $\binom{i_1,j_1}{i_2,j_2}$.
Choose the rightmost anti-parallel TM β-strands $\binom{i_2,j_2}{i_3,j_3}$ of the TM β-sheet $\binom{i_1,j_1}{i_3,j_3}$ with probability:
$p\left(\begin{smallmatrix} i_1,j_1 \\ i_2,j_2 \\ i_3,j_3 \end{smallmatrix}\right) = \frac{Q_{\text{sheet}}\binom{i_1,j_1}{i_2,j_2} \cdot Q_{\text{ap}}\binom{i_2,j_2}{i_3,j_3}}{Q_{sheet}\binom{i_1,j_1}{i_3,j_3}}$
And sample a TM β-sheet $\binom{i_1,j_1}{i_2,j_2}$.
OR:
Sample the first anti-parallel TM β-strand $\binom{i_1,j_1}{i_2,j_2}$ with probability:
$p\binom{i_1, j_1}{i_2, j_2} = \frac{Q_{\text{ap}}\binom{i_1,j_1}{i_2,j_2}}{Q_{sheet}\binom{i_1,j_1}{i_2,j_2}}$
And exit procedure.

Fig. 9 Sampling procedure: The first and last TM β-strands of the barrel are sampled (left box). Then the remaining TM β-sheet is sampled by iteratively sampling the rightmost anti-parallel β strand of the remaining sequence, until the first β-strand pair of the sheet is sampled.

5 Applications

The algorithms described in the previous section are implemented in the program *partiFold* [77]. The *partiFold* algorithms use the Boltzmann partition function to predict the ensemble of structural conformations a TMB may assume instead of predicting a single minimum energy structure. From this ensemble, experimentally testable TMB properties are computed that describe the folding landscape and suggest new hypotheses. In the following, we illustrate the flexibility of the approach and show how the method can be used for predicting individual contacts, investigating the conformational landscape and predicting Debye-Waller factors (a X-ray crystallography measure accounting for the thermal motion of the atom - a.k.a. B-factors). We finally apply whole structure sampling to demonstrate the benefits of ensemble modeling over single structure prediction and the possibilities for structural exploration provided by these techniques.

5.1 Residue contact prediction

Single contact prediction remains an important concern when reconstructing 3D models [33, 43, 57]. Several machine-learning methods have been developed for this task, among them PROFcon [57] and FOLDpro [15] (general predictors), BETApro [14] (specialized for β-structures) and TMBpro [59] (specialized for TMBs) are among the most reliable. However, it should be noted that, while some of them can provide stochastic contact map of β-strand interactions, the interaction probabilities are not related to a Boltzmann distribution of conformations, but rather based on sophisticated neural networks and graph algorithms that aim to predict a single structure. In addition, their energy models also do not appear to be common across all proteins, resulting in difficulties to interpret and compare results between different proteins.

But even conceptually, the ensemble approach is radically different from previous machine-learning methods. Indeed, while the latter first start by making individual and unrelated contact predictions and finish by reconstructing a whole structure, the ensemble method does not dissociate both aspects, since the set of TMB structures is computed first and the contact probabilities are subsequently evaluated from the folding energies of these structures.

To test the ensemble method, single contact predictions are made by selecting all pairwise contacts that have a probability greater than a given threshold p_t in the stochastic contact map, and compare those against the corresponding contacts found in X-ray crystal structures as annotated by `STRIDE` [31].

To evaluate the contact predictions, we classically rely on three standard measures: the sensitivity (or coverage), where

$$\text{sensitivity} = \frac{\text{number of correctly predicted contacts}}{\text{number of observed contacts}},$$

the positive predictive value (abbreviated PPV and also known as accuracy), where

$$\text{PPV} = \frac{\text{number of correctly predicted contacts}}{\text{number of predicted contacts}},$$

and the F-measure, where

$$\text{F-measure} = \frac{2 \cdot \text{sensitivity} \cdot \text{PPV}}{\text{Sensitivity} + \text{PPV}}.$$

To demonstrate how these metrics would apply to this type of contact prediction, we refer to Figure 10, which depicts the accuracy of contact prediction for the crystal structure of outer membrane protein X (abbreviated OmpX) [73]. The flatness of the curves further indicates a good separation between accurate, highly probable contacts, and background predictive noise. This type of result could suggest a good scaffold of likely contacts when constructing a 3D model of an unknown structure.

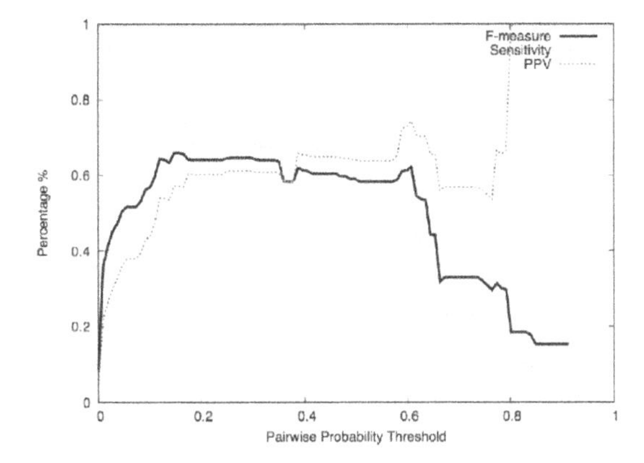

Fig. 10 Predicting residue contact probabilities in OmpX (1QJ8 [73]). The x-axis represents the threshold used to select the residue contact predictions. The graph shows the curves of the F-measure, sensitivity (or coverage) and positive predictive value (or accuracy, abbreviated PPV).

These techniques have proven to provide the state-of-the-art predictions for TMBs [77]. However, there is still room for improvement. For instance, current algorithms do not yet model bulges in β-sheets and suffer slightly in performance where bulges exist.

5.2 Representations of ensembles

The class of predictions enabled by these techniques embody whole-ensemble properties of a protein. The contact probabilities can be treated all together to represent and analyze different aspects of the folding properties of a polypeptide.

In Fig. 11, a single structure is chosen (in this case the X-ray structure of OmpX [73]), and displayed as an unrolled 2D representation of the β-barrel strands and their adjacent residue contacts. Using the stochastic contact map, residue contact pairs are then colored to indicate a high (black), a medium (dark gray) or a low (light gray) probability in the Boltzmann distributed ensemble. From this, substructures may be analyzed from their relative likelihood of pairing.

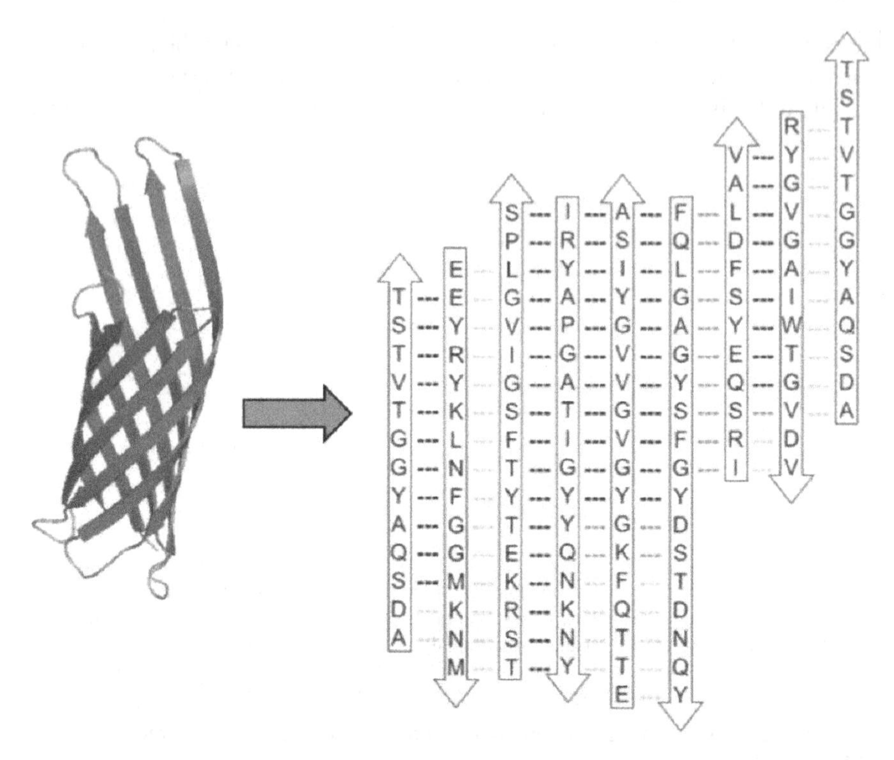

Fig. 11 Contact probabilities mapping to OmpX (1QJ8) X-ray crystal structure [73]. 2D representation (unrolled β-barrel) showing only those residues involved in β-strands (shown vertically and successively numbered) and their associated, in-register H-bonding partners. Computed contact probabilities indicated by color hue (highly probable in black, medium probability in dark gray and low probability in light gray). The leftmost β-strand is repeated on the right to allow the barrel to close.

In Fig. 12, the inter-strand residue contact probabilities are merged in the upper triangle of a single matrix called a *stochastic contact map*. This reflects the likelihood of two β-strand amino acids pairing in the (estimated) Boltzmann distribution of conformations, and not one single minimum folding energy structure. This graphical representation provides an intuitive way to depict the variety of structures that can be found in the conformational landscape. We can also compare with these maps the contacts of a given structure (in this case, the contact found in the X-ray structure are plotted in the lower triangle) to estimate its adequacy with the conformational landscape suggested by the high contact probabilities (gray regions of the stochastic contact maps).

While the mapping allows analysis of the likelihood of a given structure from the ensemble perspective, the stochastic contact maps enable us to investigate the folding landscape and estimate the variety of folds of a given polypeptide.

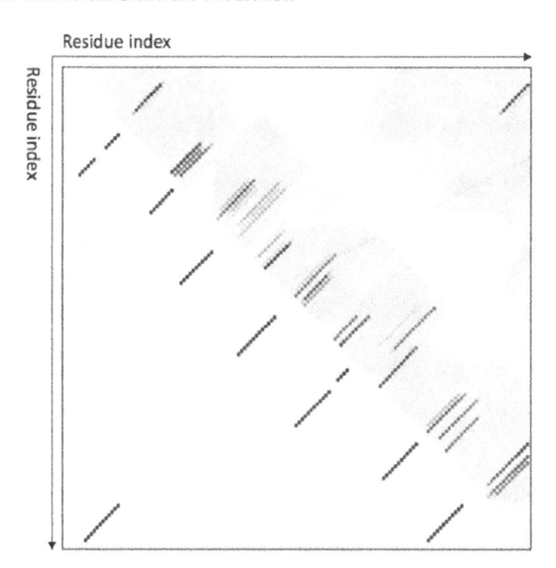

Fig. 12 Stochastic contact map for Neisserial Surface Protein A (NspA). Horizontal and vertical axes represent residue indices in sequence (indices 1 to 155 from left to right and top to bottom), and points on the map at location (i, j) in the upper triangle represent the probability of contact between residues i and j (where darker gray implies a higher probability). The X-ray crystal structure contacts of 1P4T [71] are shown in the lower triangle.

A striking example of the biological relevance of these techniques is shown in Fig. 13. The stochastic contact map of the Outer Membrane Enzyme PagP protein is computed. It contains the contacts found on the X-ray crystal structure 1THQ [2] (in black in the lower triangle), and those of the minimum folding energy structure (in gray in the lower triangle). Here, we note that (i) it is clear that the native conformation (black, lower triangle) differs radically from the minimum energy structure (gray, lower triangle), and (ii) the stochastic contact map reveals alternate β-strand pairs with high probabilities (in gray in the upper triangle).

These discrepancies may be explained through the lens of a recent experimental study. Indeed, Huysmans *et al.* [41] showed that the N-terminal α-helix found in the native structure is essential for the stability of the native β-barrel structure. If we constrain the corresponding α-helical regions of the contact map to not fold into a barrel (peach regions in Fig. 13), this prevents the protein from folding as the minimum free energy structure and thus allows it to adopt one of the other conformations suggested by the stochastic contact map (gray regions in Fig. 13), which coincides with the native structure (in black in the lower triangle). This example illustrates how the contact maps can suggest alternate folds.

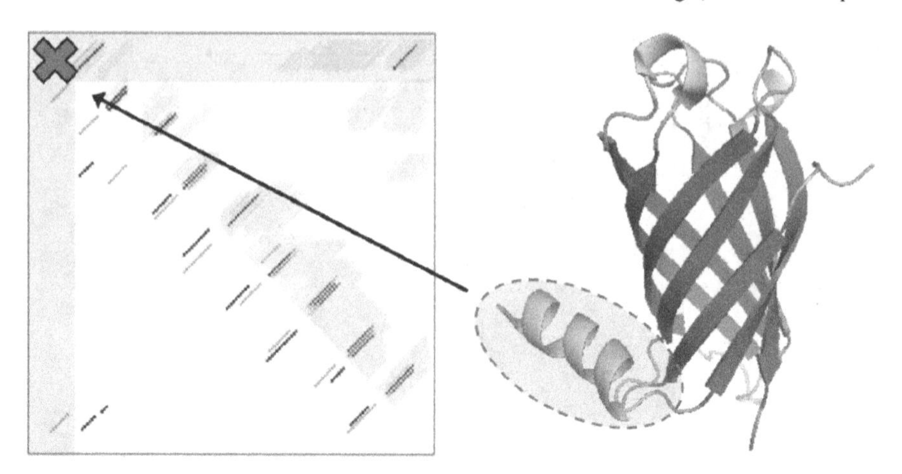

Fig. 13 Multiple conformations of PagP proteins. The stochastic contact map (gray regions in the upper triangle) is compared with the contacts found on the X-ray crystal structure 1THQ [2] (in black in the lower triangle), and those of the minimum folding energy structure (in gray in the lower triangle). If we constrain the corresponding α-helical regions of the contact map to not fold into a barrel (gray stripes), this prevents the protein from folding as the minimum free energy structure and thus allows it to adopt one of the other conformations suggested by the stochastic contact map (gray regions in the upper triangle), which coincides with the native structure (in black in the lower triangle).

5.3 Prediction of residue flexibility

We now show how the contact probabilities can be used to predict per-residue flexibility and entropy. To a first approximation, this flexibility correlates with the *Debye-Waller factor* (a.k.a. B-factor) found in X-ray crystal structures [60]. This demonstrates an important purpose for computing the Boltzmann partition function: to provide biologically-relevant grounds for the prediction of experimentally testable macroscopic and microscopic properties.

Predicting residue B-factors is important because it roughly approximates the local mobility of flexible regions, which might be associated with various biological processes, such as molecular recognition or catalytic activity [62]. In this context, flexible regions are strong candidates for loop regions connecting anti-parallel TM β-strands that extend either into the extracellular or intracellular milieu.

Classical B-factor predictors use machine learning approaches [62]. However, as is the case for contact predictions (cf. Section 5.1) these techniques do not provide a comprehensive framework facilitating the understanding of these results in a larger context. Indeed, previous methods were specifically designed to make only these predictions, while in ensemble approaches, B-factors are just one of the multiple characteristics that can be extracted from an ensemble model.

We define the *contact probability profile* of every amino acid index i in a TM β-barrel to be $P_c(i) = 2 - \sum_{j=1}^{n} p_{i,j}$, and compare this against the normalized B-factor.

Since a residue may be involved in two contacts in a β-sheet, the value of $P_c(i)$ can range between 0 and 2, where higher values indicate greater flexibility. Similarly, residues with a positive B-factor are considered flexible or disordered, while others are considered rigid. In Figure 14, we illustrate this method by comparing the curves of X-ray B-factors and contact profiles of OmpX (1QJ8) [73] and NspA (1P4T) [71] proteins.

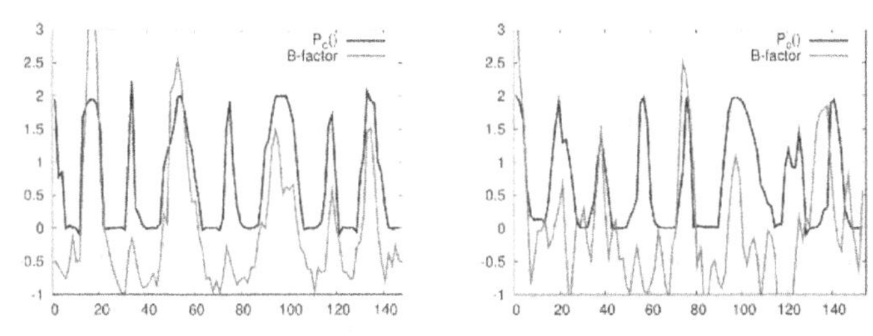

Fig. 14 Comparison of B-factors and contact probability profiles of OmpX (1QJ8) [73], left, and NspA (1P4T) [71], right, proteins.

Computing the cross-correlation coefficient between the P_c and B-factor of test proteins reveals that this method provides state-of-the-art predictions [77]. But the real purpose of this work actually goes much beyond that. The direct predictions of experimental measures are of fundamental importance. It enables biologists to directly compare computational predictions to experimental measures and avoid any misleading interpretations. These methods also can be efficiently used to tune the theoretical folding model to fit experimental data.

5.4 Whole structure prediction through Boltzmann sampling

Finally, we show how ensembles of structures can characterize protein structure better than the minimum folding energy (m.f.e.) structure. We perform stochastic conformational sampling (cf. Section 4.4) to map the landscape defined by the Boltzmann partition function. This also illustrates how the approach can be used to rigorously explore the space of all possible TMB structures. By clustering a large set of full TMB structure predictions, a small distinguishable collection of unique conformations are exposed.

Waldispühl *et al.* [77] sampled 1,000 TMB structures and grouped them into 10 clusters according to hierarchical clustering. Similar to prior methods developed for RNAs [25], for each cluster one can designate a centroid representative conformation that is chosen as the structure with the minimum total distance to all other structures in the set. To facilitate this clustering, a metric named *contact distance* is introduced: $d_c(S_1,S_2) = |\mathscr{C}_1| + |\mathscr{C}_2| - 2 \cdot |\{\mathscr{C}_1 \cap \mathscr{C}_2\}|$, where $\mathscr{C}_1$ and $\mathscr{C}_2$ are the sets of contacts in S_1 and S_2 (which represents the minimum number of contacts to be removed and added to pass from S_1 to S_2 or vice versa). Other metrics could be defined but the latter seemed to provide the best results.

The results showed that the centroid of the largest cluster usually provides a better solution than the minimum folding energy structure [77]. It has also been found than in some cases a centroid of another cluster provides significantly better structure predictions. However, the identification of the "best" cluster, as well as the the robustness of the clusters to the distance used, remains to be investigated.

6 Sampling the local neighborhood of 3D structures

We conclude this chapter by addressing a different but related problem. Instead of sampling the global folding landscape of a protein sequence of unknown structure, we aim to sample the local neighborhood of a given 3D structure. In other words, we seek to estimate the stability of a structure and explore the variations of specific folds at a precision not achieved by *partiFold*.

Unlike the methods described in previous sections, we no longer restrict our conformational space to TMBs. In the following, we overview the principal aspect of this approach. In Section 6.1, we introduce a structural modeling approach that suits the problem well. In Section 6.2, we describe the sampling procedure. Finally, in Section 6.3, we give an application of this approach.

6.1 Structure modeling

Since we aim to sample in the local neighborhood of a given 3D structure, the size of the explored conformational landscape is drastically smaller than in previous sections. Thus, we can afford to use a more detailed description of the structure. Of the many different representations of protein structure, one that has gained popularity is that of the torsion angle representation. This representation makes use of the fact that bond lengths and bond angles show little variation across structures [34], and hence can be assumed to be fixed (cf. Fig. 15). The flexibility of protein molecules can thus almost entirely be described by rotation about covalent bonds.

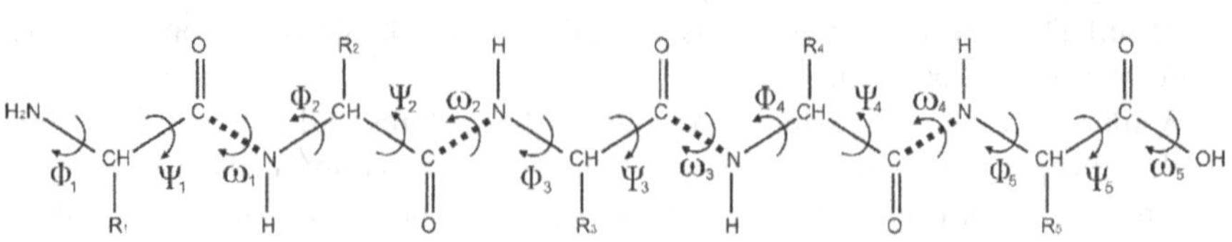

Fig. 15 Torsion angle representation of a polypeptide.

Such a simplified model has the advantage of not having to enforce regular geometry constraints. But, at the same time, non-bonded interactions are non-trivial to calculate in this reduced representation.

6.2 Sampling in the Torsion space

One of the biggest advantages of using this reduced model of a protein is the speed with which one is able to sample the conformation space. By discretizing the dihedral angle space (i.e. the Ramachandran plot [58], see Fig. 16(a)) and biasing solutions lying within specified regions, one can sample protein conformations in the neighborhood of a native structure (see Fig. 16(b)). By efficiently exploiting various algorithms developed in the Inverse Kinematics community, the algorithm ChainTweak [67] was shown to be capable of exploring a much larger conformational space than previous methods [10, 12, 35, 39, 47, 55, 70].

ChainTweak iteratively perturbs the base conformation using the torsion (a.k.a. dihedral angle) representation. A sliding window approach is used to successively move some atoms by 0-2 Å, while keeping all others fixed (see Fig. 17(a)). Inside the window, loop closure methods are used to generate such perturbations [20, 49, 80]. Moreover, residue specific Phi-Psi angle preferences, given by a Phi-Psi priority scheme (Fig. 16(b)) inherited from a Ramachandran plot [58] (Fig. 16(a)), can be used to choose a perturbation. The loop closure problem was informally discussed by Robert Diamond and M. Levitt and formally defined by Go and Scheraga [32]. The input to such a problem is the relative position of two fixed residues (anchors) at each end and the goal is to find different possible conformations for a polypeptide chain of length *m* joining the fixed ends.

Rather than being closely tied to some search strategy (or an energy function), ChainTweak is a stand-alone method that can be used by researchers as a black-box, allowing them to focus on other parts of the search problem (e.g., energy function design [48]). Unlike classical molecular dynamic simulations, which are constrained by folding trajectories, it allows an unbiased sampling of the conformational landscape in the neighborhood of a given structure.

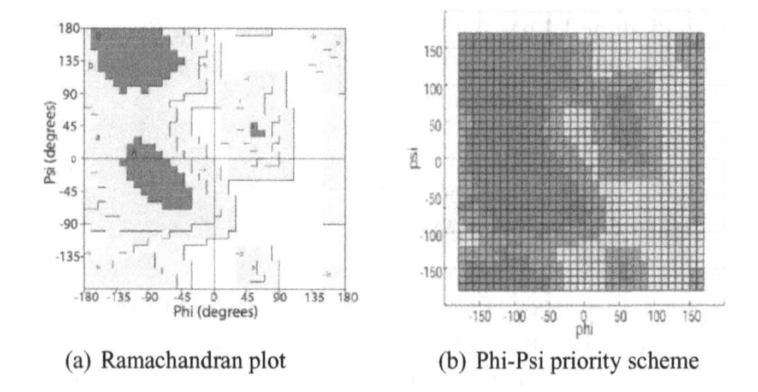

(a) Ramachandran plot (b) Phi-Psi priority scheme

Fig. 16 (a) Reference Ramachandran plot and (b) phi-psi priority scheme: (dark gray: highest priority)>(medium gray: medium priority)>(light gray: lowest priority) [67].

We show in Fig. 17(b) an application of this program and sample ten conformations from the neighborhood of a 32-residue protein structure (PDB:1CLV, chain I [56]). Using the LoopClsr [67] algorithm iteratively (Fig. 17(a)) on the backbone of a protein, we generated conformations in the neighborhood of this structure [1-4Å] within a few seconds. The size of the neighborhood explored and topology of the backbone can be constrained depending on the context of the simulations. ChainTweak is purely geometric in nature and does not inherently depend on any energy function. This makes it a useful standalone sampling algorithm, which can then be combined with existing energy functions. Such a methodology completely eliminates the dependence of sampling on the limitations of the energy function.

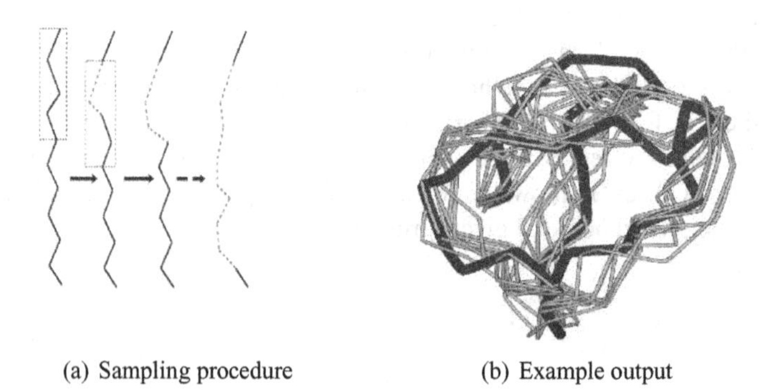

(a) Sampling procedure (b) Example output

Fig. 17 (a) Iteratively modifying the backbone of a protein. Sliding window formulation implemented in ChainTweak [67]. (b) Example output from ChainTweak. Ten conformations from the neighborhood of a 32-residue protein structure (PDB:1CLV, chain I [56]) were sampled and aligned with the original. The original structure is in black, the others are in gray [67].

6.3 Structure Determination

Structure prediction and determination are still significant bottlenecks to the goals of the Structural Genomics initiative [9]. Due to great advances in sequencing technologies and algorithms for analyzing sequence data, the gap between the number of genomes known and number of structures known is even increasing. In order to close this gap, significant advances need to be made in two areas: 1) accurately and efficiently determine structures from incomplete experimental data [9], and 2) develop accurate energy functions that can filter out native structures from a set of decoys. The first problem can be set up as an optimization problem, for which the sampling algorithm is critical for exploring diverse structures and thus maximizing the likelihood of the observed data. Furthermore, as we move away from a "static" picture of a protein to a more dynamic one, the ability to exhaustively sample all degrees of freedom becomes critical to our understanding of the structure. ChainTweak is ideally suited for such an analysis, as we have demonstrated by modeling the heterogeneity in crystal structures solved at medium to low resolutions [38]. More importantly from a biological standpoint, such an analysis potentially provides a mechanism for understanding the protein structure-function relationship [45].

7 Exercises

1. In Section 4, we add an energy term $\mathscr{L}(n)$ to the loop connecting the β-strand pairs, where n is the number of residues in the loop. Modify the recursive equations of Section 4 to account for this change.
2. Write a backtracking algorithm for computing the Boltzmann probability of closing β-strand pairs of length n (i.e. n inter-strand residue contacts).
3. We assume that the sampling algorithm in Section 4 returns the energy $E(S)$ of the sampled TMB structure S. Let $\mathscr{Z}$ be the partition function value. Then, the Boltzmann probability of a structure S is $P(S) = \frac{e^{-E(S)/RT}}{\mathscr{Z}}$. Write an iterative procedure for sampling TMBs until a ratio ρ $(0 < \rho \leq 1)$ of the folding landscape has been covered.
4. Write pseudo-code for the ChainTweak algorithm illustrated in Fig. 17(a).

8 Further reading

Many studies provide reliable techniques for sampling structures from sequence. The popular Rosetta [66] uses multiple sequence alignments to select small protein backbone fragments that are assembled together using a simulated annealing procedure. More recently, novel approaches have been proposed for overcoming the difficulty of designing a reliable energy function required to perform a simulated annealing procedure. Indeed, lattice-based techniques [82], HMMs [36], and more general Conditional Random Fields (CRFs) [84, 85] have been successfully applied for this purpose. However, it is worth noting that the methods cited here generate decoys but do not sample from a rigorously defined distribution of structures.

To conclude this chapter, we note that the program *partiFold* detailed in Section 4 has recently been extended to perform structural *ensemble* comparisons [78] and generate accurate sequence alignments of proteins with low sequence identity.

Acknowledgements The authors would like to thank Charles W. O'Donnell and Raghavendra Hosur for useful comments and suggestions.

References

1. Abe, N., Mamitsuka, H.: Predicting protein secondary structure using stochastic tree grammars. Machine Learning **29**(2-3), 275–301 (1997)
2. Ahn, V.E., Lo, E.I., Engel, C.K., Chen, L., Hwang, P.M., Kay, L.E., Bishop, R.E., Prive, G.G.: A hydrocarbon ruler measures palmitate in the enzymatic acylation of endotoxin. EMBO J **23**(15), 2931–2941 (2004 Aug 4)
3. Amato, N., Dill, K., Song, G.: Using motion planning to map protein folding landscapes and analyze folding kinetics of known native structures. Journal of Computational Biology **10**(3-4), 239–255 (2003)
4. Bartlett, A.I., Radford, S.E.: An expanding arsenal of experimental methods yields an explosion of insights into protein folding mechanisms. Nat Struct Mol Biol **16**(6), 582–588 (2009)
5. Bayrhuber, M., Meins, T., Habeck, M., Becker, S., Giller, K., Villinger, S., Vonrhein, C., Griesinger, C., Zweckstetter, M., Zeth, K.: Structure of the human voltage-dependent anion channel. Proc Natl Acad Sci U S A **105**(40), 15,370–15,375 (2008 Oct 7)
6. Berg, O.G., von Hippel, P.H.: Selection of DNA binding sites by regulatory proteins. statistical-mechanical theory and application to operators and promoters. J Mol Biol **193**(4), 723–750 (1987 Feb 20)
7. Berger, B., Leighton, T.: Protein folding in the hydrophobic-hydrophilic (HP) model is NP-complete. J Comput Biol **5**(1), 27–40 (1998)
8. Berman, H., Westbrook, J., Feng, Z., Gilliland, G., Bhat, T., Weissig, H., Shindyalov, I., Bourne, P.: The Protein Data Bank. Nucleic Acids Research **28**, 235–242 (2000)
9. Bourne, P., Weissig, H.: Structural Bioinformatics. Wiley-Liss (2003)
10. Bradley, P., Chivian, D., Meiler, J., Misura, K.M.S., Rohl, C.A., Schief, W.R., Wedemeyer, W.J., Schueler-Furman, O., Murphy, P., Schonbrun, J., Strauss, C.E.M., Baker, D.: Rosetta predictions in CASP5: Successes, failures, and prospects for complete automation. Proteins **53 Suppl 6**, 457–468 (2003)
11. Bradley, P., Cowen, L., Menke, M., King, J., Berger, B.: Betawrap: Successful prediction of parallel beta-helices from primary sequence reveals an association with many microbial pathogens. Proceedings of the National Academy of Sciences **98**(26), 14,819–14,824 (2001)
12. Cahill, M., Cahill, S., Cahill, K.: Proteins wriggle. Biophys J **82**(5), 2665–2670 (2002 May)
13. Chandler, D.: Introduction to Modern Statistical Mechanics. Oxford University Press (1987)
14. Cheng, J., Baldi, P.: Three-stage prediction of protein beta-sheets by neural networks, alignments and graph algorithms. Bioinformatics **21 Suppl 1**, i75–84 (2005 Jun)
15. Cheng, J., Baldi, P.: Improved residue contact prediction using support vector machines and a large feature set. BMC Bioinformatics **8**, 113 (2007)
16. Chiang, D., Joshi, A.K., Dill, K.: A grammatical theory for the conformational changes of simple helix bundles. Journal of Computational Biology **13**(1), 27–42 (2006)

17. Chotia, C.: The nature of the accessible and buried surfaces in proteins. J Mol. Biol. **105**(1), 1–14 (1975)
18. Clote, P., Backofen, R.: Computational Molecular Biology: An Introduction. John Wiley & Sons (2000). 279 pages
19. Clote, P., Waldispühl, J., Behzadi, B., Steyaert, J.M.: Energy landscape of *k*-point mutants of an RNA molecule. Bioinformatics **21**(22), 4140–4147 (2005)
20. Coutsias, E.A., Seok, C., Jacobson, M.P., Dill, K.A.: A kinematic view of loop closure. J Comput Chem **25**(4), 510–528 (2004)
21. Cowen, L., Bradley, P., Menke, M., King, J., Berger, B.: Predicting the beta-helix fold from protein sequence data. J of Computational Biology **9**, 261–276 (2002)
22. Dill, K., Bromberg, S.: Molecular Driving Forces. Garland Science, Taylor & Francis (2003). New York
23. Dill, K., Phillips, A., Rosen, J.: Protein structure and energy landscape dependence on sequence using a continuous energy function. J Comput Biol. **4**(3), 227–39 (1997)
24. Dill, K.A., Ozkan, S.B., Shell, M.S., Weikl, T.R.: The protein folding problem. Annu Rev Biophys **37**, 289–316 (2008)
25. Ding, Y., Lawrence, C.: A statistical sampling algorithm for RNA secondary structure prediction. Nucleic Acids Res. **31(24)**, 7280–7301 (2003)
26. Dobson, C.M.: Protein folding and misfolding. Nature **426**(6968), 884–890 (2003)
27. Dyson, H.J., Wright, P.E.: Intrinsically unstructured proteins and their functions. Nat Rev Mol Cell Biol **6**(3), 197–208 (2005 Mar)
28. Fain, B., Levitt, M.: A novel method for sampling alpha-helical protein backbones. J Mol Biol. **305**(2), 191–201 (2001)
29. Fain, B., Levitt, M.: Funnel sculpting for in silico assembly of secondary structure elements of proteins. Proc. Natl. Acad. Sci. USA **100**(19), 10,700–5 (2003)
30. Foat, B.C., Morozov, A.V., Bussemaker, H.J.: Statistical mechanical modeling of genome-wide transcription factor occupancy data by MatrixREDUCE. Bioinformatics **22**(14), e141–9 (2006 Jul 15)
31. Frishman, D., P., A.: Knowledge-based protein secondary structure assignment. Proteins **23**, 566–579 (1995)
32. Go, N., Scheraga, H.A.: Ring closure and local conformational deformations of chain molecules. Macromolecules **3**(2), 178–187 (1970)
33. Grana, O., Baker, D., MacCallum, R., Meiler, J., Punta, M., Rost B.and Tress, M., Valencia, A.: CASP6 assessment of contact prediction. Proteins **61**(7), 214–224 (2005)
34. Grosberg, A., Khokhlov, A.: Statistical Physics of Macromolecules. AIP Press (1994)
35. Guntert, P., Mumenthaler, C., Wuthrich, K.: Torsion angle dynamics for NMR structure calculation with the new program DYANA. J Mol Biol **273**(1), 283–298 (1997 Oct 17)
36. Hamelryck, T., Kent, J.T., Krogh, A.: Sampling realistic protein conformations using local structural bias. PLoS Comput Biol **2**(9), e131 (2006 Sep 22)
37. Hockenmaier, J., Joshi, A., Dill., K.: Routes are trees: The parsing perspective on protein folding. PROTEINS: Structure, Function, and Bioinformatics **66**, 1–15 (2007)
38. Hosur, R., Singh, R., Berger, B.: Personal communication
39. Huang, E.S., Subbiah, S., Tsai, J., Levitt, M.: Using a hydrophobic contact potential to evaluate native and near-native folds generated by molecular dynamics simulations. J Mol Biol **257**(3), 716–725 (1996 Apr 5)
40. Hubner, I.A., Deeds, E.J., Shakhnovich, E.I.: Understanding ensemble protein folding at atomic detail. Proc Natl Acad Sci U S A **103**(47), 17,747–17,752 (2006 Nov 21)
41. Huysmans, G.H.M., Radford, S.E., Brockwell, D.J., Baldwin, S.A.: The N-terminal helix is a post-assembly clamp in the bacterial outer membrane protein PagP. J Mol Biol **373**(3), 529–540 (2007 Oct 26)
42. Istrail, I.: Statistical mechanics, three-dimensionality and NP-completeness: I. Universality of intractability of the partition functions of the Ising model across non-planar lattices. In: A. Press (ed.) Proceedings of the 32nd ACM Symposium on the Theory of Computing (STOC00), pp. 87–96 (2000)

43. Izarzugaza, J.M.G., Grana, O., Tress, M.L., Valencia, A., Clarke, N.D.: Assessment of intramolecular contact predictions for CASP7. Proteins **69 Suppl 8**, 152–158 (2007)
44. King, J., Haase-Pettingell, C., Gossard, D.: Protein folding and misfolding. American Scientist **90**(5), 445–453 (2002)
45. Knight, J.L., Zhou, Z., Gallicchio, E., Himmel, D.M., Friesner, R.A., Arnold, E., Levy, R.M.: Exploring structural variability in X-ray crystallographic models using protein local optimization by torsion-angle sampling. Acta Crystallogr D Biol Crystallogr **64**(Pt 4), 383–396 (2008 Apr)
46. Koebnik, R.: Membrane assembly of the *Escherichia coli* outer membrane protein OmpA: Exploring sequence constraints on transmembrane β-strands. J. Mol. Biol. **285**, 1801–1810 (1999)
47. Kolodny, R., Koehl, P., Guibas, L., Levitt, M.: Small libraries of protein fragments model native protein structures accurately. J Mol Biol **323**(2), 297–307 (2002 Oct 18)
48. Krishnamoorthy, B., Tropsha, A.: Development of a four-body statistical pseudo-potential to discriminate native from non-native protein conformations. Bioinformatics **19**(12), 1540–1548 (2003 Aug 12)
49. Manocha, D., Zhu, Y., Wright, W.: Conformational analysis of molecular chains using nano-kinematics. Comput Appl Biosci **11**(1), 71–86 (1995)
50. McCaskill, J.: The equilibrium partition function and base pair binding probabilities for RNA secondary structure. Biopolymers **29**, 1105–1119 (1990)
51. McDonnell, A.V., Menke, M., Palmer, N., King, J., Cowen, L., Berger, B.: Fold recognition and accurate sequence-structure alignment of sequences directing beta-sheet proteins. Proteins **63**(4), 976–985 (2006 Jun 1)
52. Miller, D., Dill, K.: Ligand binding to proteins: the binding landscape model. Protein Sci. **6**(10), 2166–79 (1997)
53. Mirny, L., Shakhnovich, E.: Protein folding theory: from lattice to all-atom models. Annu Rev Biophys Biomol Struct. **30**, 361–96 (2001)
54. Morozov, A.V., Havranek, J.J., Baker, D., Siggia, E.D.: Protein-DNA binding specificity predictions with structural models. Nucleic Acids Res **33**(18), 5781–5798 (2005)
55. Park, B., Levitt, M.: Energy functions that discriminate X-ray and near native folds from well-constructed decoys. J Mol Biol **258**(2), 367–392 (1996 May 3)
56. Pereira, P.J., Lozanov, V., Patthy, A., Huber, R., Bode, W., Pongor, S., Strobl, S.: Specific inhibition of insect alpha-amylases: yellow meal worm alpha-amylase in complex with the amaranth alpha-amylase inhibitor at 2.0 A resolution. Structure **7**(9), 1079–1088 (1999 Sep 15)
57. Punta, B., Rost, B.: Profcon: novel prediction of long-range contacts. Bioinformatics **21**(13), 2960–2968 (2005)
58. Ramachandran, G., Sasisekharan, V.: Conformation of polypeptides and proteins. Adv. Protein. Chem. **23**, 283–437 (1968)
59. Randall, A., Cheng, J., Sweredoski, M., Baldi, P.: TMBpro: secondary structure, beta-contact and tertiary structure prediction of transmembrane beta-barrel proteins. Bioinformatics **24**(4), 513–520 (2008 Feb 15)
60. Rhodes, G.: Crystallography Made Crystal Clear, 2nd edn. Academic Press: San Diego (2000)
61. Rumbley, J., Hoang, L., Mayne, L., Englander, S.W.: An amino acid code for protein folding. Proc Natl Acad Sci U S A **98**(1), 105–112 (2001)
62. Schlessinger, A., Rost, B.: Protein flexibility and rigidity predicted from sequence. Proteins **61**(1), 115–126 (2005)
63. Schultz, C.: Illuminating folding intermediates. Nature Structural Biology **7**, 7–10 (2000)
64. Schulz, G.: β-barrel membrane proteins. Current Opinion in Structural Biology **10**, 443–447 (2000)
65. Shorter, J., Lindquist, S.: Prions as adaptive conduits of memory and inheritance. Nat Rev Genet **6**(6), 435–450 (2005 Jun)
66. Simons, K.T., Kooperberg, C., Huang, E., Baker, D.: Assembly of protein tertiary structures from fragments with similar local sequences using simulated annealing and bayesian scoring functions. J Mol Biol **268**(1), 209–225 (1997 Apr 25)

67. Singh, R., Berger, B.: ChainTweak: Sampling from the neighbourhood of a protein conformation. Proceedings of the 10th Pacific Symposium on Biocomputation pp. 52–63 (2005)
68. Sippl, M.J.: Calculation of conformational ensembles from potentials of mean force. Journal of Molecular Biology **213**, 859–883 (1990)
69. Thomas, S., Song, G., Amato, N.M.: Protein folding by motion planning. Phys Biol **2**(4), S148–55 (2005 Nov)
70. Ulmschneider, J.P., Jorgensen, W.L.: Polypeptide folding using monte carlo sampling, concerted rotation, and continuum solvation. J Am Chem Soc **126**(6), 1849–1857 (2004 Feb 18)
71. Vandeputte-Rutten, L., Bos, M.P., Tommassen, J., Gros, P.: Crystal structure of neisserial surface protein A (NspA), a conserved outer membrane protein with vaccine potential. J Biol Chem **278**(27), 24,825–24,830 (2003 Jul 4)
72. Voelz, V., Dill, K.: Exploring zipping and assembly as a protein folding principle. Proteins: Structure Function and Bioinformatics **66**, 877–888 (2007)
73. Vogt, J., Schulz, G.E.: The structure of the outer membrane protein OmpX from *escherichia coli* reveals possible mechanisms of virulence. Structure **7**(10), 1301–1309 (1999 Oct 15)
74. Wagner, G.P., Otto, W., Lynch, V., Stadler, P.F.: A stochastic model for the evolution of transcription factor binding site abundance. J Theor Biol **247**(3), 544–553 (2007 Aug 7)
75. Waldispühl, J., Berger, B., Clote, P., Steyaert, J.M.: Predicting transmembrane β-barrels and inter-strand residue interactions from sequence. Proteins: Structure, Function and Bioinformatics **65**, 61–74 (2006). Doi:10.1002/prot.2146
76. Waldispühl, J., Berger, B., Clote, P., Steyaert, J.M.: transfold: A web server for perdicting the structure of transmembrane proteins. Nucleic Acids Research (Web Server Issue) **34**, W189–W193 (2006). Doi:10.1093/nar/glk205
77. Waldispühl, J., O'Donnell, C.W., Devadas, S., Clote, P., Berger, B.: Modeling ensembles of transmembrane beta-barrel proteins. Proteins **71**(3), 1097–1112 (2008 May 15)
78. Waldispühl, J., O'Donnell, C.W., Will, S., Devadas, S., Backofen, R., Berger, B.: Simultaneous alignment and folding of protein sequences. In: S. Batzoglou (ed.) Research in Computational Molecular Biology, *Lecture Notes in Computer Science*, vol. Volume 5541/2009, pp. 339–355. Springer Berlin / Heidelberg (2009)
79. Waldispühl, J., Steyaert, J.M.: Modeling and predicting all-alpha transmembrane proteins including helix-helix pairing. Theor. Comput. Sci. **335**(1), 67–92 (2005)
80. William J. Wedemeyer, H.A.S.: Exact analytical loop closure in proteins using polynomial equations. J Comput Chem **20**(8), 819–844 (1999)
81. Wimley, W.C., White, S.H.: Reversible unfolding of β-sheets in membranes: A calorimetric study. Journal of Molecular Biology **342**, 703–711 (2004)
82. Xia, Y., Huang, E.S., Levitt, M., Samudrala, R.: Ab initio construction of protein tertiary structures using a hierarchical approach. J Mol Biol **300**(1), 171–185 (2000 Jun 30)
83. Y., Z., J., S.: SPICKER: A clustering approach to identify near-native protein folds. Journal of Computational Chemistry **25**, 865–871 (2004)
84. Zhao, F., Li, S., Sterner, B.W., Xu, J.: Discriminative learning for protein conformation sampling. Proteins **73**(1), 228–240 (2008 Oct)
85. Zhao, F., Peng, J., DeBartolo, J., Freed, K.F., Sosnick, T.R., Xu, J.: A probabilistic graphical model for ab initio folding. In: S. Batzoglou (ed.) Research in Computational Molecular Biology, *Lecture Notes in Computer Science*, vol. Volume 5541/2009, pp. 59–73. Springer Berlin / Heidelberg (2009)

Stochastic Simulation for Biochemical Systems

Yang Cao

Abstract Biochemical systems are often modeled with ordinary differential equations that are continuous and deterministic by nature. In recent years, with the development of new techniques to collect wet-lab data in a single cell, there are increasing concerns on the stochastic effect in cellular systems, where the small copy numbers of some reactant species in the cell may lead to deviations from the predictions of the deterministic differential equations of classical chemical kinetics. In this chapter, we will review important algorithms for stochastic modeling and simulation of biochemical systems.

1 History of the Problem

Biochemistry studies the chemical processes in living organisms. Typical areas of biochemistry include the gene (DNA, RNA) network, protein synthesis, cell membrane transport, and signal transduction. Traditionally, biochemical systems have been studied with mathematical modeling and simulation, based on ordinary differential equations (ODEs) , in which biochemical processes are represented using Reaction Rate Equations (RREs). The general form of the RREs can be formulated as

$$\frac{dx_i}{dt} = f_i(x_1, \cdots, x_n), \quad i = 1, \cdots, n, \tag{1}$$

where the state variables x_i represent the concentrations of involved species, and functions f_i are inferred from various chemical reactions in the system. In typical RREs, f_i's are generated from mass action kinetics [29], Mechalias-Menton or Hill equations [37, 31]. Other types of functions have also been used to model biochemical systems, such as the S-system [46, 41, 42, 43] in biochemical systems theory, a mathematical modeling framework based on ODEs that has been developed since

Yang Cao
Department of Computer Science, Virginia Tech, Blacksburg, VA, USA, e-mail: ycao@cs.vt.edu

L.S. Heath and N. Ramakrishnan (eds.), *Problem Solving Handbook in Computational Biology and Bioinformatics*, DOI 10.1007/978-0-387-09760-2_10,

the 1960s. With the development of systems biology, there have appeared more and more successful biochemical models [5, 15, 16] using ODEs. On the other hand, a model represented by ODEs is naturally continuous and deterministic. This feature limits its application to certain cellular systems [1, 17, 34] where the molecular populations of some important reactant species are small, from one to thousands. Typically, the concerned systems involve copy numbers of one or two for the number of genes of a given protein, on the order of tens to hundreds for the corresponding RNAs and on the order of thousands for regulatory proteins and enzymes. For those systems, the discreteness and stochasticity may play important roles in the dynamics of the system. The traditional modeling techniques based on ODEs cannot be used to describe the discrete and stochastic kinetics. Thus discrete stochastic modeling and simulation has become a hot area in recent years.

To include discreteness and randomness, the most accurate way to model and simulate the time evolution of a system of chemically reacting molecules is molecular dynamics (MD) simulation, which tracks the positions and velocities of all the molecules and the occurrence of all chemical reactions when molecules physically collide with each other. MD simulation was originally developed in theoretical physics in the late 1950s [2, 26] and gained popularity in material science, biochemistry, and biophysics in the 1970s. But MD simulations are generally too expensive to be practical except in the case of a relatively small number of molecules and even then only for very short time scales. To handle this difficulty, in 1970s people started to consider special cases where the dynamics of biochemical systems can be approximated by assuming that the reactant molecules are "well-stirred" such that their positions become randomized and need not be tracked in detail. When that is true, the state of the system can be defined simply by the instantaneous molecular populations of the various chemical species. The chemical reactions can be defined as events that modify the state of the system following certain biochemical rules, changing the molecular populations by integer numbers. Based on the well-stirred assumption, a practical discrete and stochastic modeling and simulation technique for biochemical systems, now known as the stochastic simulation algorithm (SSA) or Gillespie algorithm, was proposed by Dan Gillespie in two classical publications [20, 21][1]. Gillespie's pioneering work was originally proposed for chemically reacting systems and had been criticized for many years, since there were no practical chemically reacting systems known at that time that really needed this discrete stochastic modeling and simulation technique, until in the 1990s, the rapid development of biological experiment technologies enabled observations of the dynamics in a single cell and people's attentions were drawn to the biochemical models in cells. In the famous discrete stochastic model [1] of lambda phage affected E. coli published by Adam Arkin et al. in 1998, Gillespie's SSA was successfully used to simulate this discrete stochastic model and demonstrate that the randomness in

[1] Interestingly, a mathematically-equivalent simulation algorithm, the BKL method [4], was proposed independently in the 70s, for the Monte Carlo simulation of ising spin systems. Gillespie's work focuses more on chemical systems, while the BKL method focuses more on physical systems. Their similarity was noticed only recently. Here we will discuss these more from the chemical systems point of view.

the cell may cause phenotype differences. It was a milestone for the application of discrete stochastic algorithms in the modeling of cellular systems. After that, Gillespie's SSA has become very popular in research in computational biology and systems biology. It is widely accepted as a classical modeling and simulation method in this area.

With more and more applications of the SSA, it finally attracts more computational scientists to make further development. Different implementation strategies and approximation algorithms were proposed in the past ten years. For the implementation of SSA, following the original direct method (DM) and the first reaction method (FRM) proposed by Dan Gillespie [20, 21] in 1976 and 1977, the next reaction method (NRM) [18] was proposed in 2000. The optimized direct method (ODM) [13] was proposed in 2004 and the sorted direct method (SDM) [35] was proposed in 2006. For all these implementation strategies, the efficiency gains over the original DM in computation time are usually less than 20%, which are still considered very slow for many applications. This low efficiency comes from the algorithm itself. As the SSA is a procedure simulating every reaction event individually, the computational cost is inevitably high. Thus approximation methods have been developed. The first approximation method, called the tau-leaping method [23], was proposed by Dan Gillespie in 2001. Following that, new tau-leaping methods, such as the binomial tau-leaping method [14, 45], implicit tau-leaping method [40] and trapezoidal tau-leaping method [7] have been proposed. The efficient implementation of the tau-leaping method have also been discussed [10, 12, 24]. More and more robust and efficient stochastic simulation algorithms are still under research.

2 Algorithm Description

2.1 Problem Definition and the Chemical Master Equation

Consider a system of N molecular species $\{S_1, \ldots, S_N\}$ interacting through M elemental chemical reaction channels $\{R_1, \ldots, R_M\}$. We assume that the system is confined to a constant volume Ω and is well-stirred, or, in other words, is in thermal (but not chemical) equilibrium at a constant temperature. Under these assumptions, the state of the system can be represented by the populations of the species involved. We denote these populations by $\mathbf{X}(t) \equiv (X_1(t), \ldots, X_N(t))$, where $X_i(t)$ is the number of molecules of species S_i in the system at time t. The well-stirred condition is crucial. When this condition is broken, the spatial information of each species becomes important and the population information for the species will not be enough alone to determine the system dynamics. In cases where the well-stirred condition does not hold, the required simulation techniques will be different from what we discuss in this chapter. The so-called elemental reactions only include monomolecular and bimolecular reactions. Generalizations can be made to include more complicated reaction types such as the commonly used Michaelis-Menten reaction [8, 11, 38].

We note that modeling these higher-order reaction types using discrete stochastic methods are still under research and are not addressed in this chapter.

For a well-stirred system, each reaction channel R_j can be characterized by a *propensity function* a_j and a *state change vector* $\nu_j \equiv (\nu_{1j}, \ldots, \nu_{Nj})$. The propensity function is defined by the statement:

$a_j(\mathbf{x})dt \equiv$ the probability, given $\mathbf{X}(t) = \mathbf{x}$, that one R_j reaction will occur in the next infinitesimal time interval $[t, t+dt)$.

All state change vectors ν_j form a matrix ν, which is also known as the stoichiometric matrix. ν_{ij}, the element in the stoichiometric matrix ν, is the change in the molecular population S_i induced by one reaction R_j. The propensity function $a_j(x)$ reflects the fundamental characteristics of the stochastic chemical kinetics. Its value depends on the populations of the reactant populations and a reaction propensity rate constant c_j, which is defined so that

$c_j dt \equiv$ the probability that a randomly chosen combination of R_j reactant molecules will react in the next infinitesimal time dt.

Then a_j is the product of c_j and the number of all possible combinations of R_j reactant molecules.

The following are three simple examples of basic reactions and their propensity functions and state change vectors.

$$\text{For } S_1 \xrightarrow{c_1} S_2, \qquad a_j(x) = c_1 x_1, \text{ and } \nu_j = (-1, 1, 0, \cdots, 0). \tag{2}$$

$$\text{For } S_1 + S_2 \xrightarrow{c_1} S_3, \qquad a_j(x) = c_1 x_1 x_2, \text{ and } \nu_j = (-1, -1, 1, 0, \cdots, 0). \tag{3}$$

$$\text{For } S_1 + S_1 \xrightarrow{c_1} S_2, \qquad a_j(x) = \frac{c_1}{2} x_1 (x_1 - 1), \text{ and } \nu_j = (-2, 1, 0, \cdots, 0). \tag{4}$$

It is easy to see that the form of the propensity function is similar to the mass action terms in deterministic RREs. The value of c_j is similar to its counterpart, the reaction rate constant k_j in the RREs. And indeed there is a connection between c_j and k_j depending on the reaction type. For a monomolecular reaction such as in the example of equation (2), $c_1 = k_1$. For a bimolecular reaction between different species such as in (3), $c_1 = \frac{k_1}{A\Omega}$, where A is Avogadro's number and Ω is the constant volume. For a bimolecular reaction between the same species, the forms of the propensity function and the reaction rate function have a slight difference, but when x_1 is large the difference will be negligibly small and we will have $c_1 \approx \frac{2k_1}{A\Omega}$.

Once the propensity functions and stoichiometric matrix are determined, the dynamics of the system obeys the *chemical master equation* (CME):

$$\frac{\partial P(\mathbf{x}, t | \mathbf{x}_0, t_0)}{\partial t} = \sum_{j=1}^{M} [a_j(\mathbf{x} - \nu_j) P(\mathbf{x} - \nu_j, t | \mathbf{x}_0, t_0) - a_j(\mathbf{x}) P(\mathbf{x}, t | \mathbf{x}_0, t_0)], \tag{5}$$

where $P(\mathbf{x},t|\mathbf{x}_0,t_0)$ denotes the probability that $\mathbf{X}(t)$ will be $\mathbf{x}$ given that $\mathbf{X}(t_0) = \mathbf{x}_0$. In principle, the CME completely determines the dynamics of $P(\mathbf{x},t|\mathbf{x}_0,t_0)$. But the CME is essentially an ODE whose dimension is given by the number of all possible combinations of states of $\mathbf{x}$. Consider the example of a small reaction network of 5 species and assume that the population of each species is in the range from 0 to 99. The dimension of the corresponding CME will then be $100^5 = 10^{10}$. As the number of species increases, the dimension of the corresponding CME increases exponentially. The rapidly increasing dimension presents a great challenge for the numerical solution of CMEs. This challenge is known as the "curse of dimension". It is easy to see that the CME is both theoretically and computationally intractable for all but the simplest models. In recent years, there has been some interesting research [36, 47] trying to reduce the dimension of the CME or to provide an approximate numerical solution of the CME. Progress has been made but so far these methods still can only be applied practically to simple models.

2.2 The Stochastic Simulation Algorithm (SSA)

Another way to study the dynamics of a chemically reacting system is to construct realizations of $\mathbf{X}(t)$ through numerical simulation. In numerical simulation, the key is not to directly compute the probabilities $P(\mathbf{x},t|\mathbf{x}_0,t_0)$ but to generate a single trajectory (a realization) that the system may undergo. The most important simulation method in this direction is Gillespie's stochastic simulation algorithm (SSA) [20, 21]. Instead of following the time evolution of the probabilities, the SSA generates a trajectory of the system step by step. In each step, the SSA starts from a current state $x(t) = \mathbf{x}$ and asks two questions:

- When will the next reaction occur? We denote this time interval by τ.
- When the next reaction occurs, which reaction will it be? We denote the chosen reaction by the index $\mathbf{j}$.

To answer the above questions, one needs to study the joint probability density function $p(\tau, j|\mathbf{x},t)$, which is defined by

$$p(\tau, j|\mathbf{x},t)dt = \begin{array}{l}\text{the probability, given } \mathbf{X}(t) = \mathbf{x}, \text{ that the next reaction will} \\ \text{occur in the infinitesimal time interval } [t+\tau, t+\tau+dt), \\ \text{and will be an } R_j \text{reaction.}\end{array} \quad (6)$$

It can be derived [20, 21] that

$$p(\tau, j|\mathbf{x},t) = a_j(\mathbf{x})\exp(-a_0(\mathbf{x})\tau), \quad (7)$$

where $a_0(\mathbf{x}) \equiv \sum_{j=1}^{M} a_j(\mathbf{x})$. Equation (7) is the theoretical foundation for the SSA. It implies that the time τ to the next occurring reaction is an exponentially distributed random variable with mean $1/a_0(\mathbf{x})$, and that the index j of that reaction is the integer random variable with point probability $a_j(\mathbf{x})/a_0(\mathbf{x})$. To advance the system from state $\mathbf{x}$ at time t, the SSA generates two uniform random numbers r_1 and r_2 over the unit interval, and then takes the time of the next reaction to be $t+\tau$ where

$$\tau = \frac{1}{a_0(\mathbf{x})} \ln\left(\frac{1}{r_1}\right), \tag{8}$$

and the index for the next reaction to be the smallest integer $\mathbf{j}$ satisfying

$$\sum_{j'=1}^{\mathbf{j}} a_{j'}(\mathbf{x}) > r_2 a_0(\mathbf{x}). \tag{9}$$

The system state is then updated according to $\mathbf{X}(t+\tau) = \mathbf{x} + \nu_j$, and this process is repeated until the simulation final time or until some other terminating condition is reached.

2.3 The Tau-Leaping Method

The tau-leaping method (Gillespie, 2001) was designed to speed up a stochastic simulation by *leaping* over many reactions in one time step. This idea is illustrated in Figure 1. The tau-leaping method makes the leap by answering the following

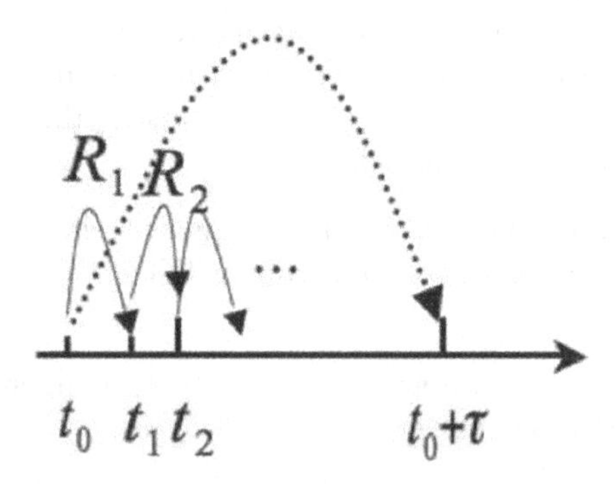

Fig. 1 The comparison between the SSA and the tau-leaping method. The tau-leaping method leaps over many reactions in one time step.

question: How many times will each reaction channel fire in the next *specified* time interval τ? More precisely, let

$$K_j(\tau;\mathbf{x},t) \triangleq \text{the number of times, given } \mathbf{X}(t)=\mathbf{x}\text{, that reaction channel } R_j \text{ will fire in the time interval } [t,t+\tau) \quad (j=1,\ldots,M). \tag{10}$$

For arbitrary values of τ it will be about as difficult to compute $K_j(\tau;\mathbf{x},t)$ as to solve the CME. The tau-leaping method chooses a small τ value to satisfy the following **Leap Condition**: *For the current state x, require τ to be small enough that the change in the state during $[t,t+\tau)$ will be so small that no propensity function will suffer an appreciable change in its value.* Under the Leap Condition, a good *approximation* to $K_j(\tau;\mathbf{x},t)$ will be provided by $P(a_j(\mathbf{x}),\tau)$, the Poisson random variable with mean (and variance) $a_j(\mathbf{x})\tau$. So if $\mathbf{X}(t)=\mathbf{x}$ and we choose τ to satisfy the Leap Condition, we can update the state to time $t+\tau$ according to the *approximate* formula

$$\mathbf{X}(t+\tau)\doteq\mathbf{x}+\sum_{j=1}^{M}v_jP(a_j(\mathbf{x})\tau), \tag{11}$$

where $P(a_j(\mathbf{x})\tau)$, for each $j=1,...,M$, denotes an independent sample of the Poisson random variable with mean and variance $a_j(\mathbf{x})\tau$. This computational procedure is the *tau-leaping approximation.*

The tau-leaping method makes a natural connection between the SSA and the deterministic RREs. When τ is chosen very small such that in every time step there is at most one reaction occurring, the tau-leaping method reduces to a linear approximation of the SSA. When τ is allowed to be large such that

$$a_j(\mathbf{x})\tau\gg 1,\ \text{for all } j=1,\ldots,M, \tag{12}$$

the Poisson random number $P(a_j(\mathbf{x})\tau)$ can be approximated by the Normal random number with mean and variance $a_j(\mathbf{x})\tau$, denoted by $N(a_j(\mathbf{x})\tau,a_j(\mathbf{x})\tau)$. Then the formula (11) reduces to the forward Euler method for the chemical Langevin equation (CLE) (Gillespie, 2001). Moreover, when the values $a_j(\mathbf{x})\tau$, for all $j=1,...,M$, are even larger, the standard deviation is then negligible compared to the mean value. The Poisson random number $P(a_j(\mathbf{x})\tau)$ can then be simply replaced by its mean value $a_j(\mathbf{x})\tau$. Then the equation (11) becomes

$$\mathbf{X}(t+\tau)\doteq\mathbf{x}+\sum_{j=1}^{M}v_ja_j(\mathbf{x})\tau, \tag{13}$$

which is the forward Euler method for the corresponding RREs. Note that here the merger of the tau-leaping method into the forward Euler method is seamless. One does not need to check the condition (12) for all j's. The idea of using the normal random number or just the mean value to approximate the Poisson random number can be applied for any individual j. This procedure can be wrapped in a Poisson random number approximation procedure. Choose two threshold values: $\mathbf{M}_1$, for which the Poisson random number $P(\mathbf{M}_1)$ can be safely approximated by a normal random number $N(\mathbf{M}_1,\mathbf{M}_1)$, and $\mathbf{M}_2$, for which $P(\mathbf{M}_2)$ can be safely approximated by $\mathbf{M}_2$.

3 Available Implementations

There have been different implementations for the SSA and the tau-leaping methods. In this section, we introduce a straightforward implementation of the SSA called the direct method (DM). Other implementation strategies can be found in Section 5. The focus is on the more complicated implementation of the tau-leaping method.

The SSA can be simply implemented as in the following algorithm.

The DM implementation of SSA

Starting from initial condition $t = t_0$ and $x = x_0$,

1. With $x(t) = x$, calculate all $a_j(x)$ and $a_0(x)$.
2. If $a_0(x) = 0$, terminate the simulation. Otherwise generate two uniform random numbers r_1 and r_2. Calculate τ and $\mathbf{j}$ according to (8) and (9) respectively.
3. Update the system by $t = t + \tau$ and $\mathbf{x} = \mathbf{x} + \nu_{\mathbf{j}}$.
4. If t reaches the end time, stop. Otherwise, go to step 1.

The implementation of the tau-leaping method is much more complicated. Any implementation of the tau-leaping method has to solve two important problems:

- Avoid negative populations. Negative populations resulting from the original tau-leaping method have been observed in the simulation of certain systems in which some consumed reactant species are present in small numbers.
- Dynamically select τ values. The key point here is to make sure that the Leap Condition is satisfied and the simulation is still efficient.

3.1 Avoid Negative Populations

The negative population problem arises often from multiple reaction channels consuming the same reactant. We made an observation [10] that most negative populations were related to species with a low population. Based on this observation, an adaptive hybrid SSA/tau-leaping implementation strategy was proposed [10], which seems to resolve the negativity problem satisfactorily.

Since negative populations typically arise from multiple firings of reactions that are only a few reaction events away from consuming all the molecules of one of their reactants, the hybrid SSA/tau-leaping strategy defines these reaction channels as the *critical* reactions. The hybrid strategy introduces a second control parameter n_c, a positive integer that is usually set somewhere between 2 and 20. Any reaction channel with a positive propensity function that is currently within n_c firings of exhausting one of its reactants is classified as a *critical* reaction. The hybrid strategy chooses τ in such a way that no more than one firing of *all* the critical reactions can occur during the leap. Essentially, the algorithm simulates the *critical* reactions using an adapted (and thus not quite exact) version of the SSA, and the remaining *non-critical* reactions using the tau-leaping method. Since no more than one firing of a critical reaction can occur during a leap, the probability of producing a negative population is reduced to nearly zero. On those rare occasions when a negative population does arise (from firings of some non-critical reaction), that step can simply be rejected and repeated with τ reduced by half, or else the simulation can be started over using a larger value for n_c.

There are still some important implementation details. First, how do we decide whether or not a reaction is *critical* with the parameter n_c? This is done by first estimating for each reaction R_j with $a_j(\mathbf{x}) > 0$ the maximum number of times L_j that R_j can fire before exhausting one of its reactants [45, 14]:

$$L_j = \min_{i \in [1,N];\, \nu_{ij} < 0} \left[\frac{x_i}{|\nu_{ij}|} \right]. \tag{14}$$

Here the minimum is taken over only those index values i for which $\nu_{ij} < 0$, and the brackets denote "greatest-integer-in". After L_j is calculated, it is compared with n_c. If $L_j < n_c$, R_j is considered as a *critical* reaction and should be simulated by the SSA part. Otherwise, R_j is noncritical and can be simulated by the tau-leaping part.

The next step is to decide how to implement the SSA part and the tau-leaping part together. To solve this problem, in every simulation step we first generate a τ' from a τ-selection procedure and a τ'' from the SSA part. If τ' is even smaller than a few fold of the expected stepsize of a pure SSA method, $\frac{1}{a_0(\mathbf{x})}$, we will stick with the pure SSA method. Otherwise, we use the tau-leaping method to simulate the non-critical reactions and the SSA method to simulate the critical reactions. The real simulation timestep τ is chosen to be the smaller value between τ' and τ''. If τ'' is smaller, the critical reaction fires. Otherwise, no critical reaction should fire before τ. In both cases, the numbers of noncritical reaction firings are calculated using the Poisson tau-leaping method. The τ'' for the SSA part can simply follow the SSA procedure limited to only critical reactions. The τ' for the tau-leaping part will be discussed below.

3.2 The Tau-Selection Formula

The simulation formula for the tau-leaping method is quite simple. The key point is how to select the τ value so that the Leap Condition is satisfied. There have been several tau-selection formulae proposed in the literature. Gillespie [23] originally proposed that the Leap Condition could be considered satisfied if the expected change in each propensity function $a_j(\mathbf{x})$ during the leap were bounded by $\varepsilon a_0(\mathbf{x})$, where ε is an error control parameter $(0 < \varepsilon \ll 1)$. Later, this condition is refined by Gillespie and Petzold [24]. Denoting the change in propensity function a_j from time t to time $t+\tau$, given $\mathbf{X}(t) = \mathbf{x}$, by $\Delta_\tau a_j(\mathbf{x})$, the requirement by the two papers

[23, 24] can be stated as

$$|\Delta_\tau a_j(\mathbf{x})| \leq \varepsilon a_0(\mathbf{x}), \quad j = 1, \ldots, M. \tag{15}$$

Although this formula does indeed limit the changes in the propensities during a leap as required, it does not fully accomplish the task with a proper scaling. The Leap Condition requires that every propensity function remains "practically constant" during a τ time period, since that is what allows the number of reaction events R_j during τ to be accurately approximated by a statistically independent Poisson random variable with mean $a_j(\mathbf{x})\tau$. If $a_j(\mathbf{x})$ for reaction channel R_j happens to be very small compared to $a_k(\mathbf{x})$ for reaction channel R_k, $a_j(\mathbf{x})$ will then be much smaller than $a_0(\mathbf{x})$. The condition (15) may allow a large *relative* change in $a_j(\mathbf{x})$, and that could result in simulation inaccuracies. To allow the formula for the Leap Condition to reflect the relative scales, we can change the condition (15) by

$$|\Delta_\tau a_j(\mathbf{x})| \leq \varepsilon a_j(\mathbf{x}), \quad j = 1, \ldots, M. \tag{16}$$

Although this formula is an improvement to the condition (15), its calculation needs the Jacobian function of the propensity functions, which could be very time-consuming, especially if both M and N are large. A new τ-selection formula [12] was then proposed to avoid this computational burden. Here we introduce a simplified version of this new formula.

The underlying strategy of this new τ-selection procedure is to bound the relative changes in the *molecular populations* by a specified value ε $(0 < \varepsilon \ll 1)$. Let

$$\Delta_\tau X_i \equiv \Delta_\tau X_i(\mathbf{x}) \triangleq X_i(t+\tau) - x_i, \quad \text{given } \mathbf{X}(t) = \mathbf{x}. \tag{17}$$

Instead of basing the τ-selection on condition (16), we base it on the condition

$$\Delta_\tau X_i \leq \max\{\varepsilon x_i, 1\}, \quad \forall i \in I_{rs}, \tag{18}$$

where I_{rs} denotes the set of indices of all reactant species (so $i \in I_{rs}$ if and only if x_i is an argument of at least one propensity function). Condition (18) evidently requires the relative change in X_i to be bounded by ε, except that X_i will never be required to change by an amount less than 1. From the tau-leaping formula, we see that the quantity defined in (17) will essentially be given by

$$\Delta_\tau X_i = \sum_{j=1}^{M} \nu_{ij} \mathscr{P}(a_j(\mathbf{x})\tau), \quad \forall i \in I_{rs}. \tag{19}$$

Since the Poisson random variables (or the corresponding approximations) $\mathscr{P}(a_j(\mathbf{x})\tau)$ on the right-hand side of Eq.(19) are statistically independent and have means and variances $a_j(\mathbf{x})\tau$, the mean and variance of that linear combination can be straightforwardly computed:

$$\langle \Delta_\tau X_i \rangle = \sum_{j=1}^{M} \nu_{ij}[a_j(x)\tau], \quad \forall i \in I_{rs}, \tag{20a}$$

$$\mathrm{var}\{\Delta_\tau X_i\} = \sum_{j=1}^{M} \nu_{ij}^2[a_j(x)\tau], \quad \forall i \in I_{rs}. \tag{20b}$$

Using the same reasoning that was used in deriving the Gillespie-Petzold τ-selection procedure [24], we may consider the bound (18) on $\Delta_\tau X_i$ to be "substantially satisfied" if it is simultaneously satisfied by the absolute mean and the standard deviation of $\Delta_\tau X_i$:

$$|\langle \Delta_\tau X_i \rangle| \le \max\{\varepsilon x_i, 1\}, \quad \sqrt{\mathrm{var}\{\Delta_\tau X_i\}} \le \max\{\varepsilon x_i, 1\}, \quad \forall i \in I_{rs}. \tag{21}$$

Substituting formulas (20) into conditions (21), we obtain the following bounds on τ:

$$\tau \le \frac{\max\{\varepsilon_i x_i, 1\}}{|\sum_{j=1}^{M} \nu_{ij} a_j(\mathbf{x})|}, \quad \tau \le \frac{\max\{\varepsilon x_i, 1\}^2}{\sum_{j=1}^{M} \nu_{ij}^2 a_j(\mathbf{x})}, \quad \forall i \in I_{rs}. \tag{22}$$

$$\hat{\mu}_i(\mathbf{x}) \triangleq \sum_{j=1}^{M} \nu_{ij} a_j(\mathbf{x}), \quad \forall i \in I_{rs}, \tag{23a}$$

$$\hat{\sigma}_i^2(\mathbf{x}) \triangleq \sum_{j=1}^{M} \nu_{ij}^2 a_j(\mathbf{x}), \quad \forall i \in I_{rs}, \tag{23b}$$

where I_{rs} is the set of indices of all reactant species, and then taking

$$\tau = \min_{i \in I_{rs}} \left\{ \frac{\max\{\varepsilon x_i, 1\}}{|\hat{\mu}_i(\mathbf{x})|}, \frac{\max\{\varepsilon x_i, 1\}^2}{\hat{\sigma}_i^2(\mathbf{x})} \right\}. \tag{24}$$

The τ-selection procedure of formulas (23) and (24) is simpler to program and faster to execute than the τ-selection procedure of those formulas in [23, 24]. Particularly, the required number of computational operations increases quadratically with the number of reaction channels in the old formulas, but only linearly with the number of species in the new formulas. Since τ-selection has to be performed prior to every tau-leap, using these new formulas leads to substantially faster simulations when the system has many reaction channels and species.

The formulas (23) and (24) are for the original tau-leaping method. In order to apply them to the hybrid SSA/tau-leaping method, they need a little modification. The calculation should not be extended to critical reactions since they are handled by the adapted SSA part. Thus we let J_{ncr} denote the set of indices of the *non-critical reactions*. If J_{ncr} is empty (i.e., there are no non-critical reactions), we simply take $\tau' = \infty$ (practically this can be a large stepsize, for example the whole simulation time interval). Otherwise, the $\hat{\mu}_i$ and $\hat{\sigma}_i$ are calculated with the following formula:

$$\hat{\mu}_i(\mathbf{x}) \triangleq \sum_{j \in J_{ncr}} \nu_{ij} a_j(\mathbf{x}), \quad \forall i \in I_{rs}, \tag{25a}$$

$$\hat{\sigma}_i^2(\mathbf{x}) \triangleq \sum_{j\in J_{ncr}} \nu_{ij}^2 a_j(\mathbf{x}), \quad \forall i \in I_{rs}. \tag{25b}$$

The formula of τ' remains the same as τ in (24) but the calculation of $\hat{\mu}$ and $\hat{\sigma}$ are replaced by (25). Notice that the difference between (23) and (25) is that in (25) only non-critical reactions are considered, while in (23) all reactions are included.

The full description of the hybrid SSA/Tau-leaping strategy is given as follows.

The Hybrid SSA/Tau-Leaping Method

1. In state $\mathbf{x}$ at time t, identify the currently critical reactions. We calculate L_j according to the formula (14). Any reaction R_j with $a_j(\mathbf{x}) > 0$ is deemed *critical* if $L_j < n_c$. Otherwise, it is *non-critical*. (We normally take $n_c = 10$ as a practical value.)
2. Let J_{ncr} denote the set of indices of the non-critical reactions. If J_{ncr} is empty, we take $\tau' = \infty$ (or the final simulation time). Otherwise, with a value chosen for ε (we normally take $\varepsilon = 0.03$), compute a candidate time leap τ' from the τ-selection formula (24) and (25). Thus τ' tentatively estimates the time to the next *non-critical* reaction.
3. If τ' is less than some small multiple (which we usually take to be 10) of $1/a_0(\mathbf{x})$, abandon tau-leaping temporarily, execute some modest number (which we usually take to be 100) of single-reaction SSA steps, and return to step 1. Otherwise, proceed to step 4.
4. Compute the sum $a_0^c(\mathbf{x})$ of the propensity functions of all the *critical* reactions. Generate a second candidate time leap τ'' as a sample of the exponential random variable with mean $1/a_0^c(\mathbf{x})$. As thus computed, τ'' tentatively estimates the time to the next *critical* reaction.
5. Take the actual time leap τ to be the smaller of τ' and τ'', and set the number of firings k_j of each reaction R_j accordingly:

    a. If $\tau' < \tau''$, take $\tau = \tau'$. For all critical reactions R_j set $k_j = 0$ (no critical reactions will occur during this leap). For all non-critical reactions R_j, generate k_j as a sample of the Poisson random variable with mean $a_j(\mathbf{x})\tau$.
    b. If $\tau'' \le \tau'$, take $\tau = \tau''$. Generate j_c as a sample of the integer random variable with point probabilities $a_j(\mathbf{x})/a_0^c(\mathbf{x})$, where j runs over the index values of the *critical* reactions only. (The value of j_c identifies the next critical reaction, the *only* critical reaction that will occur in this leap.) Set $k_{j_c} = 1$, and for all other critical reactions R_j set $k_j = 0$. For all the non-critical reactions R_j, generate k_j as a sample of the Poisson random variable with mean $a_j(\mathbf{x})\tau$.

6. If there is a negative component in $\mathbf{x} + \sum_j k_j \nu_j$, reduce τ' by half, and return to step 3. Otherwise, leap by replacing $t \leftarrow t + \tau$ and $\mathbf{x} \leftarrow \mathbf{x} + \sum_j k_j \nu_j$; then return to step 1, or else stop.

3.3 StochKit: a Stochastic Simulation ToolKit

The above hybrid method has been fully implemented in the package STOCHKIT [33], a software toolkit for discrete stochastic and multiscale simulation of chemically reacting systems. STOCHKIT is an efficient, extensible stochastic simulation toolkit developed in C++ that aims to make state of the art stochastic simulation algorithms accessible to biologists and chemists, while remaining open to extension via new stochastic and multiscale algorithms. STOCHKIT consists of a suite of software applications for stochastic simulation. The STOCHKIT core implements the simulation algorithms. Additional tools are provided for the convenience of simulation and analysis. A typical simulation process of STOCHKIT is shown in Figure 2.

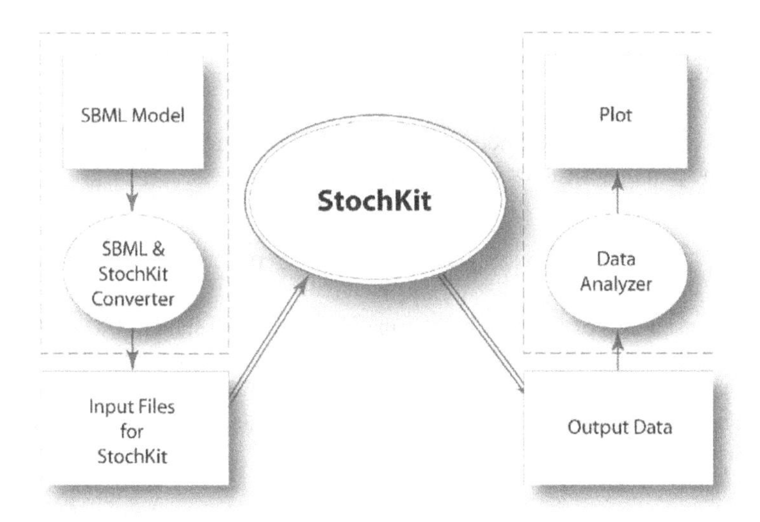

Fig. 2 Simulation Process of STOCHKIT.

A more detailed introduction to STOCHKIT is given in Reference [33]. The STOCHKIT package is freely available for download at www.engr.ucsb.edu/∼cse. The User's Guide is also available from that link.

4 Examples

To demonstrate the application of the SSA and the hybrid SSA/tau-leaping method, we apply both methods to the Schlögl model [22] and the LacZ/LacY model [30, 45]. Simulation files for both models are available in the STOCHKIT package.

4.1 The Schlögl Model

This model is famous for its bistable steady-state distribution. The reactions are

$$
\begin{aligned}
B_1 + 2X &\underset{c_2}{\overset{c_1}{\rightleftharpoons}} 3X, \\
B_2 &\underset{c_4}{\overset{c_3}{\rightleftharpoons}} X,
\end{aligned}
\tag{26}
$$

where B_1 and B_2 denote buffered species whose respective molecular populations N_1 and N_2 are assumed to remain essentially constant over the time interval of interest. There is only one time-varying species, X; the state change vectors are $\nu_1 = \nu_3 = 1$, $\nu_2 = \nu_4 = -1$; and the propensity functions are

$$
\begin{aligned}
a_1(x) &= \tfrac{c_1}{2} N_1 x(x-1), \\
a_2(x) &= \tfrac{c_2}{6} x(x-1)(x-2), \\
a_3(x) &= c_3 N_2, \\
a_4(x) &= c_4 x.
\end{aligned}
\tag{27}
$$

For some values of the parameters this model has two stable states, and that is the case for the parameter values we have chosen here:

$$
\begin{gathered}
c_1 = 3 \times 10^{-7}, \quad c_2 = 10^{-4}, \quad c_3 = 10^{-3}, \quad c_4 = 3.5, \\
N_1 = 1 \times 10^5, \quad N_2 = 2 \times 10^5.
\end{gathered}
\tag{28}
$$

We made ensembles of 10^5 simulation runs from the initial state $X(0) = 250$ to time $t = 4$ using the SSA and the hybrid SSA/tau-leaping method, the latter for a range of ε-values. Fig. 3 shows the histogram distance or "error" between the SSA ensemble and the tau-leaping ensembles as a function of ε. We can see that the errors increase roughly linearly with ε.

4.2 The LacZ/LacY Model

This model was first proposed by Kierzek [30] and later used for an efficiency test in [45]. This model has 22 reactions, 19 species, and an extremely multiscale nature. A detailed description of this model is omitted here. Interested readers can refer to the two references above and a list of the reaction channels and reaction rates of this model are given in Table 1. It was reported in [45] that negative populations were observed many times in their simulation using the original tau-leaping method. In our numerical experiments for this model on a 1.4Ghz Pentium IV Linux workstation, a single simulation from $t = 0$ to $t = 2100$ by SSA took 3,359 seconds CPU time. With an error tolerance of $\varepsilon = 0.03$, a single simulation by the hybrid SSA/tau-leaping method took 113.77 seconds CPU time with no negative population observed during the simulation.

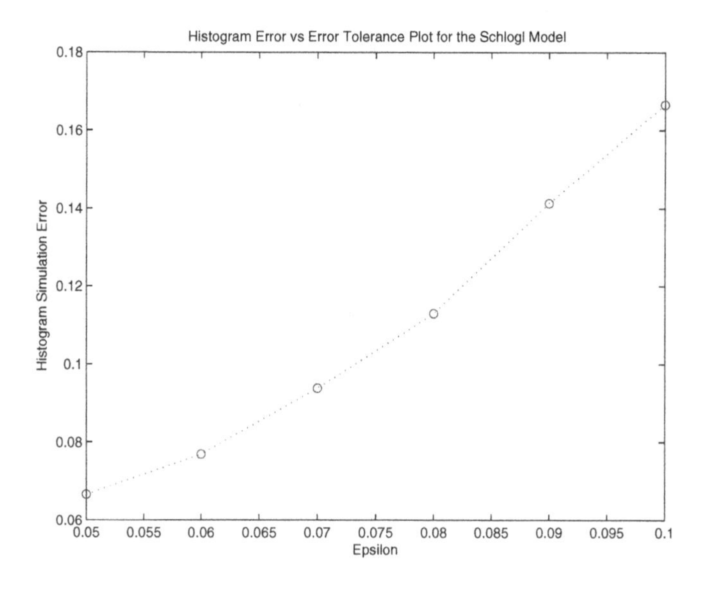

Fig. 3 Plot of histogram distance errors corresponding to different ε values for the Schlögl model. Histogram distance errors are measured by 10^5 samples generated from the SSA method and the hybrid SSA/tau-leaping method using different τ-selection formulas.

Since a single SSA simulation from $t = 0$ to $t = 2100$ took about an hour on our computer, obtaining a large number of SSA samples posed a challenge. We ran the SSA from time $t = 0$ to time $t = 1000$ to obtain an "initial" state; then we made 10^5 SSA runs from time $t = 1000$ to time $t = 1001$ (which required about 3.5 hours of computer time) and histogrammed the resulting populations. Finally, we made the same number of the SSA/tau-leaping runs over the same time interval for a range of values for ε. Fig. 4 shows the plot of histogram distance or "error" as a function of ε. We note again that the error increases roughly linearly with ε.

5 Advanced Topics

5.1 Different Implementation Strategies for SSA

Although the SSA is quite simple, due to its importance, there are several different implementation strategies proposed in the literature for the SSA. They are the direct method (DM) [21], the first reaction method (FRM) [21], the next reaction method (NRM) [18], the optimized direct method (ODM) [13], the sorted direct method (SDM) [35], and the Logarithmic Direct Method (LDM) [32].

	Reaction channel	Reaction rate
R1	PLac+RNAP $\longrightarrow$ PLacRNAP	0.17
R2	PLacRNAP $\longrightarrow$ PLac+RNAP	10
R3	PLacRNAP $\longrightarrow$ TrLacZ1	1
R4	TrLacZ1 $\longrightarrow$ RbsLacZ+PLac+TrLacZ2	1
R5	TrLacZ2 $\longrightarrow$ TrLacY1	0.015
R6	TrLacY1 $\longrightarrow$ RbsLacY+TrLacY2	1
R7	TrLacY2 $\longrightarrow$ RNAP	0.36
R8	Ribosome+RbsLacZ $\longrightarrow$ RbsRibosomeLacZ	0.17
R9	Ribosome+RbsLacY $\longrightarrow$ RbsRibosomeLacY	0.17
R10	RbsRibosomeLacZ $\longrightarrow$ Ribosome+RbsLacZ	0.45
R11	RbsRibosomeLacY $\longrightarrow$ Ribosome+RbsLacY	0.45
R12	RbsRibosomeLacZ $\longrightarrow$ TrRbsLacZ+RbsLacZ	0.4
R13	RbsRibosomeLacY $\longrightarrow$ TrRbsLacY+RbsLacY	0.4
R14	TrRbsLacZ $\longrightarrow$ LacZ	0.015
R15	TrRbsLacY $\longrightarrow$ LacY	0.036
R16	LacZ $\longrightarrow$ dgrLacZ	6.42×10^{-5}
R17	LacY $\longrightarrow$ dgrLacY	6.42×10^{-5}
R18	RbsLacZ $\longrightarrow$ dgrRbsLacZ	0.3
R19	RbsLacY $\longrightarrow$ dgrRbsLacY	0.3
R20	LacZ+lactose $\longrightarrow$ LacZlactose	9.52×10^{-5}
R21	LacZlactose $\longrightarrow$ product+LacZ	431
R22	LacY $\longrightarrow$ lactose+LacY	14

Table 1 A full list of reaction channels and deterministic reaction rates for the LacY/LacZ model.

The First Reaction Method is theoretically equivalent to the Direct Method but is quite different in the implementation details. The FRM generates a potential reaction time for each reaction and chooses the "first" reaction channel that has the earliest firing time to occur. In the FRM implementation, one generates M uniform random numbers $r_1, \ldots, r_M$ in every step and calculates a time τ_k for each reaction channel R_k by

$$\tau_k = \frac{1}{a_j(\mathbf{x})} \ln\left(\frac{1}{r_k}\right). \tag{29}$$

Then τ and j are given by

$$\begin{aligned} \tau &= \min_{1 \leq k \leq M} \tau_k, \\ \mathbf{j} &= \text{ the index for the smallest } \tau_k \end{aligned} \tag{30}$$

It can be proved that the τ and $\mathbf{j}$ generated from (30) follow the same distributions as in (8) and (9). Thus the DM and the FRM are statistically equivalent. However, in every step the FRM generates M τ_k values but uses only one of them. Thus the FRM is much less efficient than the DM.

Gibson and Bruck [18] have made remarkable progress improving the implementation efficiency of the FRM. Their method is the Next Reaction Method (NRM). The NRM uses a dependent graph to record the influence of each reaction channel on the other reaction channels. It records the absolute time $t + \tau_k$ as the expected firing time for the R_k reaction. If the firing of one reaction channel does not change

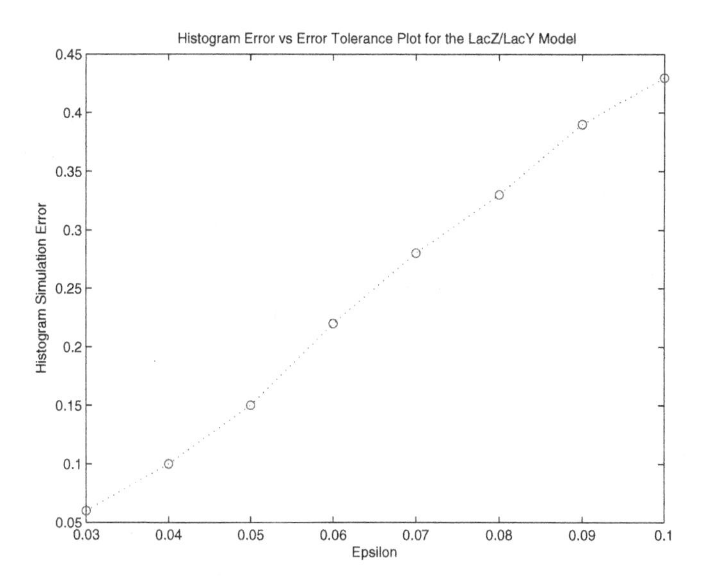

Fig. 4 Plot of histogram distance errors corresponding to different ε values for the LacZ/LacY model. Histogram distance errors are measured between the population distributions of LacZlactose in 10^5 runs of the SSA and the hybrid SSA/tau-leaping method using different τ-selection formulas.

the propensity of another reaction channel, the expected firing time for the latter reaction remains the same. In this way the NRM avoids unnecessary updates of the propensity function and expected firing time. For a reaction channel R_k whose reactants have been changed by the firing reaction, the NRM uses a cleverly designed formula to reuse the uniform random number r_k generated in the previous step. As a result, in every step there is only one uniform random number generated. The NRM turns out to be much more efficient than the FRM. However, using a detailed numerical analysis, it has been shown [13] that the NRM still has a higher computational cost than the Direct Method except for simple systems where the reactions are almost totally independent of each other.

To decrease the computational cost the Optimized Direct Method (ODM) [13] adopts the dependent graph to avoid the unnecessary recalculation of propensity functions and rearranges the indices of the reaction channels so that the more frequent reaction channels are always indexed before the less frequent ones. With these two improvements over the DM, the ODM becomes one of the most efficient SSA implementation strategies currently in use.

The re-index technique of the ODM requires one or a few sample runs using the SSA to collect the necessary information. This is not convenient in many applications. In order to dynamically adjust the indices of the reaction channels, the Sorted Direct Method (SDM) was proposed [35]. In the SDM, a bubble-up sorting method was applied to the indices of reaction channels. In the simulation, every time one reaction occurs, its reaction index decreases by one so that in the next step it is found more quickly. Then, after a certain initial simulation time, the index list will be sorted close to the optimal one. The SDM is a little less efficient than the ODM but its adaptive feature makes it a good strategy, particular in simulation of oscillation systems where a fast reaction in one time period may become slow in another time period. In that case, the dynamic indexing of this method is very useful.

Recently the Logarithmic Direct Method (LDM) was proposed [32], which applies a binary search method to the direct method. When the number of reaction channels, M, is large, the LDM can complete the search for the index j within $O(\log(M))$ time. Thus the LDM has advantages for large biochemical systems.

5.2 Multiscale Problems and Advanced Algorithms

Multiscale behavior appears in a wide range of problems. The multiscale problem in biochemical simulation has two aspects. The first is the timescale. Some reactions are much faster than others. Often the fast reactions quickly reach a stable state and the dynamics of the system are driven by the slow reactions. The SSA simulates every reaction and thus puts a great deal of effort into the more frequently occurring fast reactions, even though they do not contribute much to the dynamics and stochasticity of the system. This multiscale problem in time is known in the deterministic regime as *stiffness* [6]. Second, the populations of different species are of widely different magnitude. Some species are present with a large population while other species have very few copies in a cell. Species with a small population should be modeled by a discrete stochastic process, whereas species with a large population can be efficiently modeled by a deterministic ordinary differential equation (ODE). SSA treats all of the species as discrete stochastic processes. Overall, the SSA is not an efficient algorithm to solve a multiscale problem.

The tau-leaping method works better when the populations of all species are moderate (hundreds or thousands) or even larger, if the system is not stiff. The current challenge in research on the tau-leaping method is how to handle stiffness. While implicit tau-leaping methods [7, 40] have been proposed for stiff problems, the implementation details for the implicit tau-leaping methods are still under research. The current implementation is not as adaptive as the implementation of explicit methods. The large error in the variance of fast variables caused by the implicit methods, known as the *damping effect* [40], is still a challenge for implicit methods. Meanwhile, practical error estimation and control for tau-leaping methods still remain as an open question. There have been some initial progress in this direction. The recently developed tau-selection formula [12] gave *a priori* error control mechanism for the explicit tau-leaping method, while the *a posteriori error* control mechanism has been given by David Anderson [3].

Another big group of multiscale methods is the partition-based hybrid methods. The general idea is to partition the whole system into a deterministic part and a stochastic part. Naturally numerical methods for ODEs are applied to the deterministic part, and SSA is applied to the stochastic part. Many different methods [8, 11, 28, 27, 25, 38, 44] in this group have been proposed and the major differences among them are focused on the partition criteria and implementation details.

6 Exercises

1. Write down the Chemical Master Equation for the Schlögl model (26).
2. The Lotka-Volterra system consists of three reaction channels and two species:

$$\begin{aligned} S_1 &\xrightarrow{c_1} S_1 + S_1, \\ S_1 + S_2 &\xrightarrow{c_2} S_2 + S_2, \\ S_2 &\xrightarrow{c_3} \mathit{Decayed}. \end{aligned} \tag{31}$$

   with rate constants: $c_1 = 10,\ c_2 = 0.01,\ c_3 = 10$ and initial conditions $x_1(0) = x_2(0) = 1000$. Write down the reaction rate equations and simulate the equation in matlab. Run an SSA simulation using StochKit and compare the result with the reaction rate equation result.
3. Consider a set of chemical reactions given below

$$\begin{aligned} S_1 &\xrightarrow{c_1} \mathit{Decayed} \\ S_1 + S_1 &\underset{c_2}{\overset{c_3}{\rightleftharpoons}} S_2 \\ S_2 &\xrightarrow{c_4} S_3. \end{aligned} \tag{32}$$

   a. Write down the propensity functions and stoichiometric matrix.
   b. Write an SSA code to simulate this system from $t = 0$ to $t = 10$ with the following parameters:

$$c_1 = 1, \quad c_2 = 0.002, \quad c_3 = 0.5, \quad c_4 = 0.04,$$

   and initial state $x_1(0) = 10,000$, $x_2(0) = x_3(0) = 0$.
   c. Use your SSA code to simulate the same system with the same initial state from $t = 0$ to $t = 2$ with a different set of parameters:

$$c_1 = 1, \quad c_2 = 10, \quad c_3 = 1000, \quad c_4 = 0.1,$$

   Pay special attention to the difference in computational time and the number of reactions for each reaction channel.
   d. (Advanced) This example is a test problem in StochKit. Try it with the tau-leaping method in StochKit for the two parameter sets given in b) and c).

You will find that for the parameter set in c), the tau-leaping method is even slower than the SSA. That is the effect of stiffness.

7 Further Reading

It is always worthwhile to read Gillespie's three milestone papers [20, 21, 23] and his book [22]. Interested readers can also find a detailed review of stochastic chemical kinetics by Dan Gillespie in [19]. For a detailed introduction to molecular dynamics simulation, readers will find Haile's book [26] very useful. For people who are looking for interesting research topics, there are plenty of challenges on the implementation strategies of SSA [35, 13, 18, 32], the efficient implementation of tau-leaping methods [10, 12], the implicit tau-leaping methods[7, 40, 39] and the SSA method for multiscale problems [8, 9, 11, 28, 27, 25, 38, 44].

Acknowledgements This work was supported by the University of California Santa Barbara under award KK9103, the National Science Foundation under award CCF-0726763, and the National Institutes of Health under award GM073744.

References

1. A, A., J, R., H, M.: Stochastic kinetic analysis of developmental pathway bifurcation in phage λ-infected E. Coli cells. Genetics **149**, 1633–1648 (1998)
2. Alder, B.J., Wainwright, T.E.: Studies in molecular dynamics. i. general method. J. Chem. Phys. **31**(2), 459–466 (1959)
3. Anderson D: Incorporating postleap checks in tau-leaping. J. Chem. Phys. **128**, 054,103 (2008)
4. BORTZ, A., KALOS, M., LEBOWITZ, J.: New algorithm for Monte-Carlo simulation of ising spin systems. J. Comput. Phys. **17**(1), 10–18 (1975)
5. Bray, D, Bourret, R, and Simon, M: Computer simulation of the phosphorylation cascade controlling bacterial chemotaxis. Mol. Bio. Cell **4**, 469–482 (1993)
6. Brenan, K.E., Campbell, S.L., and Petzold, L.R.: Numerical Solution of Initial-Value Problems in Differential-Algebraic Equations. SIAM, Philadelphia, PA (1996)
7. Cao, Y. and Petzold, L.: Trapezoidal tau-leaping formula for the stochastic simulation of chemically reacting systems. Proceedings of Foundations of Systems Biology in Engineering (FOSBE 2005) pp. 149–152 (2005)
8. Cao, Y., Gillespie, D. and Petzold, L.: The slow-scale stochastic simulation algorithm. J. Chem. Phys. **122**, 014,116 (2005)
9. Cao, Y., Gillespie, D. and Petzold, L.: Accelerated stochastic simulation of the stiff enzyme-substrate reaction. J. Chem. Phys. **123**, 144,917 (2005)
10. Cao, Y., Gillespie, D. and Petzold, L.: Avoiding negative populations in explicit tau leaping. J. Chem. Phys. **123**, 054,104 (2005)
11. Cao, Y., Gillespie, D. and Petzold, L.: Multiscale stochastic simulation algorithm with stochastic partial equilibrium assumption for chemically reacting systems. J. Comput. Phys. **206**, 395–411 (2005)
12. Cao, Y., Gillespie, D. and Petzold, L.: Efficient stepsize selection for the tau-leaping method. J. Chem. Phys. **124**, 044,109 (2006)

13. Cao, Y., Li, H. and Petzold, L.: Efficient formulation of the stochastic simulation algorithm for chemically reacting systems. J. Chem. Phys. **121**, 4059–67 (2004)
14. Chatterjee, A., Vlachos, D. and Katsoulakis, M.: Binomial distribution based tau-leap accelerated stochastic simulation. J. Chem. Phys. **122**, 024,112 (2005)
15. Chen K, Calzone L, Csikasz-Nagy A, Cross F, Novak B and Tyson J: Integrative analysis of cell cycle control in budding yeast. Mol. Biol. Cell **15**, 3841–3862 (2004)
16. Drew E, Lingchong Y, John Y and Ian M: Computation, prediction, and experimental test of fitness for bacteriophage T7 mutants with permuted genomes. PNAS pp. 5375–5380 (1997)
17. Fedoroff, N. and Fontana, W.: Small numbers of big molecules. Science **297**, 1129–1131 (2002)
18. Gibson, M. and Bruck, J.: Efficient exact stochastic simulation of chemical systems with many species and many channels. J. Phys. Chem. A **104**, 1876 (2000)
19. Gillespie, D.: Stochastic Chemical Kinetics (Sec. 5.11). Springer, Dordrecht (2005)
20. Gillespie, D.: A general method for numerically simulating the stochastic time evolution of coupled chemical reactions. J. Comput. Phys. **22**, 403–434 (1976)
21. Gillespie, D.: Exact stochastic simulation of coupled chemical reactions. J. Phys. Chem. **81**, 2340–61 (1977)
22. Gillespie, D.: Markov Processes: An Introduction for Physical Scientists. Academic Press (1992)
23. Gillespie, D.: Approximate accelerated stochastic simulation of chemically reacting systems. J. Chem. Phys. **115**, 1716 (2001)
24. Gillespie, D. and Petzold, L.: Improved leap-size selection for accelerated stochastic simulation. J. Chem. Phys. **119**, 8229–34 (2003)
25. Griffith M, Courtney T, Peccoud J and Sanders W: Dynamic partitioning for hybrid simulation of the bistable hiv-1 transactivation network. Bioinformatics **22**, 2782–2789 (2006)
26. Haile J: Molecular Dynamics Simulation: Elementary Methods. Wiley-Interscience (1997)
27. Harris L and Clancy P: A partitioned leaping approach for multiscale modeling of chemical reaction dynamics. J. Chem. Phys. **125**, 144,107 (2006)
28. Haseltine, E. and Rawlings, J.: Approximate simulation of coupled fast and slow reactions for stochastic chemical kinetics. J. Chem. Phys. **117**, 6959–6969 (2002)
29. Horn F and Jackson R: General mass action kinetics. Archive for Rational Mechanics and Analysis **47**, 81–116 (1972)
30. Kierzek, A.: STOCKS: STOChastic Kinetic Simulations of biochemical systems with Gillespie algorithm. Bioinformatics **18**, 470–81 (2002)
31. L. Michaelis and M. L. Menten: The kinetics of invertase activity. Biochem. Z **49**, 333–369 (1913)
32. Li H and Petzold L: Logarithmic Direct Method for DiscreteStochastic Simulation of Chemically Reacting Systems (2006). technical report, computer science department, UCSB
33. Li, H., Cao, Y., Petzold, L. and Gillespie, D.: Algorithms and software for stochastic simulation of biochemical reacting systems. Biotechnology Progress **24**, 56–61 (2008)
34. McAdams, H., Arkin, A.: Stochastic mechanisms in gene expression. PNAS **94**, 814–819 (1997)
35. McCollum JM, Peterson GD, Cox CD, Simpson ML and Samatova NF: The sorting direct method for stochastic simulation of biochemical systems with varying reaction execution behavior. Computational Biology AND Chemistry **30**, 39–49 (2006)
36. Munsky B and Khammash M: The finite state projection algorithm for the solution of the chemical master equation. J. Chem. Phys. **124**, 044,101 (2006)
37. Plowman KM: Enzyme Kinetics. New York, McGraw-Hill (1971)
38. Rao, C. and Arkin, A.: Stochastic chemical kinetics and the quasi steady-state assumption: application to the Gillespie algorithm. J. Chem. Phys. **118**, 4999–5010 (2003)
39. Rathinam, M., Petzold, L., Cao, Y. and Gillespie, D.: Consistency and stability of tau leaping schemes for chemical reaction systems. SIAM Multiscale Modeling **4**, 867–895 (2005)
40. Rathinam, M., Petzold, L., Cao, Y., Gillespie, D.: Stiffness in stochastic chemically reacting systems: the implicit tau-leaping method. J. Chem. Phys. **119**, 12,784–94 (2003)

41. Savageau MA: Biochemical systems analysis: I. Some mathematical properties of the rate law for the component enzymatic reactions. J. Theor. Biol. **25**, 365–369 (1969)
42. Savageau MA: Biochemical systems analysis: A study of function and design in molecular biology. Reading, MA, AddisonWesley (1976)
43. Savageau MA: Development of fractal kinetic theory for enzyme-catalysed reactions and implications for the design of biochemical pathways. Biosystems **47**, 9–36 (1998)
44. T. Kiehl, R. Mattheyses and M. Simmons: Hybrid simulation of cellular behavior. Bioinformatics pp. 316–322 (2004)
45. Tian, T. and Burrage, K.: Binomial leap methods for simulating stochastic chemical kinetics. J. Chem. Phys. **121**, 10,356–64 (2004)
46. Voit EO: Computational Analysis of Biochemical Systems. A Practical Guide for Biochemists and Molecular Biologists. Cambridge University Press, Cambridge, U.K. (2000)
47. Zhang J. and Watson L.: A modified uniformization method for the chemical master equation. Proc. 7th IEEE Internat. Conf. on Bioinformatics and Bioengineering, Boston, MA pp. 1429–1433 (2007)

Part IV
Networks

This fourth part of the book broadens our study of biological networks into cellular response networks and modules hidden in interaction networks.

Cellular Response Networks

Christopher D. Lasher, Christopher L. Poirel, and T. M. Murali

Abstract Complex networks of interactions between genes, proteins, and other molecules choreograph cellular processes. The interactions that are active in the cell change over time, both as a natural outcome of the cell's natural life cycle and in response to external signals. The set of active interactions, called the *response network*, are likely to be significantly different between a normally-functioning cell and a diseased cell. The wide availability of DNA microarray data and experimentally-determined interaction networks has made it possible to automatically compute response networks. This chapter surveys algorithms that have been developed to compute response networks.

1 History of the Problem

Genes carry genetic information that is used to synthesize essential components of the living cell. These components are called gene products, typically RNA molecules or proteins. Coordinated interactions among gene products comprise and control many fundamental cellular processes such as the formation of protein complexes, the metabolism of food by biochemical pathways, and signaling pathways triggered by external signals. Gene products also control and modulate the synthesis and activity of other gene products. These interactions constitute an intricate

Christopher D. Lasher
Genetics, Bioinformatics, and Computational Biology Program, Virginia Polytechnic Institute and State University, Blacksburg, VA, USA, e-mail: lasher@vt.edu

Christopher L. Poirel
Department of Computer Science, Virginia Polytechnic Institute and State University, Blacksburg, VA, USA, e-mail: poirel@vt.edu

T. M. Murali
Department of Computer Science, Virginia Polytechnic Institute and State University, Blacksburg, VA, USA, e-mail: murali@cs.vt.edu

L.S. Heath and N. Ramakrishnan (eds.), *Problem Solving Handbook in Computational Biology and Bioinformatics*, DOI 10.1007/978-0-387-09760-2_11,

network that dynamically changes in response to a myriad of cues. Therefore, discovering *response networks*, the set of molecular interactions that are active in a given cellular context, and understanding how normal response networks may be perturbed in a disease are fundamental biological questions [12].

Gene expression is the process by which a gene is first transcribed to messenger RNA (mRNA). The *expression level* of a gene is the number of copies of its mRNA that are present in a cell. DNA microarrays have allowed biologists to simultaneously measure the average expression level of each gene in a set of cells. DNA microarrays offer a powerful experimental platform to study diverse contexts, since they capture a snapshot of the activity of all genes in the cells in the sample. However, DNA microarrays measure levels of the *nodes* (genes) and do not directly provide any information on the *edges* (interactions). Data regarding edges are available from datasets of physical and functional interactions between genes and proteins that are now widely available. Integrated analysis of gene expression data and protein-protein interaction (PPI) networks is emerging as a powerful technique for computing response networks. This chapter surveys several algorithms that are available to perform this type of analysis.

We stress that this type of analysis is distinct from methods that find modules in PPI networks alone (see next chapter). Such analysis is usually performed on protein interaction networks integrated from a variety of different experimental sources and public repositories. However, an experiment that reports an interaction often does not yield information on the conditions under which that interaction takes place in the cell. In many situations, the experimental context in which an interaction happens is lost when the interaction is recorded in a database. In other cases, the context may simply not be apparent. For instance, an interaction between two human proteins may be detected by a yeast 2-hybrid experiment [6]. Since such an experiment is performed in *Saccharomyces cerevisiae* (baker's yeast), it simply cannot produce any information on when the detected interaction may take place in a human cell. As a consequence, protein interaction networks typically represent the *universe* of interactions that take place in multiple, different contexts within the cell. Integrating them with measurements of molecular levels, such as DNA microarray data, is necessary for computing response networks.

2 Algorithm Descriptions

We divide response network algorithms into two broad classes, depending on the design of the experiment used to collect DNA microarray data:

1. A very common experimental design partitions the set of samples into two subsets, with one subset corresponding to an experimental treatment and another subset corresponding to a control. Numerous methods have been developed to assess to what degree each gene is differentially expressed when comparing the treatment to the control. Using a hypothesis testing framework, for each gene g, these methods yield a p-value $0 \leq p_g \leq 1$ representing the statistical significance

of the difference between the two sets of expression levels of the gene. These p-values form the starting point of response network computations. We call such datasets *treatment-control* data and examine these methods in Section 2.1.
2. Another common experimental design yields a gene expression dataset consisting of measurements from multiple samples under a particular experimental condition; the samples can correspond to multiple time-points after exposing cells to a particular treatment or stimulus or to multiple patients diagnosed with a particular disease. The complete gene expression data is part of the input to an algorithm to compute response networks. Analysis of such datasets usually starts by computing co-expression or similarity values for gene pairs. We discuss co-expression-based techniques in Section 2.2.

2.1 Detecting Response Networks from Treatment-Control Data

Experiments for analyzing gene expression often produce treatment-control data. The treatment samples offer measurements of the expression of different genes under a certain experimental condition or phenotype (e.g., after a gene knock-out or for a specific disease). The control samples measure gene expression without the influence of the experimental condition (e.g., wild-type cells or normal cells).

We discuss three algorithms that integrate this type of gene expression data with molecular interaction networks. The ActiveModules algorithm of Ideker *et al.* [11] and the algorithm of Dittrich *et al.* [4] are methods to estimate the differential expression of each node in the protein-protein interaction network and subsequently find high-scoring subnetworks, i.e., subgraphs that have large differential expression in total. The DEGAS algorithm of Ulitsky and Shamir [28] uses a different approach: for each gene, the method computes a separate p-value in every sample in the treatment. After combining this expression data with a protein-protein interaction network, the algorithm searches for a minimally connected subnetwork of genes that respond to the experimental condition for at least some specified number of samples in the treatment.

The inputs to the algorithms discussed in this section are an undirected protein-protein interaction network $G = (V, E)$ and two sets of gene expression data $V_T = \{g_T \mid g \in V\}$, where T is the set of samples in the treatment and $g_T : T \rightarrow \mathbb{R}$ denotes the expression values of gene g in each of the samples in T, and $V_C = \{g_C \mid g \in V\}$, where C is the set of samples in the control and we define g_C analogously to g_T. Informally, the goal of these methods is to compute the connected subgraph of G such that the genes in the subgraph show the most differential expression between the samples in T and the samples in C.

The ActiveModules Algorithm

Ideker *et al.* [11] introduce the ActiveModules algorithm for computing highly-perturbed response networks from treatment-control data. For each gene g in G, they compute a p-value p_g based on the expression values g_T and g_C of that gene in the treatment samples and the control samples. Many tools are available to calculate such p-values [5]. For instance, a simple approach is to apply the t-test to g_T and g_C. In general, the p-value represents the statistical significance of the observed difference between the expression levels of a gene in T and in C. A smaller p-value indicates a more statistically significant difference. Ideker *et al.* [11] convert each value p_g to a z-score z_g using the inverse normal cumulative distribution function evaluated at $1 - p_g$, i.e., $z_g = \Phi^{-1}(1 - p_g)$, where Φ is the cumulative normal distribution function. This transformation converts small p-values to large z-scores. Consequently, connected subnetworks composed of genes with high z-scores are desirable.

The authors do not simply discard genes with low z-scores. Instead, they develop a method for scoring any subgraph of G based on the z-scores of all the genes in the subgraph. For a subgraph A of G on a k-node set B, define the Liptak-Stouffer z-score z_A as

$$z_A = \frac{\sum_{g \in B} z_g}{\sqrt{k}}.$$

Clearly if A is a subgraph of genes with high z-scores, then it will have a large aggregate Liptak-Stouffer z-score and may possess some biological significance. The final step in scoring a subgraph lies in determining whether or not z_A is statistically significant. Ideker *et al.* [11] compute the statistical significance empirically: they compute the aggregate Liptak-Stouffer z-scores for multiple subgraphs induced by k randomly selected genes, and estimate the mean μ_k and standard deviation σ_k of these random subgraphs of size k. They define the *corrected subgraph score* s_A as follows,

$$s_A = \frac{z_A - \mu_k}{\sigma_k}.$$

This transformation adjusts the z-score z_A so that a randomly-selected subgraph on k nodes will have a corrected subgraph score with mean 0 and standard deviation 1.

With a function to score subgraphs in hand, the authors proceed to discover highly-scoring subgraphs. They demonstrate that a similar problem is NP-complete (we describe this problem in more detail in Section 2.1, page 238). Thus, it is unlikely that an efficient (polynomial time) algorithm exists that computes the subgraph that maximizes s_A. Ideker *et al.* [11] resort to simulated annealing [16], a heuristic method often used to solve computationally intractable combinatorial optimization problems. The following algorithm demonstrates the simulated annealing technique. The algorithm has three user-determined parameters: n, the number of iterations, a starting temperature T_s and an ending temperature $T_e < T_s$.

1: Label each node in V either 'in' or 'out' with equal probability, and let I be the set of all nodes labeled 'in'.

2: Compute s_I.
3: $T \leftarrow T_s$
4: **for** $i = 1 \ldots n$ **do**
5: $s \leftarrow s_I$.
6: Select a node $v \in V$ uniformly at random and switch its label.
7: Compute s_I.
8: **if** $s_I > s$ **then**
9: Keep the new label for v.
10: **else**
11: Keep the new label for v with probability $e^{(s_I - s)/T}$
12: $T \leftarrow T \times (\frac{T_e}{T_s})^{\frac{1}{n}}$
13: Return the subgraph of G induced by I.

The variable T represents a temperature that decreases geometrically with each iteration, by a factor of $(T_e/T_s)^{1/n}$. The algorithm always accepts a modification that increases the corrected subgraph score. However, when $s_I < s$, the algorithm accepts the change with a probability $0 < p = e^{(s_I - s)/T} < 1$. For a fixed value of T, the closer s_I is to s, the closer p is to 1. For a fixed value of $s_I - s$, the probability p decreases as T decreases, indicating that the algorithm is more liberal in earlier iterations, being more likely to keep changes that lower the corrected score. Since the returned graph induced by I is not guaranteed to be connected, the authors simply take the highest-scoring connected component as the result. Note that this approach will not necessarily find the optimal solution, but operates under the belief that high-scoring networks are likely to have some biological significance.

The Algorithm of Dittrich *et al.*

Dittrich *et al.* [4] build on the Ideker *et al.* [11] approach by developing a new scoring function and a different method for discovering high-scoring subgraphs. First, they follow Pounds and Morris [23] to model the distribution of p-values over all genes in V as a mixture of noise and signal components. Let $B(a,b)$ denote the beta distribution, where a and b are the two parameters that define the shape of the beta distribution function. The probability density function of $B(a,b)$ is

$$f(x) = \frac{\Gamma(a+b)}{\Gamma(a)\Gamma(b)} x^{a-1}(1-x)^{b-1},$$

where $\Gamma(x) = \int_0^\infty t^{x-1} e^{-x} dt$ is the gamma function. Dittrich *et al.* assume that the signal component of the distribution of p-values has a $B(a,1)$ distribution, i.e., given that a p-value x is generated by the signal component, its probability distribution function is ax^{a-1}. Similarly, they assume that if a p-value is generated by the noise component, then the p-value is $B(1,1)$ or uniformly distributed on (0,1). Therefore, if λ (respectively, $1-\lambda$) is the probability that a p-value is generated by the noise (respectively, signal) component of the mixture, then the probability distribution function for a p-value x can be rewritten as

$$f(x|a,\lambda) = \lambda + (1-\lambda)ax^{a-1}, \qquad 0 < x \leq 1; 0 < \lambda < 1.$$

where λ and a are mixture and shape parameters, respectively. Given the p-values $P_V = \{p_v | v \in V\}$, define the likelihood of these values as

$$\mathscr{L}(\lambda, a; P_v) = \prod_{v \in V} \left(\lambda + (1-\lambda)ap_v^{a-1}\right).$$

The authors use numerical optimization methods to estimate the mixture and shape parameters $[\lambda^*, a^*] = \arg\max_{\lambda,a} \mathscr{L}(\lambda, a; P_V)$ that maximize the likelihood of the p-values.

The ultimate goal of this approach is to develop a scoring function that associates a p-value arising primarily from the signal component with a positive score and a p-value generated by background noise with a negative score. The following scoring function captures this property:

$$s_x = \log \frac{ax^{a-1}}{a\tau^{a-1}} = (a-1)\left(\log(x) - \log(\tau)\right),$$

where τ is a p-value threshold that yields a user-specified false discovery rate[1] (see Pounds and Morris [23] for details). A p-value is deemed significant when it is smaller than τ, thus the corresponding node is assigned a positive score. Conversely, if a p-value is larger than τ, the corresponding node is assigned a negative score. Dittrich *et al.* define the score s_A for a subgraph A as

$$s_A = \sum_{g \in A} s_{p_g},$$

that is, s_A is simply the sum of the scores for each of the nodes in A. Dittrich *et al.* aim to compute the subgraph of G with the largest score. This problem is known as the *maximum-weight connected subgraph* (MWCS) problem: given a graph $G = (V, E)$ and node weight $w_v \in \mathbb{R}$ for each $v \in V$, the MWCS problem asks for the connected subgraph $G' = (V', E')$ of G that maximizes $w_{G'} = \sum_{v \in V'} w_v$. The MWCS problem has been proven to be NP-complete [11]. Notice that the MWCS problem is trivial if all weights are positive, since the entire graph G would clearly be the optimal solution.

Dittrich *et al.* convert an instance of the MWCS problem into an instance of the prize-collecting Steiner tree (PCST) problem. While the PCST problem is also NP-complete, Ljubić *et al.* [19] provide an elegant algorithm based on formulating the PCST problem as an integer linear program (ILP). They propose a branch-and-cut heuristic to solve this ILP. The algorithm does not have a running time that is polynomial in the size of the input. However, Dittrich *et al.* show that this approach finds provably-optimal solutions in a reasonable amount of time for biologically-relevant network sizes.

[1] The false discovery rate is the ratio of false positives (values incorrectly identified to be significant) and the total number of values deemed to be significant.

The DEGAS algorithm

Ulitsky and Shamir [28] develop the DEGAS[2] algorithm for identifying disease-related pathways within the cell. In contrast to the algorithms discussed previously in this section, they calculate multiple p-values for each gene, one for each sample in T. The main goal of the algorithm is to discover subgraphs containing several genes that are differentially expressed in multiple samples in T. The authors call such subgraphs *dysregulated pathways* (DPs). The process is two-fold. First, discover minimal connected subgraphs that have at least k differentially-expressed genes, where k is a parameter to the algorithm. Second, they find those minimal connected subgraphs that are statistically significant.

Recall that T is the set of n treatment samples. Associate with each node $v \in V$ a set of treatment samples $S_v \subseteq T$ in which v is differentially expressed (in comparison to the expression of v in the control samples C). For every node $v \in V$, Ulitsky and Shamir compute the set S_v by (i) estimating a p-value in each sample $t \in T$ that represents the differential expression of g in t, (ii) applying a user-specified cutoff on the p-values, and (iii) including a treatment sample t in S_v if the p-value is below the cutoff. The authors construct a bipartite graph $B = (V, T, E^B)$, where $E^B = \{(t, v) | t \in S_v\}$. The graph B is simply a bipartition between genes and samples, with an edge between gene v and sample t when $t \in S_v$. Now define a subset $C \subseteq V$ of genes to be a *connected* (k,l)*-cover* $CC(k,l)$ if the following two conditions hold:

1. C induces a connected subgraph in G.
2. There exists a set of $n - l$ treatment samples $T' \subseteq T$ such that for every sample $t' \in T'$, $|N(t') \cap C| \geq k$, where $N(t')$ is the set of genes that are adjacent to t' in the graph B.

The second property of a connected (k,l)-cover C states at least k genes in C are differentially expressed in all but l samples in T. This notion ties together a connected subgraph of G with a set of treatment samples, in each of which a sufficiently large number of genes in the subgraph are perturbed. By not requiring all genes in C to be perturbed in all samples in T', a connected (k,l)-cover is able to accommodate inter-sample variation and noise.

Given integers k and l, the *minimum connected* (k,l)*-cover* problem $MCC(k,l)$ is to find the connected (k,l)-cover with the fewest number of nodes. Since discovering minimal connected subgraphs is NP-hard, Ulitsky *et al.* develop approaches that offer provably good results. They propose Covering Using Shortest Paths (CUSP), an algorithm which provides a $k(n-l)$-approximation for $MCC(k,l)$. Define the *distance* between two nodes $d(u,v)$ as the minimum number of edges in any path connecting u and v in G. The algorithm proceeds in four major steps:

1. Find the k shortest paths from each node $r \in V$ to each sample $u \in T$. More specifically, for each node $r \in V$, for each sample $u \in T$, and for each $1 \leq i \leq k$, let $P[r,u]_i$ be the ith closest node to r in G that is a neighbor of u, and let $D[r,u]_i = d(r, P[r,u]_i)$. Compute $D[r,u]_i$ and $P[r,u]_i$ for $1 \leq i \leq k$.

---

[2] This name for the algorithm appears only in the software implementation; see Table 1.

2. Find a set of $n-l$ samples in T for which the k shortest paths from r are not very long. Specifically, compute S_r, the set of $n-l$ samples in T that have the smallest values for $m[r,u] = max_q\{D[r,u]_q, 1 \leq q \leq k\}$. In other words, compute $m[r,u]$ for each sample $u \in T$ and include the $n-l$ samples that have the smallest values of $m[r,u]$ in the set S_r.
3. Extract the shortest paths between r and the samples in S_r, i.e., compute X_r, the union of the paths to the nodes in G that neighbor the samples in S_r. The authors claim that X_r is a $CC(k,l)$ in G. Indeed, it induces a connected component in G. Furthermore, each of the $n-l$ samples in $S_r \subseteq T$ is covered once for each $P[r,u]_i$ where $1 \leq i \leq k$.
4. Output the smallest X_r, i.e., return the $CC(k,l)$ instance $X = \arg\min_{v \in V} |X_v|$, which is designated to be a DP.

To assess the statistical significance of the DP returned by the CUSP algorithm, the authors generate multiple random networks with the same number of nodes as G, using degree preserving randomization [20]. They apply the CUSP algorithm on the original graph B with different values of k. They also run the CUSP algorithm on each random network with different values of k, and compute a distribution of DP sizes for each value of k. The p-value of each DP computed in B is the fraction of DPs with a larger size computed in the random graphs. They return the most statistically-significant DP computed from the original network that corresponds to this k.

Notice that the CUSP algorithm computes only one DP. In practice, we want to return multiple DPs and test each of them for statistical significance, since any significant DP may be biologically interesting. The authors describe a method for discovering multiple DPs. Suppose X is the first DP returned by the CUSP algorithm. For each node $v \in X$ remove all edges adjacent to v from E^B in the graph B and call the resulting graph B'. Then run CUSP on B' to produce a new DP. Continue this procedure until CUSP no longer returns a statistically significant DP.

2.2 Co-Expression-Based Methods for Detecting Response Networks

Gene expression data sets with many samples per condition or phenotype, or with samples from many conditions or phenotypes, afford the opportunity to calculate co-expression, or similarity values for gene pairs. These datasets give rise to another class of algorithms that use similarity values to detect response networks. We discuss four different approaches to integrating expression and network data in order to determine response networks. The first approach, by Hanisch *et al.* [10], simultaneously uses gene expression similarity and distances in the PPI network, and then clusters genes after combining these distances. The second approach, by Murali and Rivera [22], overlays expression-based similarities on the edges of the PPI network as edge weights, and then detects heavy subgraphs in the weighted network. The final pair of closely-related approaches, by Ulitsky and Shamir [29, 30], indirectly leverage the PPI network to constrain the actions for refining sets of similarly expressed genes.

The algorithms discussed in this section use as inputs one undirected protein-protein or protein-reaction interaction network $G = (V, E)$ and one gene expression dataset $V_S = \{g_S \mid g \in V\}$, where S is the set of samples and $g_S : S \to \mathbb{R}$ denotes the expression values of gene g in the samples S. Informally, these methods strive to compute a connected subgraph of G such that the genes in the subgraph show the most similar expression across the samples in S.

The Algorithm of Hanisch *et al.*

Hanisch *et al.* [10] present an algorithm that clusters genes using distances between their expression profiles in combination with distances between their gene products in a PPI network. They begin by converting curated metabolic pathways from KEGG [14] into a bipartite graph $G = (V, E)$ with biological molecules (e.g., enzymes and metabolites) as one set of nodes and reactions as the other set. Edges in G connect molecules to reactions in which they participate. In order to emphasize the relationships between the biological molecules, and to disfavor paths through ubiquitous molecules which take part in many reactions (e.g., ATP), each edge e receives a weight w_e equal to the degree of the incident biological molecule. The distance $d_{\text{net}}(u, v)$ between two nodes u and v in this bipartite graph is the minimum sum of all edge weights along some path connecting the two nodes. Hanisch *et al.* then calculate the distance in expression $d_{\exp}(g, h)$ for all pairs of genes g and h, as $1 - c(g, h)$, where $c(g, h)$ is the Pearson's correlation coefficient between g_S and h_S.

After mapping genes in the gene expression data to the enzymes they code for in G, the authors combine the computed distances $d_{\exp}$ and d_{net} into a joint distance $\Delta(u, v)$ as follows:

$$\Delta(u, v) = 1 - \frac{\lambda_{\exp}(u, v) + \lambda_{\text{net}}(u, v)}{2},$$

where the *logistic regression* function $\lambda_\Psi(u, v)$, $\Psi \in \{\exp, \text{net}\}$, is

$$\lambda_\Psi(u, v) = \frac{1}{1 + e^{-s_\Phi(\delta_\Phi(u, v) - v_\Phi)}}.$$

User defined parameters s_Φ and v_Φ control the shape of the logistic curve, giving the slope of the curve and the point at which the curve reaches $\frac{1}{2}$, respectively. Hanisch *et al.* set the values of these parameters empirically.

Finally, Hanisch *et al.* use agglomerative hierarchical clustering to partition the genes into a user-defined number of groups. This aspect distinguishes their algorithms from the others presented below, which do not require pre-defining the number of computed response networks. To assist with choosing an appropriate number of clusters, Hanisch *et al.* plot silhouette values [24], which measure the separation and tightness of clusters, for different cut points, and heuristically select appropriate points. As a final observation, although this algorithm does not directly compute response networks, by taking distances in the bipartite graph into account, it indirectly discovers those metabolic pathways perturbed in an experiment.

The ActiveNetworks Algorithm

The ActiveNetworks algorithm presented by Murali and Rivera [22] projects co-expression values as edge weights onto an interaction network and casts the problem of finding response networks as one of finding dense subgraphs within the weighted interaction network. First, they remove all genes with little variation in expression and their incident edges from G. Next, they compute the weight w_e of each edge $e = (g,h)$ in E as the absolute value of Pearson's correlation coefficient of g_S and h_S. Murali and Rivera then assess the statistical significance of the weight of each edge in the PPI network using a permutation test and remove edges with insignificant weights from G.

Given a subgraph $H = (V', E')$ of G, they define its *density* as

$$w_H = \frac{\sum_{e \in E'} w_e}{|V'|},$$

i.e., the total weight of the edges in H divided by the number of nodes in H. Computing the subgraph of maximum density can be solved in polynomial time [7] or by using linear programming [2]. In practice, Murali and Rivera use a greedy algorithm that guarantees a 2-approximation, i.e., the subgraph computed by the algorithm has density at least half as much as that of the most dense subgraph in G. Define the weight of a node to be the total weight of the edges incident on it. The algorithm repeatedly deletes the node of smallest weight till G becomes empty. It reports the most dense subgraph encountered during this process. Murali and Rivera embed this algorithm in a heuristic to find all "dense pockets" in G: apply the greedy algorithm to G, delete the edges of the computed subgraph from G, and repeat this process, until the density of G falls below its initial density. They return the union of all dense subgraphs computed as the response network. See Section 4 for an application of this method to data for *S. cerevisiae*.

The MATISSE and CEZANNE Algorithms

Ulitsky and Shamir [29] present an algorithm called MATISSE that seeks to find sets of genes (called *modules*) with high expression similarity, but with the additional constraints that (i) each set must induce a connected subgraph in the interaction network, and (ii) no gene appears in more than one set. In the context of this chapter, we consider the union of these modules to comprise a response network.

Ulitsky and Shamir begin by computing a likelihood ratio for each pair of genes from their similarity (measured as the value of Pearson's correlation coefficient): this likelihood ratio compares the probability that such similarity would be observed under the assumption that the two genes respond to the experiment versus the assumption the two genes have no relation. Large positive values of the logarithm of the likelihood ratio indicate greater support for the hypothesis that the two genes have related expression patterns and respond to the experimental condition. Conversely, large negative values of the logarithm indicate greater support that the two genes have unrelated expression patterns.

Ulitsky and Shamir then construct a complete similarity graph $X = (V, E, w)$, where the set V of nodes is the set of all genes, the set E of edges consists of all pairs of genes, and $w : E \rightarrow \mathbb{R}$ is a function specifying the log-likelihood for every edge in E. Given a set V' of genes, they define the score $s_{V'}$ of this set of genes as the sum of the log-likelihoods of all pairs of genes in V'; they define the score for a set of gene sets as the sum of the scores for all gene sets in the set. They address the problem of finding multiple disjoint gene sets in X such that each set of genes induces a connected subgraph in the interaction network G, and the total score of the gene sets is as large as possible. The MATISSE algorithm finds these subgraphs in three stages: identification of small subgraph "seeds", improving subgraphs from the seeds, and, finally, identifying statistically significant subgraphs.

To detect seeds, Ulitsky and Shamir settle on a "best-neighbors" heuristic that operates as follows. First, rank all nodes in X by the sum of their edge weights. Next, take the subgraph induced by the top ranked node and all the nodes connected to it in X by edges with positive weight as a seed. Remove this subgraph from X. Repeat the process with the next remaining highest-ranked node until X is empty.

They proceed to simultaneously refine all seeds using a greedy algorithm. At each step, they add a node to a module, remove a node from a module, reassign a node from one module to another, or merge two modules. They proceed with an action if it increases the overall score and maintains the connectivity of the subgraph induced by each module in G. This procedure terminates when no action meets these criteria.

In the final stage, the algorithm reports modules that are statistically significant. The authors use the following approach that is standard in the literature. Given a module, sample sets of genes of the same size from X, and compute the score of each set of genes. Next, compute the rank of the module's score among the scores of these random gene sets, and set the statistical significance of the module to be its score's rank divided by the number of sampled gene sets.

Ulitsky and Shamir further extend this algorithm to accommodate the situation when each edge in G has a weight that indicates the probability that it is an interaction that does take place in the cell [30].[3] They restate their objective as one of finding disjoint modules of strongly co-expressed genes in X that have a high probability of connectedness in G. More formally, given a user-specified probability q,

[3] Numerous methods have been developed to assess such confidence values. See Suthram *et al.* [26] for a comparison of such methods.

the algorithm detects subsets of genes which induce node-disjoint modules in X that are *q-connected* in G, i.e., have a probability of connectedness of at least q.

For each edge $e \in E$, let p_e denote the probability that e is a true interaction. The authors assign a confidence value $-\log(1-p_e)$ to the edge e. Let G_U be the subgraph of G induced by the set of genes $U \subseteq V$. Consider any *cut* of G_U, i.e., a set of edges in G_U that partition U into two non-empty subsets. The *weight* of this cut is the sum of the confidence values of the edges in this cut. With these definitions, a subset U of genes is *q-connected* if, for each possible cut in G_U, the weight of the cut is at least $-\log(1-q)$. (Note that $\log(1-q)$ represents the probability the subgraph is not connected.) Formally, a subset U is q-connected if, for all $W \subset U$

$$\sum_{e=(x,y), x \in W, y \in U-W} -\log(1-p_e) \geq -\log(1-q).$$

To determine if a subset is q-connected, it is sufficient to check if the weight of the minimum cut in G_U exceeds $-\log(1-q)$.

The steps for identifying disjoint modules that are q-connected closely follow those of MATISSE: seed identification, module optimization, and filtering for significant modules. CEZANNE starts with the modules computed by MATISSE; recall that MATISSE does not take edge weights into account. The authors identify q-connected seeds by recursively splitting the modules into smaller subgraphs along the minimum cut, until the weight of the cut is at least $-\log(1-q)$. The computations required to refine these seeds must satisfy the constraint that any modification must preserve q-connectedness (as opposed to connectivity). Ulitsky and Shamir employ several heuristics for performing the optimizations within acceptable running times; we refer the reader to their paper for details. Finally, Ulitsky and Shamir report only statistically-significant modules as follows: they create an empirical distribution of 100 scores by shuffling each gene's expression values among the samples, applying the CEZANNE algorithm, and recording the highest similarity score for each run. They assess the p-value for a module by ranking it within this distribution of similarity scores for randomized gene expression data.

3 Available Implementations

Table 1 summarizes the algorithms discussed in this chapter. We provide the name of the software package containing an implementation of each algorithm and a reference to available source code. In the column titled "Description", the phrase "Treatment-control data" refers to algorithms discussed in Section 2.1, "co-expression based" refers to algorithms discussed in Section 2.2, and "Comparing response networks" refers to the algorithms that we discuss below in Section 5.

Table 1 Implementations of algorithms to compute response networks.

Software	Algorithm	Description
http://www.cytoscape.org/tut/modules.complexes.php		
Cytoscape Plugin	ActiveModules [11]	Treatment-control data
https://www.mi.fu-berlin.de/w/LiSA/Heinz		
LiSA	Dittrich *et al.* [4]	Treatment-control data
http://acgt.cs.tau.ac.il/matisse		
MATTISE	DEGAS [28] MATISSE [29] CEZANNE [30] Hanisch *et al.* [10]	Treatment-control data Co-expression based Co-expression based Co-expression based
http://bioinformatics.cs.vt.edu/∼murali/software/network-lego		
NetworkLego	ActiveNetworks [22] Network Legos [22]	Co-expression based Comparing response networks
Source code is available upon request from Liu *et al.* [18].		
GNEA	GNEA [18]	Comparing response networks

4 Examples

In this section, we showcase the application of two co-expression-based methods for computing response networks to different stresses applied to *S. cerevisiae*.

Application of ActiveNetworks to amino acid starvation

In unpublished work, we applied the ActiveNetworks algorithm to a time-course of DNA microarray data collected upon amino acid starvation [9] and an interaction network integrated by Kelley and Ideker from multiple sources [15]. This network contains 15,429 protein-protein interactions from the Database of Interacting Proteins (DIP) [25], 5869 protein-DNA interactions (between transcription factors and their target genes) [17], and 6,306 metabolic interactions (interaction between proteins that operate on at least one common metabolite) based on the KEGG database [14]. As a negative control, this network included 4,812 genetic interactions [27]. Since genetically interacting genes are unlikely to be co-expressed, such interactions should not appear in a response network. Overall, this network contains 32,416 (27,604 physical and 4,812 genetic) interactions between 5601 proteins.

Figure 1 displays a layout of the computed response network. At the center of this network are two transcription factors PHD1 and GCN4. PHD1 is a transcriptional activator that enhances pseudohyphal growth, a pattern of cell growth that occurs in conditions of nitrogen limitation and an abundant fermentable carbon source. GNC4 is a transcriptional activator of amino acid biosynthetic genes in response to

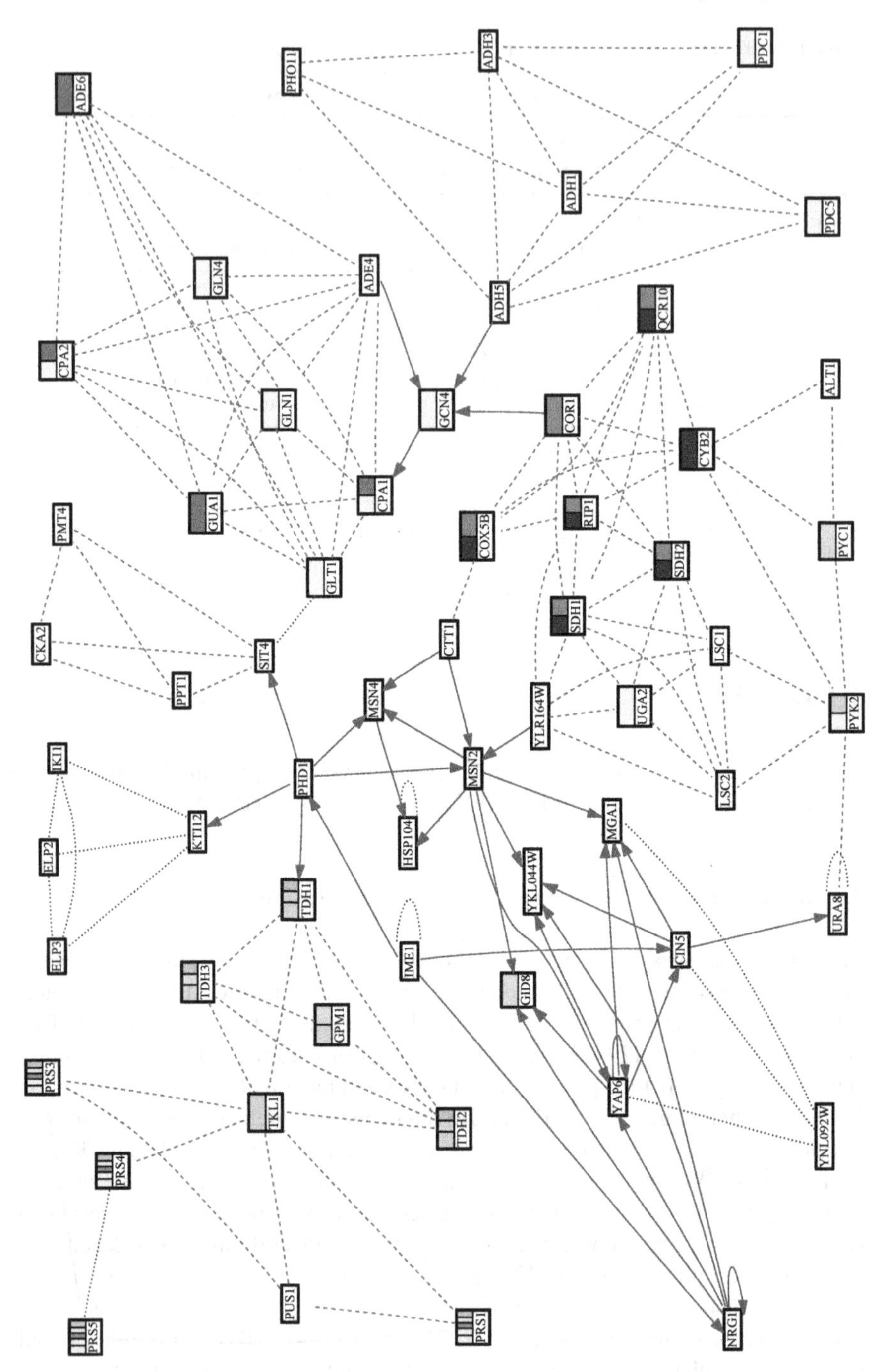

Fig. 1 Response networks in *S. cerevisiae* following amino acid starvation showing interactions between transcription factors and their target genes (solid arrows), PPIs (dotted), and metabolite-based interactions (dashed).

amino acid starvation. The Gene Ontology (GO) [1] biological processes enriched in this network include purine ribonucleoside salvage, electron transport, glucose catabolism, carboxylic acid metabolism and gluconeogenesis, pointing to the intricate network of transciptional regulatory interactions, protein complexes, signaling circuits, and metabolic pathways activated in response to the stress. The response network included only two genetic interactions (not displayed in Figure 1), indicating that genetically interacting gene pairs are not highly co-expressed in this gene expression data set.

Application of CEZANNE to DNA damage response

To detect response networks of *Saccharomyces cerevisiae* (yeast) under conditions inducing DNA damage, Ulitsky and Shamir obtained expression data from Gasch *et al.* [8] and a PPI network with confidence values derived from previous work by Collins *et al.* [3]. Ulitsky and Shamir applied and compared the CEZANNE [30], MATISSE [29], and Hanisch co-clustering [10] algorithms, as well as methods that used only expression data and not the interaction network.

CEZANNE identified a total of 14 significant response networks covering 471 of the 6167 genes in the interaction network, ranging in size from 3 to 346 genes. All response networks were enriched for at least one term in the "biological process" category, and 11 were enriched in at least one term in the "molecular function" category. The largest response network featured many proteins associated with ribosomal biosynthesis; as a whole, the module experienced down-regulation in response to DNA damage. The other modules include enrichment for genes with annotations related to function in the ribosome, proteasome, and mitochondrion. CEZANNE was able to detect both modules not detected by the other methods as well as more specific and precise modules, as assessed by F-tests. Such modules included genes correlated in the literature to cell response to stress but not previously detected by the other methods.

5 Advanced Topics: Comparing Response Networks

At times, researchers wish to identify similarities and differences between multiple response networks. These approaches are motivated by the desire to compare the cell's response to different conditions. Noting that Ideker *et al.* [11] extend their ActiveModules algorithm (Section 2.1) to incorporate expression profiles for multiple experimental conditions, we discuss two other algorithms capable of handling multiple response networks, namely, Gene Network Enrichment Analysis (GNEA) and Network Legos. With the availability of public repositories such as the National Center for Biotechnology Information's Gene Expression Omnibus that now contain thousands of gene expression datasets, we expect many novel methods for analyzing and comparing response networks to appear in the future.

5.1 Gene Network Enrichment Analysis

Motivated by the question of whether genes responsible for insulin production and uptake appeared frequently in response networks computed from contrasts of diabetic and non-diabetic patients, Liu *et al.* [18] develop a method for detecting enrichment of gene sets[4] across a collection of response networks, which they call *gene network enrichment analysis* (*GNEA*) . For each gene set F, they compute the significance of its enrichment in each response network using the one-sided version of Fisher's exact test and tally the number of response networks c_F for which the gene set has a statistically significant enrichment (p-value at most some user-defined threshold). To empirically determine the significance of c_F, Liu *et al.* construct a distribution of counts from 10,000 random gene sets of the same size as F: the p-value of c_F is the fraction of random gene sets whose counts are larger than c_F. Finally, they report all gene sets with highly significant counts.

5.2 Network Legos

Murali and Rivera [22] introduce the concept of *network legos* as a means for explicitly representing similarities and differences between response networks. They treat a response network simply as a set of edges. Given a collection $\mathscr{A}$ of response networks, each of which is a subgraph of an undirected interaction network G, they first define the notion of a *block* as a triple $(H, \mathscr{P}, \mathscr{N})$, where H is a subgraph of G, $\mathscr{P}$ and $\mathscr{N}$ are disjoint subsets of $\mathscr{A}$, and $\mathscr{P} \neq \emptyset$ such that

$$H = \Big(\bigcap_{P \in \mathscr{P}} P\Big) \bigcap \Big(\bigcap_{N \in \mathscr{N}} (G - N)\Big),$$

where "$\cap$," "$-$," and "$\cup$" respectively denote the intersection, difference, and union of the edge sets of two graphs and

1. $\mathscr{P}$ is maximal, i.e., there is no response network $P \in \mathscr{A} - \mathscr{P}$ such that $H \subseteq P$, and
2. $\mathscr{N}$ is maximal, i.e., there is no response network $N \in \mathscr{A} - \mathscr{N}$ such that $H \cap N = \emptyset$.

In other words, they form H by taking the intersection of all the response networks in $\mathscr{P}$ and removing any edge that appears in any of the response networks in $\mathscr{N}$. Informally, H represents the cellular response that is common to all the experimental conditions whose response networks are members of the "positive" set $\mathscr{P}$. It also does not incorporate any aspect of the cellular response captured in the "negative" set $\mathscr{N}$.

---

[4] In this context, a gene set is simply a set of genes that perform the same function or have been grouped together based on some common attribute.

Murali and Rivera reduce the problem of computing all blocks to that of computing all closed biclusters in an appropriately defined binary matrix representing presence and absence of interactions in each response network. They use the CHARM algorithm for this purpose [31]. To estimate σ_H, the statistical significance of an observed block $(H, \mathscr{P}, \mathscr{N})$, Murali and Rivera construct a set of blocks $R_H = \{(H', \mathscr{P}', \mathscr{N}')\}$, composed from random selections of response networks $\mathscr{P}'$ and $\mathscr{N}'$, where $|\mathscr{P}| = |\mathscr{P}'|$ and $|\mathscr{N}| = |\mathscr{N}'|$. They set σ_H to be the fraction of blocks in R_H whose subgraph has at least as many interactions as H.

Next, Murali and Rivera define a natural partial order between blocks: Given two distinct blocks $(H_1, \mathscr{P}_1, \mathscr{N}_1)$ and $(H_2, \mathscr{P}_2, \mathscr{N}_2)$, they say that $H_1 \prec H_2$ if

1. $\mathscr{P}_1 \subseteq \mathscr{P}_2$ and $\mathscr{N}_1 \subseteq \mathscr{N}_2$ or
2. $\mathscr{P}_1 \subseteq \mathscr{N}_2$ and $\mathscr{N}_1 \subseteq \mathscr{P}_2$.

Finally, Murali and Rivera define a *network lego* to be a block $(H, \mathscr{P}, \mathscr{N})$ such that $\sigma_H < \sigma_{H'}$, for every H' where $H \prec H'$ or $H' \prec H$. In other words, $(H, \mathscr{P}, \mathscr{N})$ is a network lego if it is more statistically significant than blocks formed by combining any subset of $\mathscr{P}$and $\mathscr{N}$or by combining any superset of $\mathscr{P}$and $\mathscr{N}$. They output all the blocks that satisfy this condition as the set of network legos.

6 Exercises

1. Let $z_1, z_2, \ldots, z_k$ be k independent normally distributed variables. The random variable $\sum_{i=1}^{k} z_i$ follows a Gaussian distribution. What are its mean and variance? Use these values to justify the score used by Ideker *et al.* in the ActiveModules algorithm.
2. In the ActiveModules algorithm, suppose we drop the requirement that the computed subnetwork be connected. Show that the optimal network can be computed in $O(|V| \log |V|)$ time. What is the disadvantage of not requiring that the output subnetwork be connected? Could the result be highly disconnected? What is the biological relevance of such a network?
3. Consider the following modification to the ActiveNetworks algorithm: given a subgraph $H = (V', E')$ of G, define its *density* as

$$w_H = \frac{2 \sum_{e \in E'} w_e}{|V|(|V| - 1)},$$

   i.e., the total weight of the edges in E' divided by the number of possible edges in H. With this new definition, how easy or difficult is it to compute the subgraph that has the largest density, over all subgraphs of G?
4. Assessing the statistical significance of a response network is a theme that appears repeatedly in this chapter. Summarize all the methods presented for computing statistical significance and compare and contrast them, with a focus on the different null hypotheses assumed by each method.

5. In the Network Lego algorithm, the definition of a block $(H, \mathscr{P}, \mathscr{N})$ does not require that H be connected. It is easy to add this requirement to the definition. What are the implications of this modification on the rest of the approach (the definition of the partial order and network legos, the computation of blocks, and the assessment of statistical significance)?

7 Outlook and Further Reading

The algorithms discussed in this chapter have integrated *gene expression* data with networks of physical interactions between *proteins*. They make the assumption that the expression level of a gene can be used as a surrogate for the expression or the activity of the protein produced by the gene. This assumption is simplistic, since a single gene may code for multiple proteins due to alternate splicing, and because post-transcriptional and post-translation modifications play a major role in regulating protein levels and activity. Nevertheless, this assumption is very useful in practice, since gene expression does play a major role in controlling physiological process and because DNA microarrays are the most widely-available experimental technology for genome-wide measurement of gene expression. As technologies that measure protein levels and activity (e.g., protein arrays) mature, we anticipate that the next generation of response network algorithms will directly integrate protein data with protein interaction networks.

Sharan and Ideker surveyed numerous methods that study protein interaction networks in the context of diseases [12]. Their review provides numerous entry points into the literature on response networks and related topics. There are many approaches to integrate gene expression data with interaction networks and other types of functional genomics data with the goal of finding gene modules. Such general-purpose data integration techniques have been reviewed by Joyce and Palsson [13]. For a more general survey of analyses of interaction networks, especially on connections to theoretical computer science, the reader can consult Murali and Aluru's survey [21].

References

1. Ashburner, M., Ball, C.A., Blake, J.A., Botstein, D., Butler, H., Cherry, J.M., Davis, A.P., Dolinski, K., Dwight, S.S., Eppig, J.T., Harris, M.A., Hill, D.P., Kasarskis, A., Lewis, S., Matese, J.C., Richardson, J.E., Ringwald, M., Rubin, G.M., Sherlock, G.: Gene Ontology: Tool for the unification of biology. The Gene Ontology Consortium. Nat Genet **25**(1), 25–9 (2000)
2. Charikar, M.: Greedy approximation algorithms for finding dense components in a graph. In: Proceedings of 3rd International Workshop on Approximation Algorithms for Combinatorial Optimization, pp. 84–95. Springer (2000)
3. Collins, S.R., Kemmeren, P., Zhao, X., Greenblatt, J.F., Spencer, F., Holstege, F.C.P., Weissman, J.S., Krogan, N.J.: Toward a comprehensive atlas of the physical interactome of *Saccha-*

romyces cerevisiae. Mol Cell Proteomics **6**(3), 439–450 (2007)
4. Dittrich, M.T., Klau, G.W., Rosenwald, A., Dandekar, T., Mller, T.: Identifying functional modules in protein-protein interaction networks: An integrated exact approach. Bioinformatics (Oxford, England) **24**(13), i223–231 (2008)
5. Durinck, S.: Pre-processing of microarray data and analysis of differential expression. Methods in Molecular Biology (Clifton, N.J.) **452**, 89–110 (2008)
6. Fields, S., Song, O.: A novel genetic system to detect protein-protein interactions. Nature **340**(6230), 245–6 (1989)
7. Gallo, G., Grigoriadis, M.D., Tarjan, R.E.: A fast parametric maximum flow algorithm and applications. SIAM J. Comput. **18**(1), 30–55 (1989)
8. Gasch, A.P., Huang, M., Metzner, S., Botstein, D., Elledge, S.J., Brown, P.O.: Genomic expression responses to DNA-damaging agents and the regulatory role of the yeast ATR homolog mec1p. Mol. Biol. Cell **12**(10), 2987–3003 (2001)
9. Gasch, A.P., Spellman, P.T., Kao, C.M., Eisen, M.B., Storz, G., Botstein, D., Brown, P.O.: Genomic expression programs in the response of yeast cells to environmental changes. Mol Biol Cell **11**(12), 4241–57 (2000)
10. Hanisch, D., Zien, A., Zimmer, R., Lengauer, T.: Co-clustering of biological networks and gene expression data. Bioinformatics **18**(suppl_1), S145–154 (2002)
11. Ideker, T., Ozier, O., Schwikowski, B., Siegel, A.F.: Discovering regulatory and signalling circuits in molecular interaction networks. Bioinformatics (Oxford, England) **18 Suppl 1**, S233–240 (2002)
12. Ideker, T., Sharan, R.: Protein networks in disease. Genome Res. **18**(4), 644–652 (2008)
13. Joyce, A.R., Palsson, B.O.: The model organism as a system: Integrating 'omics' data sets. Nat Rev Mol Cell Biol **7**(3), 198–210 (2006)
14. Kanehisa, M., Araki, M., Goto, S., Hattori, M., Hirakawa, M., Itoh, M., Katayama, T., Kawashima, S., Okuda, S., Tokimatsu, T., Yamanishi, Y.: KEGG for linking genomes to life and the environment. Nucleic Acids Research **36**(suppl_1), D480–484 (2008)
15. Kelley, R., Ideker, T.: Systematic interpretation of genetic interactions using protein networks. Nat Biotechnol **23**(5), 561–6 (2005)
16. Kirkpatrick, S., Gelatt, C.D., Vecchi, M.P.: Optimization by simulated annealing. Science **220**(4598), 671–680 (1983)
17. Lee, T.I., Rinaldi, N.J., Robert, F., Odom, D.T., Bar-Joseph, Z., Gerber, G.K., Hannett, N.M., Harbison, C.T., Thompson, C.M., Simon, I., Zeitlinger, J., Jennings, E.G., Murray, H.L., Gordon, D.B., Ren, B., Wyrick, J.J., Tagne, J.B., Volkert, T.L., Fraenkel, E., Gifford, D.K., Young, R.A.: Transcriptional regulatory networks in *Saccharomyces cerevisiae*. Science **298**(5594), 799–804 (2002)
18. Liu, M., Liberzon, A., Kong, S.W., Lai, W.R., Park, P.J., Kohane, I.S., Kasif, S.: Network-based analysis of affected biological processes in type 2 diabetes models. PLoS Genetics **3**(6), e96 (2007)
19. Ljubić, I., Weiskircher, R., Pferschy, U., Klau, G.W., Mutzel, P., Fischetti, M.: An algorithmic framework for the exact solution of the prize-collecting Steiner tree problem. Mathematical Programming **105**(2), 427–449 (2006)
20. Milo, R., Shen-Orr, S., Itzkovitz, S., Kashtan, N., Chklovskii, D., Alon, U.: Network motifs: Simple building blocks of complex networks. Science **298**(5594), 824–827 (2002)
21. Murali, T.M., Aluru, S.: Algorithms and Theory of Computation Handbook, second edn., chap. Computational Systems Biology, p. to appear. CRC Press (2009)
22. Murali, T.M., Rivera, C.G.: Network legos: Building blocks of cellular wiring diagrams. Journal of Computational Biology **15**(7), 829–844 (2008)
23. Pounds, S., Morris, S.W.: Estimating the occurrence of false positives and false negatives in microarray studies by approximating and partitioning the empirical distribution of p-values. Bioinformatics **19**(10), 1236–1242 (2003)
24. Rousseeuw, P.: Silhouettes: A graphical aid to the interpretation and validation of cluster analysis. Journal of Computational and Applied Mathematics **20**, 53–65 (1987)
25. Salwinski, L., Miller, C.S., Smith, A.J., Pettit, F.K., Bowie, J.U., Eisenberg, D.: The database of interacting proteins: 2004 update. Nucleic Acids Res **32 Database issue**, D449–51 (2004)

26. Suthram, S., Shlomi, T., Ruppin, E., Sharan, R., Ideker, T.: A direct comparison of protein interaction confidence assignment schemes. BMC Bioinformatics **7**, 360 (2006)
27. Tong, A.H.Y., Lesage, G., Bader, G.D., Ding, H., Xu, H., Xin, X., Young, J., Berriz, G.F., Brost, R.L., Chang, M., Chen, Y., Cheng, X., Chua, G., Friesen, H., Goldberg, D.S., Haynes, J., Humphries, C., He, G., Hussein, S., Ke, L., Krogan, N., Li, Z., Levinson, J.N., Lu, H., Ménard, P., Munyana, C., Parsons, A.B., Ryan, O., Tonikian, R., Roberts, T., Sdicu, A.M., Shapiro, J., Sheikh, B., Suter, B., Wong, S.L., Zhang, L.V., Zhu, H., Burd, C.G., Munro, S., Sander, C., Rine, J., Greenblatt, J., Peter, M., Bretscher, A., Bell, G., Roth, F.P., Brown, G.W., Andrews, B., Bussey, H., Boone, C.: Global mapping of the yeast genetic interaction network. Science **303**(5659), 808–13 (2004)
28. Ulitsky, I., Karp, R., Shamir, R.: Detecting disease-specific dysregulated pathways via analysis of clinical expression profiles. In: Proceedings of the 12th Annual International Conference on Research in Computational Molecular Biology, no. 4955 in Lecture Notes in Computer Science, pp. 347–259 (2008)
29. Ulitsky, I., Shamir, R.: Identification of functional modules using network topology and high-throughput data. BMC Systems Biology **1**, 8 (2007)
30. Ulitsky, I., Shamir, R.: Identifying functional modules using expression profiles and confidence-scored protein interactions. Bioinformatics **25**(9), 1158–1164 (2009)
31. Zaki, M.J., Hsiao, C.J.: Efficient algorithms for mining closed itemsets and their lattice structure. IEEE Trans. Knowl. Data Eng. **17**(4), 462–478 (2005)

Identification of Modules in Protein-Protein Interaction Networks

Sinan Erten and Mehmet Koyutürk

Abstract In biological systems, most processes are carried out through orchestration of multiple interacting molecules. These interactions are often abstracted using network models. A key feature of cellular networks is their modularity, which contributes significantly to the robustness, as well as adaptability of biological systems. Therefore, modularization of cellular networks is likely to be useful in obtaining insights into the working principles of cellular systems, as well as building tractable models of cellular organization and dynamics. A common, high-throughput source of data on molecular interactions is in the form of physical interactions between proteins, which are organized into protein-protein interaction (PPI) networks. This chapter provides an overview on identification and analysis of functional modules in PPI networks, which has been an active area of research in the last decade.

Proteins that make up a functional module tend to interact with each other and form a densely connected subgraph in a PPI network. Motivated by this observation, module identification is often formulated as a problem of partitioning a PPI network into dense subgraphs, which is also known as graph clustering. This chapter begins with a brief introduction to the module identification problem in PPI networks. Then, graph theoretical measures of modularity such as density, clustering coefficient and edge connectivity are introduced. Algorithmic approaches for identifying modules are then presented in a systematic manner. These clustering approaches are broadly categorized as (i) Bottom-up (ii) Top-down (iii) Iterative Improvement and (iv) Flow Based methods. Subsequently, a sample application of modularization, namely, predicting the function of uncharacterized proteins, is briefly discussed. More advanced methods to identify functional modules often integrate other data sources such as gene expression data with PPI data or use multiple networks to find

Sinan Erten
Dept. of Electrical Engineering & Computer Science, Case Western Reserve University, Cleveland, OH, USA, e-mail: sinan.erten@case.edu

Mehmet Koyutürk
Dept. of Electrical Engineering & Computer Science and Center for Proteomics & Bioinformatics, Case Western Reserve University, Cleveland, OH, USA, e-mail: koyuturk@eecs.case.edu

L.S. Heath and N. Ramakrishnan (eds.), *Problem Solving Handbook in Computational Biology and Bioinformatics*, DOI 10.1007/978-0-387-09760-2_12,

conserved regions in the networks. After an overview on these advanced methods, some exercises are presented to the reader.

1 History of the Problem

For years in the earlier century, the study of biological systems focused on understanding behaviour of single molecules [21]. Although this molecular approach had great impact, it does not adequately reflect what really is going on inside the cell. With recent advances in biotechnology, availability of high throughput biological data enables investigation of biological processes from a systems perspective. Biological processes are often orchestrated through interaction of multiple molecules. Considered together, these interactions form complex biological networks that underlie cellular organization [37]. Computational analyses of the structure of these networks provide significant insights into the mechanisms that drive complex biological systems [27].

In this chapter, we focus on protein-protein interaction (PPI) networks that model physical interactions and functional associations between proteins. These interactions are captured via a variety of experimental and computational methods [6, 23, 36]. PPI networks are often abstracted by graph models, in which the proteins are represented by nodes and the interactions (often physical) among them are represented by undirected edges. This abstraction enables application of graph theoretical approaches to the analysis of cellular organization.

One of the most common applications of PPI network analysis is the identification of functional modules. *Functional modules* are generally defined as groups of proteins with a distinct biological function [43]. For example, a protein complex is a macromolecular machine formed by multiple interacting proteins to perform specific functions (e.g., RNA polymerase I consists of 8 to 14 protein subunits and it is responsible for the transcription of ribosomal RNA from DNA in eukaryotic cells). Proteins within a module tend to interact with each other, whereas they are somewhat isolated from the rest of the network [43]. Consequently, such modular groups of proteins usually induce densely connected subgraphs in a PPI network [21]. Identification and analysis of these modules renders the study of complex biological systems tractable, and provides significant insights into the essential characteristics of these systems, including robustness, adaptability, and efficiency [21].

The problem of network-based module identification is studied extensively in recent years and various computational approaches are developed to tackle various challenges, including the following:

- Protein interaction data is not reliable. The interactions captured by high-throughput experiments provide only a generic and static representation of cellular dynamics, and there are a large number of unknown interactions, as well as false positives [17].

- It is not straightforward to unambiguously formulate mathematical properties (e.g. size, density) of a group of interacting proteins to qualify it as a biologically meaningful functional module.
- Functional modules are not rigid structures that have specific functions. Proteins may dynamically form different modules in different contexts [47]. Consequently, the methods should ideally be able to identify overlapping modules.

In the following sections, we discuss how different approaches tackle these challenges to deliver efficient and effective algorithms for network based module identification. First, we discuss several algorithmic approaches to identify functional modules in PPI networks and introduce various modularity measures. Next, we provide a list of some of the publicly available software solutions for network-based module identification. Subsequently, in the Examples section, we illustrate the use of identified functional modules. We then overview several advanced methodologies that combine other sources of information with PPI network data to enhance identification of functional modules. Next, we present a few practical exercises to deepen the readers' understanding about this subject. Finally, in the Further Reading section, we mention several other resources for interested readers who would like to broaden their knowledge in this area.

2 Algorithm Description

As discussed in the previous section, proteins that make up a functional module tend to interact with each other and form a densely connected subgraph, which is somewhat isolated from the rest of the network. Motivated by this observation, module identification is often formulated as a problem of partitioning a PPI network into dense subgraphs, which is also known as graph clustering. Many existing graph clustering methods are applied to the analysis of PPI networks, with a view to addressing the domain-specific challenges discussed in the previous section.

In this section, we first discuss various graph theoretical measures for assessing the modularity of a group of proteins. Next, we briefly go through several algorithmic approaches that tackle module identification problem based on these measures. Finally, we provide an overview of some methods for further refinement of identified modules, which usually make use of other sources of information such as known protein complexes.

2.1 Graph Theoretical Measures of Modularity

What qualifies a set of proteins as a functional module in a PPI network? Different measures are developed to provide a quantitative answer to this question. Here, we review some extensively used measures.

A PPI network, denoted by the undirected graph $G = (V, E)$, consists of a set V of proteins and a set E of interactions connecting pairs of these proteins, i.e., an edge $v_i v_j \in E$ represents an interaction between proteins v_i and v_j. In graph theory, a clique in an undirected graph is defined as a set of vertices, such that any pair of vertices in this set are connected with an edge. In the context of PPI networks, a clique is obviously a good candidate for being interpreted as a functional module, since it is maximally connected. However, it is not realistic to assume that all proteins in a functional module interact with each other; for example, a protein complex may indeed contain proteins that do not physically interact with each other, but are rather held together by another protein in the complex. Furthermore, due to the large amount of false positives and missing interactions, as well as the variability of the way experimental data is interpreted (e.g., spoke vs. matrix model [7]), searching for a perfect clique in a PPI network would not be realistic. Consequently, several measures are developed to assess the *cliquishness* of a subgraph.

For a single protein, *clustering coefficient* is defined to measure how close the neighborhood structure of that protein is to a clique [48]. In mathematical terms, clustering coefficient c_i for a node i is defined as $c_i = 2e_i/M_i(M_i - 1)$ where e_i is the total number of edges among the neighbours of node i and M_i is the number of immediate neighbours of node i. A node in a perfect clique structure therefore has a clustering coefficient of 1.

However, since many proteins may be involved in more than one module, they often have many interacting partners that do not interact with each other. Consequently, the clustering properties of these proteins may not be adequately captured by clustering coefficient (it will be close to 0 for proteins that are involved in many modules). Therefore, another metric, *core-clustering coefficient* , is defined, which is very similar to clustering coefficient, but is more flexible especially in capturing modules that involve hub-like proteins [8]. In order to define core-clustering coefficient formally, we first introduce another term, *k-core* . A k-core of a graph is simply a subgraph, in which all vertices have degree at least k. Consequently, the core-clustering coefficient of a protein i is defined as the density of the k-core with largest k of the immediate neighbourhood of i.

From the perspective of a set of proteins (as opposed to a single protein), the equivalent of clustering coefficient is density, which is another metric that is widely used as a measure of modularity of a subgraph. *Density* of a set of proteins is defined as the fraction of the number of interactions present in the induced subgraph to the maximum possible number of interactions that can potentially occur in that module. In mathematical terms, density d of a set N of proteins is defined as $d_N = 2e_N/|N|(|N| - 1)$ where e_N is the total number of interactions in the set and $|N|$ is the number of proteins in the set.

Furthermore, to measure how likely a protein belongs to a given module, one can use the connectivity of that protein to the proteins in the module. *Cluster property* , k_{iN} of a protein i with respect to a cluster N is calculated as $k_{iN} = e_{iN}/d_N|N|$ where e_{iN} is the total number of edges between protein i and the proteins in N, d_N is the density of N, and $|N|$ is the total number of nodes in N [2]. This measure quantifies

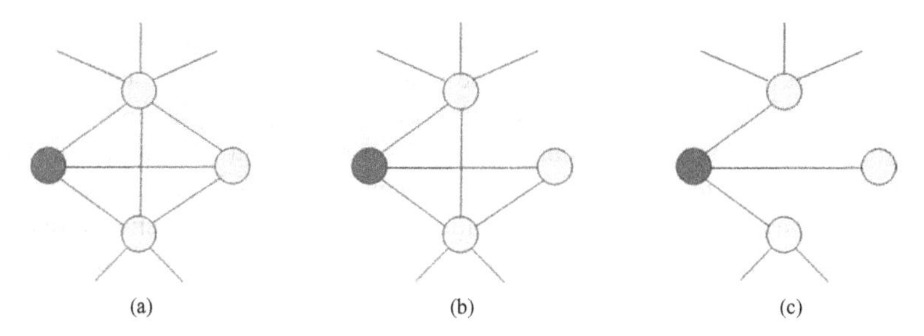

Fig. 1 Illustration of common measures of modularity in a PPI network. Each figure shows a sample subgraph of a larger network. The edges with missing incident nodes show the interactions of the proteins in these subgraphs with other proteins in the network. The clustering coefficient of the red vertex in each subgraph is (a) 1 (b) $2/3$ (c) 0. The density of each subgraph is (a) 1 (b) $5/6$ (c) $3/6$. Edge-connectivity of each subgraphs is (a) 3 (Highly Connected) (b) 2 (Not Highly Connected) (C) 1 (Not Highly Connected) .

the relative connectivity of protein i to the proteins in N with respect to the average connectivity of these proteins to each other.

Another measure of modularity of a subgraph is *edge connectivity* , which is defined as the minimum number of interactions to be removed in order to disconnect the subgraph. A set of proteins is referred to as *highly connected* if the edge connectivity is greater than half of the number of proteins in the set [20]. The concept of high connectivity is defined to reflect the idea that a subgraph can be highly connected if, for each protein in this subgraph, the number of proteins that it interacts with is larger than those it does not interact with.

Note that, many functionally associated proteins interact with each other through indirect paths as well [13]. Consequently, besides evaluating the connectivity of a set of proteins in terms of their direct interactions, it is also useful to assess the reliability of paths connecting these proteins. For example the reliability of path between a single protein and a set of proteins in a module can be used to estimate the probability of that protein being a member of that module [5, 9]. The reliability of a path is a measure that depends on the reliability of interactions that make up that path, which indeed depends on the strength of the evidence that those interactions exist in vivo. Here, the reliability of an interaction is generally scored using various statistical techniques to reflect the strength of various sources of evidence on the existence of the interaction [15].

A major problem associated with these modularity measures is that they are rather arbitrary, i.e., they are not directly associated with a quantitative model of underlying modular process. However, assessment of the significance of these measures with respect to a reference statistical model may provide a statistical basis for the potential biological relevance of a module. Commonly, statistical significance is assessed by generating many random graphs from the source PPI network by preserving the node degrees to obtain a background distribution for the module's topological characteristics (e.g., density, edge connectivity) [22, 35, 43]. This can be achieved by repeatedly switching the incident nodes of random pairs of edges. If the structure of a candidate module is observed frequently in these random graphs, this may be an indicator that the structure is likely to occur by chance and therefore it may not reflect any biological insight. On the other hand, candidate modular structures that happen to occur very rarely in random graphs are more likely to have a biological meaning. However, assessment of statistical significance via Monte Carlo simulations is often computationally expensive. Alternatively, one can explicitly formulate statistical significance of a module and use this measure directly as an objective criterion for the module identification algorithms [33].

2.2 Algorithms for the Identification of Network Modules

Here we present different approaches for searching PPI networks to identify subgraphs that are (locally) optimal in terms of the modularity measures discussed above.

2.2.1 Bottom-Up Clustering

This type of algorithms usually start with a set of seed proteins and grow clusters around them based on a variety of methods in a bottom-up manner. Whenever a cluster reaches a certain threshold for a modularity measure, that cluster is qualified as a functional module. Due to their greedy nature, these algorithms are generally quite efficient, often with time complexity linear in the number of interactions in the network. However, they are also more likely to lack global awareness, because of the greedy choices they make at each step of the search process.

One typical example for this approach is MCODE [8] . MCODE first assigns a weight to each node that is correlated to the core-clustering coefficient of the corresponding node. Then, it initiates a module composed of the protein with largest weight and iterates by adding proteins that interact with the module and have weight above some threshold in a depth first manner, until there are no unassigned nodes left. The premise here is that connected nodes with high clustering coefficients are likely to be parts of the same module.

Another bottom-up clustering method is proposed in [2]. Similar to MCODE, this algorithm also starts with a seed protein that makes up its own module, and grows it gradually by adding the neighbours to the cluster as long as the density of the cluster and the added node's cluster property are above a certain threshold value. Whenever there are no more neighbours left that may be added to the cluster, that cluster is removed from the network and whole process starts again, until all nodes are assigned to a cluster.

2.2.2 Top-Down Clustering

This type of algorithms start with the whole network in hand, and recursively partition it to the clusters, until each cluster reaches to a certain value for a defined property. While these algorithms provide more globally aware solutions by making decisions based on global network properties, they are computationally more expensive since they solve global optimization problems at each step of the algorithm.

One typical example for this type of algorithms is HCS (Highly Connected Subgraphs) [20]. The recursive step of the algorithm proceeds as follows: the network is partitioned into two disconnected regions by removing the minimum number of edges possible (a problem known as min-cut partitioning [1]). Subsequently, each subgraph is checked to be highly connected, and partitioned recursively if they are not. The recursion stops whenever a subgraph that is highly connected is found. Consequently, each final subgraph (at the leaves of recursion tree) is guaranteed to be at least half as dense as a clique of the same size, by the definition of a highly connected graph.

An important problem with the application of HCS to PPI networks is the following. Due to the nature of the PPI networks (most proteins have low degree), repeated solution of the min-cut problem tends to produce many clusters containing only a single node, which obviously do not represent a significant functional module. A statistically motivated module identification algorithm, SIDES [33], alleviates this problem of HCS by using ratio-cut instead of min-cut. Ratio-cut is the ratio of the min-cut to the size of the smaller partition to be created. In this way, more balanced partitions are favored which reduces the number of singleton clusters (clusters with a single node) created.

2.2.3 Iterative Improvement

We have covered two algorithmic approaches so far. One starts with single seed proteins and grows clusters around them, and the other starts with the whole network and partitions it recursively until "good" clusters are found. Some methods instead start with an initial set of clusters and work to improve the modularity of these clusters by making changes in the assignments of proteins into modules. Initial clusters may be assigned randomly, or they may be chosen based on biological information that is available a priori.

One such method, RNSC [28] starts with partitioning the whole network to randomly assigned clusters and improves these clusters by changing the cluster assignments of the nodes based on a cost function, which is very similar to the density measure described in the previous section. Basically, a move (i.e., change of a cluster assignment) takes place if it improves the modularity of the partition based on this cost function. Similarly, the method proposed in [5] starts with initial cluster assignments that typically depend on known protein complexes. For each cluster, the rest of the proteins are ranked based on a probability value that represent how likely they will be a member of that cluster. This probability is estimated based on

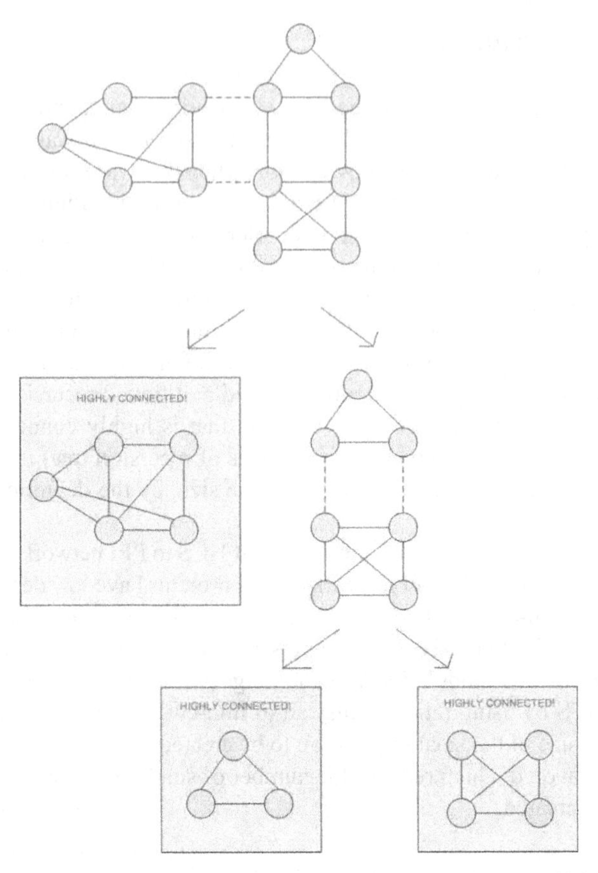

Fig. 2 A sample run of the HCS algorithm. Having an edge connectivity 2 which is smaller than half of the number of nodes, the initial graph is not highly connected. It is partitioned into two subgraphs and each partition is checked to be highly connected. Dashed lines correspond to min-cut edges between the two parts identified by the min-cut algorithm. The process continues until all partitions are highly connected. SIDES implements a similar algorithm, in which min-cut is replaced with ratio-cut and the high connectivity criterion is replaced by statistical significance of subgraph density.

the reliability of the paths between the query protein and proteins that make up the cluster. A similar approach is used in SEEDY [9]. SEEDY adds new proteins to the given seed modules if the reliability of the most reliable path from the candidate protein to the seed cluster is above a certain threshold.

2.2.4 Flow-Based Clustering

Information flow based approaches are often utilized in the analysis of biological networks [44] and they also find application in module identification. A well known flow based algorithm is MCL [18], which clusters the graphs based on random walks. MCL is based on the intuition that a random walk that starts from a node in a module is unlikely to leave that cluster without visiting most of the nodes in the module. After simulating a large number of random walks on a network, regions with high flow are most likely to be dense subgraphs and vice versa. One other algorithm for graph clustering based on network flow is proposed in [12]. Prior to applying this method, each edge is assigned a reliability score in the protein interaction network, by utilizing the semantic similarity and semantic interactivity of the nodes. The most informative proteins are selected based on the weighted degree of the node, which is the sum of the weights of all edges from this node to its immediate neighbours. In the flow simulation stage, flow starts with the most informative proteins and the influence of these proteins on others are calculated based on the weights of the edges. The intuition behind this algorithm is that, more reliable interactions carry more influence of a protein to its interacting partners. If the influence of a protein on another one is high, they are likely to be in the same functional module.

2.3 Refining Identified Modules

The module identification algorithms covered so far assess the modularity of a subgraph using topological measures such as density, clustering coefficient, etc. To address domain-specific challenges and enhance the biological relevance of identified modules, after this clustering process, identified modules may go through a post-processing stage, where they are extended [8], merged [12] or filtered [8, 28] using various strategies. As mentioned previously, due to the existence of multi-functional proteins, functional modules often overlap with each other. Traditional partitioning-based clustering algorithms (e.g., bottom-up,top-down) are not suited to identifying overlapping modules [47]. In order to have more realistic functional modules, clusters are generally extended through a post processing phase. One way to achieve this is to extend a cluster by adding the neighbours of the border proteins to it as long as the used modularity measure is satisfied [8]. One helpful resource for refining identified modules is the already available information on known protein complexes. Identified modules may be compared to protein complexes in terms of density, size and other topological metrics and be modified (or filtered out) accordingly to achieve final modules as similar as possible to known protein complexes in terms of these measures [28].

3 Available Implementations

The implementations of some of the algorithms explained in the previous section are summarized in Table 1.

Table 1 Publicly available implementations of major module identification algorithms.

Name	Algorithm	Platform
MCODE [8] http://baderlab.org/Software/MCODE	Bottom-Up	Java
SIDES [33] http://vorlon.case.edu/∼mxk331/sides/	Top-Down	C, Java
UVCLUSTER [3] http://www.uv.es/genomica/UVCLUSTER/	Iterative	C
MCL [18] http://www.micans.org/mcl/source	Flow Based	C
COMPLEXPANDER [5] http://llama.med.harvard.edu/cgi/Complexpander/Complexpander.pl	Iterative	online
SEEDY [9] available upon request	Iterative	Perl
SPC [45] http://mips.gsf.de/proj/spc	Superparamagnetic Clustering	C++
NETWORKBLAST [25] http://www.cs.tau.ac.il/∼bnet/networkblast.htm	Network Alignment	Java
MAWISH [32] http://vorlon.case.edu/∼mxk331/mawish/	Network Alignment	C
MULE [31] http://vorlon.case.edu/∼mxk331/mule/	Graph Mining	C

4 Examples

In this section, we illustrate the practical uses of identified modules in PPI networks for generating novel biological knowledge. Clustering a PPI network and exploring the topological measures of obtained clusters are not always sufficient to identify realistic functional modules. Scientists often make use of curated data that provides direct information on the function of individual molecules, which in turn can be used to assess the functional coherence of modules.

Many proteins are already assigned some known functions. A valuable resource that provides large scale information on molecular function is the publicly available Gene Ontology Database [4]. These annotations are often used to calculate the statistical significance of the enrichment of a particular function in identified functional modules [14, 3, 10, 47]. In other words, functional annotations of proteins that make up a module can be evaluated in terms of their functional coherence (or homogeneity), as well as coverage of a specific function. This process can be summarized as follows: for each GO term, the ratio of the proteins in a module that are annotated with this term is found. This ratio is compared with the ratio of the proteins annotated with that particular GO term in the whole network. Statistical models based on hypergeometric distribution are used to estimate the significance of the term in that module. If there exists any GO terms that are significantly enriched in a module, that module can be classified as functionally significant [10].

Predicting unknown functional annotations of proteins is an important problem. Many approaches are proposed so far that make use of identified functional modules for this task. The general idea is to find significantly enriched functional annotations in a module and annotate all proteins in that module with those functions. Although these methods are usually shown to be effective in function prediction, they are based on the assumption that interacting proteins tend to share functions, which may be an oversimplification (e.g., most of the metabolic enzymes perform their functions without any collaborators [41]).

5 Advanced Topics

Since different data sources capture different aspects of biological processes, integration of different 'omic datasets prove invaluable in discovering novel biological information. In this section, we overview various methods that integrate network information with other sources of biological data to identify functional modules.

5.1 Using Gene Expression and Protein Interaction Data Together

PPI data provides a generic and static picture of cellular organization in that it does not capture the temporal and spatial specificity of interactions between biomolecules. On the other hand, genome-scale measurements of mRNA expression via DNA microarrays [42] provide useful information on the expression of genes in a particular context (e.g., after introduction of a ligand, phase of cell cycle, disease). This information may be used in conjunction with molecular interaction data to investigate the dynamics of functional modules, in terms of the availability of the interacting molecules in the cell in specific conditions. Indeed, it is shown that proteins that have correlated expression profiles across a range of natural/experimental conditions have more tendency to interact with each other. Expression levels of proteins that are identified to be in the same functional module are also shown to be significantly correlated [24, 46]. This co-expression is especially apparent among elements of permanent protein complexes, while being somewhat weak for transient interactions. Transient interactions occur between proteins that come together in only certain cellular states to undertake a biomolecular function. Based on these observations an algorithm [34] is developed to identify groups of proteins with (i) high connectivity in PPI networks and (ii) significant correlation of gene expression. This algorithm is shown to perform better in terms of capturing biologically relevant functional modules when compared to using these data sources separately.

5.2 Using Network Alignment

Availability of interaction data for multiple species can also be utilized to enhance the identification of functional modules, through identification of conserved subgraphs across multiple networks , a problem also known as network alignment. The intuition here is that, conserved regions in PPI networks often correspond to biologically functional subgraphs [26, 29]. Since the mapping of orthologs between different species (networks) is not one-to-one, these algorithms generally construct a Cartesian-product graph, in which each node represents a group of orthologs, one from each species [30, 39, 40]. Subsequently, by assigning weights to edges based on their conservation (e.g., based on maximum likelihood [40] or match score-mismatch penalty [30], these algorithms reduce the problem into one of finding dense subgraphs in the product graph.

Since the size of such a product graph grows exponentially with the number of networks, these algorithms do not scale well to large number of networks. This problem is alleviated by summarizing PPI networks through contraction of nodes that correspond to ortholog proteins and using dedicated frequent subgraph mining algorithms on the resulting uniquely-labeled graphs [32]. Furthermore, by formulating multiple graph alignment problem as one of assigning nodes into equivalence classes, the complexity of the problem can be rendered linear in the number of networks [19].

6 Exercises

1. Find the core-clustering coefficient of the red vertices in each graph given in Figure 1. Compare this value to the clustering coefficient of those vertices. Which of these two metrics do you think is more appropriate to be used as a modularity measure of vertices in a PPI network?
2. Provide an example to illustrate that the density of a subgraph is not necessarily proportional to its edge connectivity. Namely, give two subgraphs, S_1 and S_2, such that $d_{S_1} < d_{S_2}$, while $c_{S_1} < c_{S_2}$, where d_S and c_S denote the density and edge connectivity of subgraph S, respectively.
3. Apply SIDES algorithm to partition the graph given in Figure 2. Is there a difference in the resulting partitions when using HCS or SIDES? If not, give an example graph for which these methods will partition differently.
4. Explain how the HCS algorithm guarantees that final partitions will at least be as half dense as a clique of the same size.

7 Further Reading

This chapter provides a broad overview of existing approaches to the problem of module identification in PPI networks. For details on any particular method, we recommend the reader to consult the corresponding references. [38] provides a comprehensive overview on characterization of biological processes through comparative analysis of biological networks. Furthermore, [41] provides a comprehensive review of protein function prediction through application of various module identification techniques and use of the identified modules to predict unknown functions of proteins. Finally, [11] is a review that compares various clustering algorithms on PPI networks in terms of robustness, sensitivity and performance.

References

1. Ahuja, R.K., Magnanti, T.L., Orlin, J.B.: Network Flows: Theory, Algorithms, and Applications. Prentice Hall (1993)
2. Altaf-Ul-Amin, M., Shinbo, Y., Mihara, K., Kurokawa, K., Kanaya, S.: Development and implementation of an algorithm for detection of protein complexes in large interaction networks. BMC Bioinformatics **7**, 207 (2006)
3. Arnau, V., Mars, S., Marin, I.: Iterative cluster analysis of protein interaction data. Bioinformatics **21**(3), 364+ (2005)
4. Ashburner, M., Ball, C.A., Blake, J.A., Botstein, D., Butler, H., Cherry, J.M., Davis, A.P., Dolinski, K., Dwight, S.S., Eppig, J.T., Harris, M.A., Hill, D.P., Issel-Tarver, L., Kasarskis, A., Lewis, S., Matese, J.C., Richardson, J.E., Ringwald, M., Rubin, G.M., Sherlock, G.: Gene ontology: tool for the unification of biology. the gene ontology consortium. Nature Genetics **25**(1), 25–29 (2000)
5. Asthana, S., King, O.D., Gibbons, F.D., Roth, F.P.: Methods predicting protein complex membership using probabilistic network reliability. Genome Res **14**, 1170–1175 (2004)
6. Auerbach, D., Thaminy, S., Hottiger, M., Stagljar, I.: The post-genomic era of interactive proteomics: Facts and perspectives. Proteomics **2**, 611–623 (2002)
7. Bader, G.D., Hogue, C.W.: Analyzing yeast protein-protein interaction data obtained from different sources. Nat Biotech **20**(10), 991–997 (2002)
8. Bader, G.D., Hogue, C.W.: An automated method for finding molecular complexes in large protein interaction networks. BMC Bioinformatics **4**(1) (2003)
9. Bader, J.S.: Greedily building protein networks with confidence. Bioinformatics **19**(15), 1869–1874 (2003)
10. Bauer, S., Grossmann, S., Vingron, M., Robinson, P.N.N.: Ontologizer 2.0 - a multifunctional tool for go term enrichment analysis and data exploration. Bioinformatics (Oxford, England) (2008)
11. Brohee, S., van Helden, J.: Evaluation of clustering algorithms for protein-protein interaction networks. BMC Bioinformatics **7**, 488+ (2006)
12. Cho, Y.R., Hwang, W., Zhang, A.: Identification of overlapping functional modules in protein interaction networks: Information flow-based approach. In: ICDMW '06: Proceedings of the Sixth IEEE International Conference on Data Mining - Workshops, pp. 147–152. IEEE Computer Society, Washington, DC, USA (2006)
13. Chua, H.N., Sung, W.K., Wong, L.: Using indirect protein interactions for the prediction of gene ontology functions. BMC Bioinformatics **8** (2007)
14. Database, S.G.: Saccharomyces genome database gene ontology termfinder. URL http://db.yeastgenome.org/cgi-bin/GO/goTermFinder

15. Deng, M., Sun, F., Chen, T.: Assessment of the reliability of protein-protein interactions and protein function prediction. Pac Symp Biocomput pp. 140–151 (2003)
16. Dutkowski, J., Tiuryn, J.: Identification of functional modules from conserved ancestral protein protein interactions. Bioinformatics **23**(13) (2007)
17. Edwards, A.M., Kus, B., Jansen, R., Greenbaum, D., Greenblatt, J., Gerstein, M.: Bridging structural biology and genomics: assessing protein interaction data with known complexes. Trends in Genetics **18**(10), 529–536 (2002)
18. Enright, A.J., Van Dongen, S., Ouzounis, C.A.: An efficient algorithm for large-scale detection of protein families. Nucleic Acids Res **30**(7), 1575–1584 (2002)
19. Flannick, J., Novak, A., Srinivasan, B.S., McAdams, H.H., Batzoglou, S.: Graemlin: general and robust alignment of multiple large interaction networks. Genome Res **16**(9), 1169–1181 (2006)
20. Hartuv, E., Shamir, R.: A clustering algorithm based on graph connectivity. Information Processing Letters **76**, 175–181 (2000)
21. Hartwell, L.H., Hopfield, J.J., Leibler, S., Murray, A.W.: From molecular to modular cell biology. Nature **402**(6761 Suppl) (1999)
22. Ideker, T., Ozier, O., Schwikowski, B., Siegel, A.F.: Discovering regulatory and signalling circuits in molecular interaction networks. Bioinformatics **18 Suppl 1** (2002)
23. Ito, T., Chiba, T., Ozawa, R., Yoshida, M., Hattori, M., Sakaki, Y.: A comprehensive two-hybrid analysis to explore the yeast protein interactome. Proc Natl Acad Sci U S A **98**(8), 4569–4574 (2001)
24. Jansen, R., Greenbaum, D., Gerstein, M.: Relating whole-genome expression data with protein-protein interactions. Genome Research **12**, 37–46 (2002)
25. Kalaev, M., Smoot, M., Ideker, T., Sharan, R.: Networkblast: Comparative analysis of protein networks. Bioinformatics (2008)
26. Kelley, B.P., Sharan, R., Karp, R.M., Sittler, T., Root, D.E., Stockwell, B.R., Ideker, T.: Conserved pathways within bacteria and yeast as revealed by global protein network alignment. Proc Natl Acad Sci U S A **100**(20), 11,394–11,399 (2003)
27. Kiel, C., Beltrao, P., Serrano, L.: Analyzing protein interaction networks using structural information. Annual Review of Biochemistry **77**(1) (2008)
28. King, A.D., Przulj, N., Jurisica, I.: Protein complex prediction via cost-based clustering. Bioinformatics **20**(17), 3013–3020 (2004)
29. Koyutürk, M., Grama, A., Szpankowski, W.: An efficient algorithm for detecting frequent subgraphs in biological networks. In: Bioinformatics, pp. 200–207 (2004)
30. Koyutürk, M., Grama, A., Szpankowski, W.: Pairwise local alignment of protein interaction networks guided by models of evolution. In: Lecture Notes in Bioinformatics 3500, pp. 48–65 (2005)
31. Koyutürk, M., Kim, Y., Subramaniam, S., Szpankowski, W., Grama, A.: Detecting conserved interaction patterns in biological networks. J Comput Biol **13**(7), 1299–1322 (2006)
32. Koyutürk, M., Kim, Y., Topkara, U., Subramaniam, S., Szpankowski, W., Grama, A.: Pairwise alignment of protein interaction networks. J Comput Biol **13**(2), 182–199 (2006)
33. Koyutürk, M., Szpankowski, W., Grama, A.: Assessing significance of connectivity and conservation in protein interaction networks. Journal of Computational Biology **14**(6), 747–764 (2007)
34. Maraziotis, I.A., Dimitrakopoulou, K., Bezerianos, A.: Growing functional modules from a seed protein via integration of protein interaction and gene expression data. BMC Bioinformatics **8**, 408 (2007)
35. Milo, R., Shen-Orr, S., Itzkovitz, S., Kashtan, N., Chklovskii, D., Alon, U.: Network motifs: simple building blocks of complex networks. Science **298**(5594), 824–827 (2002)
36. Pellegrini, M., Marcotte, E.M., Thompson, M.J., Eisenberg, D., Yeates, T.O.: Assigning protein functions by comparative genome analysis: protein phylogenetic profiles. Proc Natl Acad Sci U S A **96**(8), 4285–4288 (1999)
37. Ravasz, E., Somera, A.L., Mongru, D.A., Oltvai, Z.N., Barabási, A.L.: Hierarchical organization of modularity in metabolic networks. Science **297**(5586), 1551–5 (2002)

38. Sharan, R., Ideker, T.: Modeling cellular machinery through biological network comparison. Nature Biotechnology **24**(4), 427–433 (2006)
39. Sharan, R., Ideker, T., Kelley, B., Shamir, R., Karp, R.M.: Identification of protein complexes by comparative analysis of yeast and bacterial protein interaction data. J Comput Biol **12**(6), 835–846 (2005)
40. Sharan, R., Suthram, S., Kelley, R.M., Kuhn, T., McCuine, S., Uetz, P., Sittler, T., Karp, R.M., Ideker, T.: From the cover: Conserved patterns of protein interaction in multiple species. Proc Natl Acad Sci U S A **102**(6), 1974–1979 (2005)
41. Sharan, R., Ulitsky, I., Shamir, R.: Network-based prediction of protein function. Mol Syst Biol **3** (2007)
42. Slonim, D.K.: From patterns to pathways: gene expression data analysis comes of age. Nat Genet **32 Suppl**, 502–508 (2002)
43. Spirin, V., Mirny, L.A.: Protein complexes and functional modules in molecular networks. Proc Natl Acad Sci U S A **100**(21), 12,123–12,128 (2003)
44. Stojmirović, A., Yu, Y.K.: Information flow in interaction networks. J Comput Biol **14**(8), 1115–1143 (2007)
45. Tetko, I.V., Facius, A., Ruepp, A., Mewes, H.W.: Super paramagnetic clustering of protein sequences. BMC Bioinformatics **6**, 82 (2005)
46. Tornow, S., Mewes, H.W.: Functional modules by relating protein interaction networks and gene expression. Nucleic Acids Res **31**, 6283–6289 (2003)
47. Ucar, D., Asur, S., Catalyurek, U.V., Parthasarathy, S.: Improving functional modularity in protein-protein interactions graphs using hub-induced subgraphs. Knowledge Discovery in Databases: PKDD 2006 pp. 371–382 (2006)
48. Watts, D.J., Strogatz, S.H.: Collective dynamics of 'small-world' networks. Nature **393**(6684), 440–442 (1998)

Part V
Biological Data Management and Mining

This final part of the book covers miscellaneous topics pertaining to biological data management and mining.

Designing Microarray Experiments

Paola Sebastiani, Jacqui Milton, and Ling Wang

Abstract Gene expression microarrays have become an important exploratory tool in many screening experiments that aim to discover the genes that change expression in two or more biological conditions and can be used to build molecular profiles for both diagnostic and prognostic use. The still very high costs of microarrays and the difficulty in generating the biological samples are critical issues of microarray-based screening experiments, and the experimental design plays a crucial role in how informative an experiment is going to be. In this chapter, we describe some of the major issues related to the design of either randomized control trials or observational studies and discuss the choice of powerful sample sizes, the selection of informative experimental conditions, and experimental strategies that can minimize confounding. We conclude with a discussion of some of the open problems in the design and analysis of microarray experiments that need further research.

1 Designed experiments versus observational studies

Since the work of Fisher in the 1920's, randomization has been viewed as a necessary ingredient of experimental design, and the randomized control trial has been considered the ideal experimental framework that can lead to causal inference [2]. By randomization, that is, the random allocation of experimental subjects to any of the "controlled" biological conditions that need to be tested, the investigator can remove the effect of those exogenous variables that are not explicitly included in the experiment and can therefore ensure that only the differences in biological condi-

Paola Sebastiani
Boston University, Boston MA e-mail: sebas@bu.edu

Jacqui Milton
Boston University, Boston MA, e-mail: jnmilton@gmail.com

Ling Wang
Novartis Vaccines and Diagnostics, Emeryville CA, e-mail: ling-1.wang@novartis.com

L.S. Heath and N. Ramakrishnan (eds.), *Problem Solving Handbook in Computational Biology and Bioinformatics*, DOI 10.1007/978-0-387-09760-2_13,

tions are responsible for changes in the outcome. Randomization is often unusable in studies with human subjects because of ethical reasons or it is simply impracticable, and observational studies in which the experimenter has no control over the assignment of subjects to treatments are to be used instead [21].

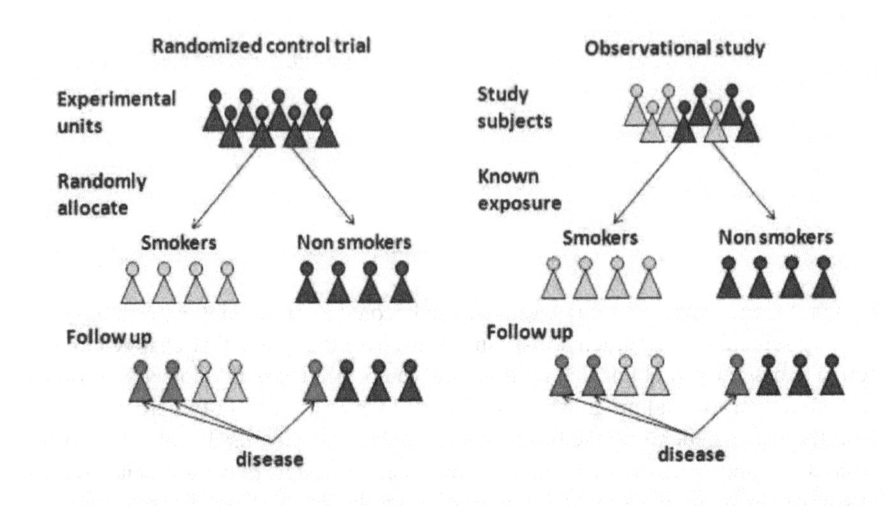

Fig. 1 *Randomized control trials versus observational studies. The panel on the left illustrates a randomized control trial in which experimental subjects are randomly allocated to one of two exposure groups and then followed. At the end of the experiment, the number of diseased subjects in each of the two groups can be attributed to the exposure because the randomization makes the effect of confounders unlikely. The panel on the right illustrates a prospective study in which the study subjects have known exposure at the time of entering the study and, after the follow up, the number of diseased subjects can be used to estimate the effect of the exposure on the disease risk. The lack of randomization, however, does not exclude that other unmeasured variables may be associated with the exposure and contribute to the disease risk. For example, smoking in the past was more frequent in male subjects and an observational study that does not control for gender may lead to confounding between smoking and gender predisposition on the disease outcome.*

As an example, suppose that an investigator wishes to discover the molecular profile of a certain type of human cancer after 20 years of exposure to active smoking. If the study design is a randomized control trial, the investigator would randomly allocate experimental subjects to one of the two groups — smoker or non smoker — and after 20 years exposure collect biological samples for the expression profiling experiment. Because the harm of smoking is well established, the experiment would not be considered ethical. An alternative solution is to use data from an observational study in which subjects either exposed to smoking or not can be followed prospectively to discover those who develop cancer in either group, and biological specimens can be collected and analyzed. See Figure 1 for a schematic.

Cohort studies that have been conducted for many years, such as the Framingham Heart Study [35], are the best approximation to a randomized control trial if the risk factors are carefully measured when subjects enter the study and are longitudinally monitored. The study design allows for estimation of the disease risk as a function of the exposure but precludes causal inference because of the lack of randomization [21]. Prospective studies however are expensive and time consuming, and a cheaper alternative is to use a case control study in which experimental subjects are chosen based on disease status (either 'case' if affected or 'control' if disease free) and the exposure to risk factors is observed retrospectively and is therefore the random outcome of the study [37]. An even simpler solution is a retrospective study in which only the total number of subjects is controlled and both disease status and exposure are random outcomes. Neither a case/control design nor a retrospective design can protect from confounding — that is the indistinguishable effect of two variables on the outcome of a study — and may lead to spurious associations but techniques such as matching can help reduce these defects [21]. Matching cases and controls by the levels of possible confounders should remove sources of confounding. For example, if gender is a possible confounder of the effect of smoking on cancer because of gender-specific detoxifying genes, then matching cases and disease free controls by gender will remove the confounding effect of gender [21]. Matching however restricts the generalizabilty of the results, so some care must be taken to avoid over-matching [8].

To design either an experiment or an observational study investigators need to answer some important questions. First, it is important to list carefully the hypotheses that one wishes to test, and have a clear analysis plan to ensure that the study design will be informative and allow for the planned analysis. For example, if the effect of two or more experimental factors or covariates are to be tested, it is important to design the study to make sure that all the effects of interest are estimable. The analysis plan may consist of using multi-way ANOVA to estimate the main effects as well as interaction effects of the factors on the outcome, and the estimable interaction effects are determined by the number of combined factor levels that are tested. The choice of experimental groups is also necessary in the design of an observational study, to make sure that the effects of covariates are not confounded with each other. The main intuition here is to allow for sufficient variation of either experimental factor levels or groups of exposure to make sure that their combined joint effects are observed: for example the experimental design that changes one-factor-at-a-time allows only for the estimation of main effects, whereas a factorial design will usually allow for the estimation of important, if not all, interaction terms. Figure 2 shows some examples. There is substantial literature about the design of factorial experiments that can be used to choose the optimal combination of experimental conditions subject to a fixed number of experimental runs and observations, and we refer to [6] for a comprehensive description of the topic.

In the remainder of this chapter, we will look at the design of microarray experiments that have been used to discovery genes that have differential expression across two or more biological conditions and to build models for molecular classification. We will also examine some of the issues related to the execution of microarray experiments that can introduce systematic bias.

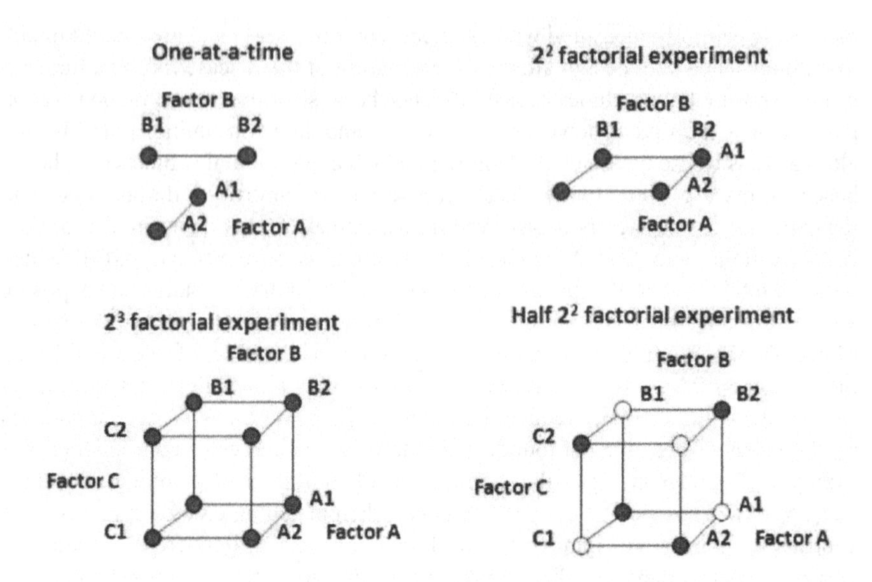

Fig. 2 *Example of different experimental designs. In each graph, each node represents an experimental run in which a combination of factor levels is tested. The one-at-a-time design (top left) examines the output of 4 different runs: in the first two runs the factor A levels (A1 and A2) are changed while factor B is not controlled. In the 3rd and 4th runs, factor B levels are changed while factor A is not controlled. This design provides no information about the interaction of the two factors. The* 2^2 *factorial experiments (top right) also uses 4 runs, however, each run tests one of the possible 4 combinations of the two level factors, (A1,B1), (A1,B2), (A2,B1), (A2,B2), and the design allows the estimation of the main effects of the two factors and their interaction. The* 2^3 *factorial experiments (bottom left) uses 8 runs to test the effect of every possible combination of factor levels on the output. This design allows the estimation of the 3 main effects, the 3 2-way interaction effects, as well as the full 3-way interaction effect of the three factors. This is a fully informative experiment that becomes quickly impracticable when the number of factors to be tested increases. There are several methods to choose fractions of the full factorial design that require a smaller number of experimental runs and will allow estimation of the effects of interest. Examples are orthogonal fractions of full factorial experiments. The schematic in the bottom right picture shows an example.*

2 Discovery of Differentially Expressed Genes

A typical microarray experiment produces the expression level of thousands of genes in two or more biological conditions with the goal to identify those genes that change expression level between two or more conditions. The collection of gene expression levels that are measured in one biological condition is called the expression profile and this is often measured multiple times because of technical and biological variability that are difficult to control. Repeated measurements of the same biological condition may be *pure replications* when the target hybridized to the microarrays is made of mRNA extracted from the same cells or *biological replications* when the target hybridized to the microarrays is made of mRNA extracted from different cells. In a cancer genomics experiment with cells extracted from human subjects, for example, the replications are often biological because of the substantial variability between subjects. In this section, we will mainly focus on biological replications and will return to the issue of pure replications in Section 4.

We will use the notation y_{kji} to describe a normalized gene expression level, where the index k specifies gene k in the microarray, $k = 1, \ldots, p$, the index i denotes sample i measured in condition j, and the index j represents one of c biological conditions. We denote by n_j the number of samples measured in condition j, so that $i = 1, \ldots, n_j$ and $\sum_j n_j = n$ is the overall sample size. We term the collection of expression levels measured for a gene across different conditions its *expression profile*, and we use the term *sample molecular profile* (or simply sample) to denote the expression level of all the genes measured with one microarray in a particular condition. Formally, the expression profile of a gene k in condition j will be described by the vector $y_{kj} = \{y_{kj1}, \ldots, y_{kjn_j}\}$, the overall expression profile of the same gene across all conditions will be the vector $y_k = \{y_{k11}, \ldots, y_{k1n_1}, \ldots, y_{kc1} y_{kcn_c}\}$, and sample i profile of condition j will be the vector $y_{ji} = \{y_{1ji}, \ldots, y_{pji}\}$.

We suppose the aim of the experiment is to discover the genes with differential expression in two or more conditions, so that the main design questions are (1) the choice of the biological conditions to be tested and (2) the number of samples to be measured per condition. Because the cost of a microarray experiment is still substantial, often the investigator needs to trade off the number of conditions c and the number of samples per condition n_j, subject to the constraint that the overall sample size n is fixed to be small. Therefore, the choice of c and n cannot be disentangled. To begin with, let us assume that there are only two biological conditions of interest, so that the problem is to choose the number of observations n_j per condition. We use this assumption to show how statisticians determine the minimum number of samples n_j that are necessary to achieve specific properties and use the results to suggest experimental designs for more complex studies in which also the number of conditions is to be chosen.

The common solution is to frame the search for differentially expressed genes as a series of independent hypothesis testing problems: for each gene the null hypothesis that the gene has the same expression level in the two conditions is tested against the alternative hypothesis that the gene changes expression level from one condition to the other. The frequentist approach to hypothesis testing weights the evidence against the null hypothesis by the p-value that, in this particular context, would be defined as the probability of observing a change of expression at least as extreme as that seen in the data under the null hypothesis of no differential expression. Although not impossible, this event is unlikely to happen if the null hypothesis is true and so a small p-value is taken as evidence that the null hypothesis is false and the alternative hypothesis is true. The probability to reject the null hypothesis when true is known as the type I error while the statistical power is the probability to accept the alternative hypothesis when true. The sample size is usually chosen to guarantee a certain level of power of the hypothesis testing procedure when the probability of the type I error is bounded to 5% or less. This upper bound on the p-value is called the significance level of the test.

It is important to consider that statistical power is a property of the analysis method and not only of the data. Therefore, different analysis strategies will require different sample sizes. One simple rule of thumb is that the more naive the statistical analysis, the larger the sample size required to achieve a desirable power. Often power calculations are based on the normal approximation of the t-statistic that is commonly used for identifying the genes with different expression in two biological conditions $j = 1, 2$:

$$t_k = \frac{\bar{y}_{k1} - \bar{y}_{k2}}{SE(\bar{y}_{k1} - \bar{y}_{k2})},$$

where $\bar{y}_{kj}$ is the mean expression level of gene k in condition j, and the standard error of the sample mean difference, $SE(\bar{y}_{k1} - \bar{y}_{k2})$, is computed assuming different variances in the two conditions. Typically the common variance is unknown, but the choice of sample size is based on a formula that assumes known variance [44]. The formula shows that the overall sample size that is necessary to achieve a statistical power $(1 - \beta)$ when the objective is to detect a difference δ between the expression level of a gene in two conditions is

$$n = 4\frac{\sigma(z_{\alpha/2} + z_\beta)^2}{\delta^2}.$$

In the formula, $z_{\alpha/2}$ and z_β are the $\alpha/2$ and β quantiles of the standard normal distribution — say Z — so that $P(Z \leq z_{\alpha/2}) = \alpha/2$ and $P(Z \leq z_\beta) = \beta$. The standard deviation in both conditions is known to be approximately σ, and the significance level of the test is α. The overall sample size is to be equally split between the conditions. Assuming the underlying conditions are true, this formula guarantees a power $(1 - \beta)$ to detect a difference in expression for one particular gene, and its derivation can be found for example in [44]. To apply this formula to microarray experiments, the investigator needs to choose an average value of δ that adequately describes changes in expression that have to be estimated across all genes and the same variability in gene expression data. It is convenient to work in a log base 2, so that a one-unit change in the log intensity corresponds to a two-fold change. For example, if $\sigma = 0.5, \delta = 2$ and $\alpha = 0.0005$, then we need approximately 20 samples, 10 for each condition, to achieve 80% power ($\beta = 0.20$). The closed-form formula makes the sample size estimation a very easy task. Nevertheless, there are several tools on the web that can be used for this calculation and that help to correct the significance level α for multiple comparisons using simple Bonferroni corrections (see for example http://bioinformatics.mdanderson.org/MicroarraySampleSize). More sophisticated approaches are based on false discovery rate rather than false positive rate [31].

The Biometric Research Branch at the National Cancer Institute (NCI) has published a series of methods for sample size estimation for high dimensional microarray studies that extend this approach to include multifactorial experiments [12]. Although these sample size estimations make simplistic assumptions that can impact sensitivity and specificity and the resulting sample size tends to be more conservative, they are helpful to suggest reasonable designs of microarray experiments.

A serious limitation of the approach based on the standard t-test or the general linear models is that the gene expression data have a type of variability that may make the assumption of normality inadequate even after a log-transformation of the data [38]. To address this issue at least in the simple case of two sample comparison, authors have suggested some forms of penalization for the denominator of the t-statistic. For example, Golub et al. [18] suggested replacing the standard error $SE(\bar{y}_{k1} - \bar{y}_{k2})$ by the quantity

$$s_{S2Nk} = \frac{s_{k1}}{\sqrt{n_1}} + \frac{s_{k2}}{\sqrt{n_2}},$$

where s_{kj} is the sample standard deviation of the gene expression in condition j. The ratio $|\bar{y}_{k1} - \bar{y}_{k2}|/s_{S2Nk}$ is termed the *signal-to-noise ratio* and, although lacking a theoretical derivation, it performs very well and has been used in a variety of applications. Other forms of penalization are justified by the fact that the standard error may be very small for genes with small expression values, thus inflating the value of the t-statistic. Based on this intuition, Tusher et al. [47] suggested to adjust the standard error by $a + SE(\bar{y}_{k1} - \bar{y}_{k2})$ where the constant a — called the fudge factor — is chosen to minimize the coefficient of variation of the t-statistic of all the genes. Efron [16] suggested to replace a by the 90th percentile of the standard error of all the genes.

These variants of the t-test statistics have the limitation that their statistical distribution no longer follows a Student's t distribution and the choice of the threshold to select the genes with a statistically significant change of expression needs to be based on either asymptotic approximations or permutation tests. The main idea of permutation tests is to compute the value of a statistic from the data in which the sample labels that represent the conditions are randomly reshuffled. By repeating this process a large number of times, it is possible to construct the empirical distribution of the statistic under the null hypothesis of no differential expression. From this distribution one can select a gene specific threshold to reject the null hypothesis with a particular significance. Authors have also developed algorithms for multiple comparison adjusted p-values [15].

Distribution free methods tend to be widely used in practice, but they often require a large sample size to detect the genes with different expression and a small false positive rate [56]. Some authors have suggested making distribution assumptions on the gene expression data, and the most popular choice is to assume that gene expression data follow a lognormal distribution [3, 20]. Another stream of work focuses on the estimation of the fold change of expression, that is, the ratio of the sample means assuming Gamma distribution for the gene expression data [10, 29]. We have shown that these distributional assumptions are often inadequate to describe gene expression data and proposed a Bayesian procedure based on model averaging that limits the false positive rate without reducing the power. Briefly, the method analyzes the differential expression by assuming both lognormal and gamma distributions, and then the results are averaged using weights that depend on the probability of each of the two models given the data. The method is implemented in the program BADGE and it is described in [39] and [40]. We showed that with a sample size of approximately 15 samples per conditions the method has more than 85% power to detect a wide range of fold changes with a negligible false positive rate. Figure 3 shows some empirical power curves.

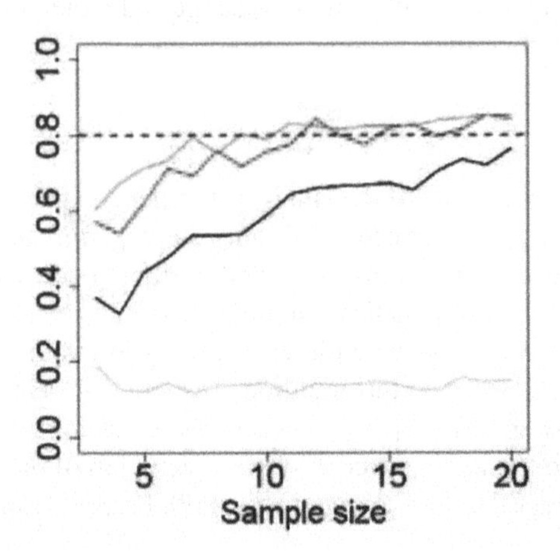

Fig. 3 *Power of BADGE for different sample sizes. The green line shows the approximate power of BADGE to detect genes that change expression by at least 2.5 fold as a function of the sample size. The red and black lines show the power curve to detect genes that change expression by at least 2 fold (red) and 1.5 fold (black). The flat line in gray is the estimated False Discovery Rate of BADGE. In all simulations, we use 0.95 as threshold on the posterior probability of a fold change of expression exceeding 1, to detect genes with differential expression. The power curves show that to achieve a power of 0.8 to detect minimally a two-fold difference in expression patterns of genes with a false positive rate below 1%, BADGE needs about 15 samples. The same sample size provides a power of 0.9 to detect a 2.5 fold difference. The power curves were generated by simulating 1,000 gene expression profiles with a fold change of at least 1.5, using a variety of parameter values generated from distributions that were built using results of previous experiments. For each set of parameters we computed the probability of differential expression and then we estimated the power by the empirical true detection rate.*

Fifteen samples per condition seems to be the minimum requirement for powerful microarray experiments, and a good heuristic to design informative microarray experiments is to limit the number of biological conditions to $n/15$ if n is the overall sample size. When this number of conditions is smaller than the overall number of combinations of factor levels, orthogonal fractions of full factorial experiments can be used to select a subset of design trials to run.

The situation is more complicated when the experiment aims at discovering genes with specific temporal expression profiles. Common temporal gene expression studies consist of a number of unevenly spaced observations that are often replicated to average out biological and sample variability. Due to the high cost of microarray experiments, replication of the same time point experiments is often chosen at the price of a reduced sampling frequency. The sampling frequency is important to capture the essential dynamics of the data generation process and increases the power to identify biologically interesting genes. We compared two different sampling schemes to uncover genes with similar expression profiles in model selection in [53]. Both schemes have the same number of observations, but one scheme consists of a number of time points without replication, and the other scheme uses replications of a subsets of these time points. The results from our simulation studies suggest that the replicated design increases the power of several model selection criteria to identify true underlying gene expression patterns such as linear, quadratic, and cubic compared to non-replicated gene expression design. The gain of power is substantial when the patterns are less distinct. However, the performance of non-replicated design is comparable to or better than the replicated designs when the underlying patterns are moderately or highly distinct. The results of this investigation were based on the assumption that the overall number of microarray experiments is fixed in advance, and the task is to decide how to trade off sampling frequency with replications. Other investigators have proposed sequential strategies to the design of temporal experiments when the objective of the analysis is to identify subsets of genes with similar temporal profiles using hierarchical clustering [51]. The design of microarray experiments in which the temporal expression profiles are observed in several biological conditions is an open question [50].

3 Building prognostic models

Gene expression data have been used very successfully to define new disease subclasses [18] and to build models for molecular classification of tumor types and prediction of disease outcome [48, 49]. A success story is *Mammaprint*, a prognostic product for predicting breast cancer recurrence developed in the Netherlands (http://usa.agendia.com/en/mammaprint.html). *Mammaprint* uses microarray technology to compute the risk of breast cancer recurrence by measuring a signature based on the expression level of 70 genes. The tool was developed through a series of studies for identification, classification, and validation of gene expressions molecular profiles [48, 49].

The planning and design of such studies is quite complex and requires careful considerations of many different aspects. The most important steps in building prognostic models are the development and validation of classifiers based on gene expression levels. Typically, such classifiers group clinical samples/subjects into several subgroups based on a probabilistic measure and the grouping is used to define a classification rule that maps the expression profile to the outcome. Then the performance of the classification rule needs to be carefully validated, preferably on another independent population. In designing such a study, sample size estimation is necessary to guarantee a sufficient prediction accuracy [42]. Mukherjee et al. [26] developed a procedure based on a learning curve to estimate the sample size needed to build a molecular classification model. Their approach was based on the idea of using cross validation to assess the classification accuracy and permutation tests to estimate the significance of the classifier. Dobbin and Simon provide formulas for sample size when the objective is to identify molecular profiles that distinguish two biological conditions [11]. These formulas essentially use the same sample size calculations given earlier. Here, we will focus on reviewing the sample size estimation when the objective is class prediction in a follow-up study. Let γ_k be the standard deviation for the $\log_2$ expression data for gene k, and let h be the hazard ratio associated with a one-unit change of the log intensity, the number of samples required is given by

$$n = \frac{(z_{\alpha/2} + z_{\beta})^2}{(\gamma_k \ln[h])^2}$$

Since the log intensities are in log base 2, a one-unit change in the log ratio or log intensity corresponds to a two-fold change. For example, if $\gamma_k = 0.5, h = 2, \alpha = 0.0005, \beta = 0.05$ then we need 219 samples.

When building a good classifier that can predict the disease status using the expression profile of a set of genes, usually the number of genes is small compared to the tens of thousands of genes being assayed. For this purpose, it is a good strategy to divide the task into a two-step process: finding the small number of genes to use in the classifier and building the prognostic model. Dobbin and Simon [12] explain that, in this two-step process, one needs to estimate the sample size needed to ensure a high expected probability of correct classification instead of a high power. They proposed an algorithm for calculating sample size in order to optimize the probability of correct classification for the classifier, and offered a web-based tool for such computation. This method considers the scenario of having two disease states, and uses three input parameters: standardized fold change, total number of genes assayed and disease prevalence in the population. Figure 4 shows a snapshot of the web page. Dobbin et al. [13] illustrates this method in a more applied, clinical context.

After a classifier (prognostic model) is developed, its performance should be evaluated. This evaluation should avoid using the same data that were used to develop the classifier. One can use cross-validation or split sample methods, but ideally a data set obtained from an independent population should be used to demonstrate that the classifier can deliver a good discrimination.

While the class predication ability of microarray data leads to the hope of better prognosis and "individualized medicine", there are still several hurdles to overcome. For instance, the planning and execution of the development of such prognostic tools is very complex, and there are many pitfalls that one needs to be cautious about [43]. Another aspect lies in the microarray technology itself. The high-throughput capability is coupled with the loss of sensitivity, compared to PCR technology. Finally, many other aspects, including tissue biopsy handling, can complicate the application in clinical use. This is one of the reasons that although the prognostic tool *Mammaprint* has been approved by the US Food and Drug Agency (FDA), it has not yet received endorsement from the American Society of Clinical Oncology [19].

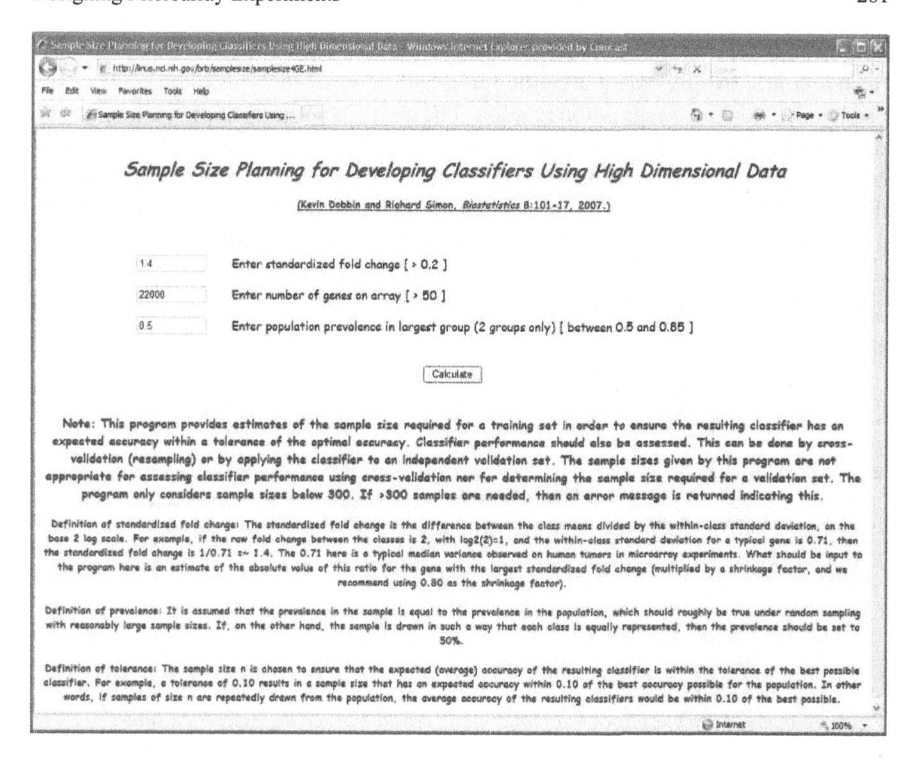

Fig. 4 *Screen shot of the NCI web page that offers sample size calculation for developing classifiers from microarray data.*

4 Running the experiment

In the previous sections, we examined issues related to the design of microarray experiments and focused attention on the choice of the conditions to be tested and the number of microarray experiments to be run for each trial. There are other sources of systematic variation that can affect microarray experiments and are related to variability of the technology, for example print tip effects, spatial patterns on the arrays, and known bias due to the fluorescent dye used to label the target mRNA that is hybridized to the array [38]. The process of normalization is used to remove some of this variation by forcing the distribution of the data to have the same median [54]. Normalization is conducted by fitting polynomial or loess regression and can introduce other sources of variation. Thus it is important that the systematic variation removed by normalization is greater than variation introduced by the normalization process. Popular methods include linear normalization, nonlinear regression, cyclic loess, contrast based method, and quantile normalization and were evaluated in [5] and [36]. Linear normalization assumes that the intensities between two or more arrays are linearly related so that the multiplication of a scaling factor is used to make the mean expression of all arrays the same. If the relationship between the arrays is nonlinear, other regression models can be used to map each array to a baseline array. Schadt et al. [36] illustrated a method in which a set of rank invariant probes was used to fit a nonlinear regression relation. The quantile normalization method makes the distribution of probe intensities for each array the same. The general principle behind this method is that a quantile-quantile plot that shows the distribution of two different probe intensities with the same distribution will be a straight diagonal line. Thus, each probe intensity distribution is transformed to be aligned to the baseline.

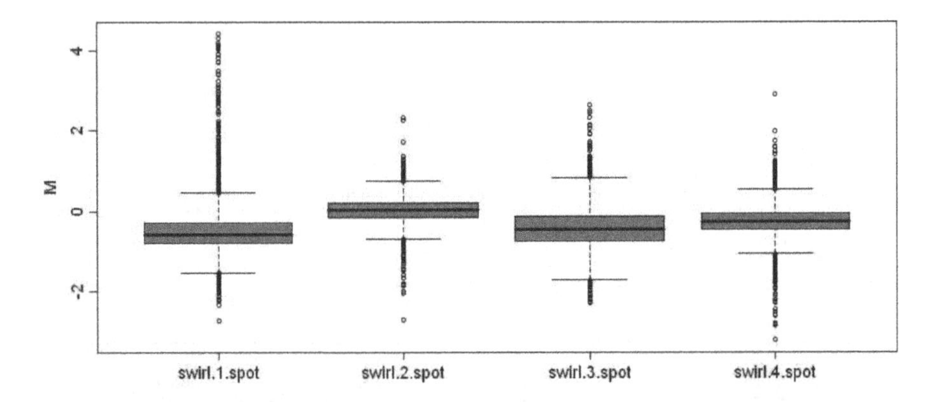

Fig. 5 *Example of dye bias. Each box plot displays the normalized log-expression values of 8,448 cDNA probes processed with a two-channel cDNA (data from the package Bioconductor). The experiment was carried out using zebra fish to study the effect of mutations in the BMP2 gene on early development in vertebrates. Swirl is a point mutant in the BMP2 gene that affects the dorsal/ventral body axis and the goal of the experiment was to identify genes with altered expression in the swirl mutant compared to wild-types. Two sets of dye-swap experiments were performed, for a total of four replicate hybridizations. In experiments 1 and 3 (labeled swirl.1.spot and swirl.3.spot) the target cDNA from the swirl mutant was labeled using the Cy3 dye while the target cDNA wild-type mutant was labeled using the Cy5 dye. In experiments 2 and 4 (labeled swirl.2.spot and swirl.4.spot) the dye assignment was reversed. The box plots show that, in the experiments 1 and 3, the expression of the control mRNA is less intense than the expression of the target mRNA, because the median expression values are negative. On the other hand, the expression of the control mRNA in the experiments 2 and 4 are more intense than the expression of the target mRNA as shown by the positive median values.*

Normalization was originally introduced to correct the dye bias of cDNA arrays in which the target and control mRNA are labeled with the Cy3 and Cy5 dyes that are known to be one more intense than the other [54]. Dye bias can easily be seen when two identical mRNA samples are labeled with different dyes and hybridized to the array [32], and Figure 5 shows an example. It has been argued that dye bias is a negligible part of the variation [25]. However, if inferences are made for individual genes, this can easily become a substantial source of bias. One can either try to avoid dye bias during the experiment or the analysis phase of the study. Normalization techniques can be introduced to remove bias during the analysis phase. A better solution is to remove bias with the experimental design and the dye swap experiment has been proposed to address this issue [55].

In a dye swap experiment, each hybridization is performed twice with the dye assignment reversed in the second experiment. This is also known as a "double reference" design which is highly effective but very expensive. The rationale of this design is that by replicating each hybridization twice, with a reversed assignment of the dye, the systematic bias cancels out and the analyst has to consider only random sources of variability rather than systematic sources of error. This effect is clearly shown in Figure 5, where the replication of the hybridizations 1 and 2 (labeled swirl.1.spot and swirl.2.spot) should remove the effect of bias. To handle dye bias during the analysis phase of the study, the following model was proposed

$$y_{kjid} = \mu + A_i + D_d + V_j + \varepsilon_{ki} + \rho_{kjid}$$

where y_{kjid} is the log ratio of the target to control intensity ratio for the ith replicate in which the target is labeled with the dye d, A_i is the array effects for the ith array, D_d is the dye effect for $d = 1,2$, V_j is the effect of condition j, ε_{ki} is the random effect of the kth gene, and ρ_{kjid} is the error term. This model was described in [23] to analyze the data correctly when they are generated using a dye swap experiment.

In addition to normalization and dye swap experiments, there are other techniques that can be used to minimize the systematic variation that results from performing a microarray experiment. Experience shows that microarray data are prone to influence by technical artifacts so randomization is important to avoid bias [23]. Arrays should be randomly chosen for each planned hybridization in case there is a systematic variation in the order in which the arrays were printed. With two-channel arrays, if one is not performing a dye swap experiment, randomization of dye assignments should be used to control for dye bias. Another important issue is whether microarray experiments should be replicated. We briefly touched this topic in Section 2 and in the context of designing temporal experiments and now we consider it more in details. It has been argued that replication is essential to estimate technical variability [25], and there are several ways in which this can be done. For example, replication can be performed by using more than one spot of the same cDNA probe on each slide. However, when creating these replicates it is important to note that adjacent spots can never be considered full replicates. The printing, hybridization, and scanning conditions will be shared by adjacent spots which creates a lack of independence. Duplicate spots should be well spaced apart. Alternatively, one can replicate the whole array hybridization and it has been estimated that three replicates of each hybridization should be conducted [25].

We wish to emphasize that technical replicates are different from biological replicates that involve hybridizations with mRNA from different biological samples. Biological replicates allow for estimation of between subjects variability, while technical replicates can be used to estimate the "pure error" that is due to the measurement instrument. Historically, this difference goes back to the theory of factorial design and the emphasis put on replication of a full factorial design to gain degrees of freedom to estimate the pure error, versus the use of different design points that can be used to estimate the lack of fit of a statistical model [6]. Consider an experiment with output y_{ji} at replication i of condition j, and suppose that a model is fitted to this data (for example a linear regression model) and denote by $\widehat{y}_{ji}$ the value estimated for y_{ji}. The distinction between pure error and lack of fit is derived from the decomposition of the residual sum of squares:

$$RSS = \sum_{ji}(y_{ji} - \widehat{y}_{ji})^2$$

that is an overall distance between observed and fitted values. The *RSS* can be decomposed into the sum of

$$\text{Pure error} = \sum_{ji}(y_{ji} - \bar{y}_{j*})^2$$

and

$$\text{Lack of Fit} = \sum_{ji}(\widehat{y}_{ji} - \bar{y}_{j*})^2$$

where $\bar{y}_{j*}$ is the average of observations in condition j. The pure error term can only be estimated when observations are replicated and provides the correct estimate of the error variance. The lack of fit term describes the departure of the fitted model from the means per condition that represent the best fit [6]. In principle, replication is needed only to estimate pure error, and therefore given the high cost of microarray experiments, it seems unnecessary to replicate each measurement three times as suggested for example in [25], and only replication of a few "crucial" experimental trials should be performed. For example, in the context of temporal experiments, replication of the baseline observations seems to be crucial to obtain good estimates of the reference values that will be used for normalization of the observations. The identification of a few crucial points to replicate during the course of microarray experiments is an important issue that needs further investigations. Work done in the area of response surface, see for example [7], could suggest valuable and more affordable design of microarray experiments.

There are practical considerations that often limit the number of samples that can be amplified and hybridized at the same time, including the amount of mRNA to be extracted and pre-processed, costs, and time. Thus at times it may be necessary to perform runs of a microarray experiment over the course of several days or months. This approach can introduce a form of systematic bias known as batch effects [22]. For each created batch, different conditions including the reagents used or the time of day could have an effect on the results making different batches not directly comparable [14]. Often, scientists will want to combine these batches to increase the sample size and consequently increase the power of their study. However, it is inappropriate to combine data sets without first adjusting for batch effects. Several methods have been suggested to adjust for batch effects. The best method to use depends on the researcher's particular study design. Benito et al. [4] and Alter et al. [1] have each proposed methods to adjust for batch effects, but they require many samples (>25) and may remove real biological variation. Johnson et al. [22] proposed an Empirical Bayes method that is robust for adjusting for batch effects in data with small batch sizes. The Empirical Bayes method borrows information across genes and experimental conditions to calculate better estimates and create more stable inferences.

Batch effects can also be controlled through a randomized block design. Arrays can be considered experimental blocks with two dye colors used to compare two categories of a factor. Blocking is the arranging of experimental units (in this case, arrays) into groups (blocks) that are similar to one another. Blocking can be used to reduce or eliminate the contribution to experimental error. The objective is to create homogeneous blocks and allow the factor of interest to vary. Thus by blocking, we create batches that are homogeneous and eliminate or minimizing batch effects. See Figure 6 for an example.

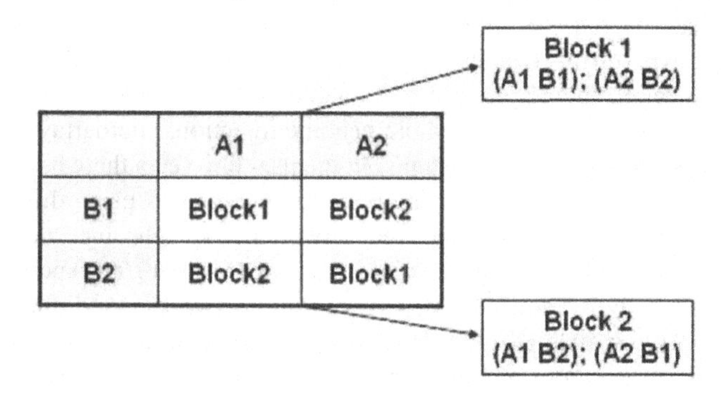

Fig. 6 *Example of a randomized block design with two experimental factors, with levels $A1, A2$ and $B1, B2$. The four design trials are allocated into two blocks in such a way that none of the two factors is confounded with the block allocation. In this case, the randomized block design is based on a Latin square.*

5 Advanced topics and further reading

With the ability to observe genome-wide expression in a target cell, microarray technology has opened new research avenues and the opportunities for ambitious experiments never thought before. When the first microarray experiments made their appearance in the scientific world, Eric Lander stated the following:

> *The challenge is no longer in the expression arrays themselves, but in developing experimental designs to exploit the full power of a global perspective. The issues are both technical and conceptual.* [24]

After almost 10 years of microarray experiments, this is still very much the case, and well designed experiments continue to be very important, because the costs of microarray experiments are prohibitive for many laboratories. Optimal experimental design should play a crucial role, and there are many areas of research that need substantial development [27]. Many microarray experiments are conducted to discover new disease taxonomies and use cluster analysis as the analytic tool for the discovery process. With the exception of a few manuscripts [52], there are no general guidelines about the optimal design of microarray experiments when the goal is cluster analysis for the discovery of either new sets of functionally similar genes or new disease taxonomies. In principle, one could use simulations to estimate the effect of different sample sizes on the power of different clustering algorithms to discover the true grouping of the data. We used a similar procedure to evaluate the accuracy of the "Bayesian clustering by dynamics" algorithm that we introduced in [33] to cluster time series. More extensive evaluations would provide valuable insight about the proper design of microarray experiments for class discovery. We conjecture that one can use prior information about the expected number of clusters to design experiments that allows for the discovery of reproducible results. The Bayesian framework should prove to be valuable to this endeavor.

By providing snapshots of the whole genome in action, microarray data can be used to try dissecting biological systems. In the past few years there have been several attempts to integrate data from microarray experiments using different types of network models, including scale-free networks [34], relevance networks [9], Boolean networks [41], and Bayesian networks [17]. Designing an experiment when the goal of the study is knowledge discovery using networks is a difficult problem and few results are available. Active learning has been suggested as a way to reduce the number of observations and experimental trials that are necessary to achieve a specific power to induce a Bayesian network from data [45]. Designs of informative temporal experiments to induce Dynamic Bayesian networks of gene expression data are presented in [30]. The review article by Needham et al. [28] contains some suggestive examples about the effect of sample size on the power to induce Bayesian networks from gene expression data. Trinh et al. [46] examine the effect of sample size on the power of standard algorithms for learning Bayesian networks classifiers with gene expression data. Their work suggests that at least 50 samples per condition are necessary to achieve sufficient accuracy. This is an important topic that needs further investigation and even simple exploratory studies based on well designed simulations could produce useful insights.

6 Exercise

1. A microarray experiment is conducted to identify genes with differential expression between endothelium cells of 30 healthy donors and 30 patients with sickle cell anemia. Comments on these four proposed designs:
   a) The 30 healthy donors comprise 20 males and 10 females. The sickle cell anemia patients comprise 10 males and 20 females.

b) The age range of the 30 healthy donors is 45 to 70 years. The age range of the sickle cell anemia patients is 15 to 50 years.
c) Both healthy donors and sickle cell anemia patients comprise 15 males and 15 females. Both the ages of sickle cell anemia patients and healthy donors range between 15 and 50 years.
d) The sickle cell anemia patients comprise 15 males and 15 females, with age ranging between 15 and 50 years. Each healthy donor was chosen to match gender and age of a sickle cell anemia patient.

2. Design a microarray experiment that can detect genes that change expression by more than 2 folds between two conditions, with at least 70% power and 5% significance in each test.
3. The family wise error rate is the probability of making one or more type I error when testing a set of N hypotheses. Compute the family wise error rate in testing 100 hypotheses, each with a significance level of 5%.
4. Compute the expected number of type I error in testing 100 hypotheses, when each hypothesis is tested with a significance level of 5%; and 1%.
5. The Bonferroni correction attempts to limit the number of type I errors in multiple testing by reducing the individual significance of each test so that the overall number of expected false positive associations is fixed. Compute the significance level of each individual test that is sufficient to guarantee that the family wise error rate is 5%.
6. Suppose you need to run an experiment comparing 8 microarray samples of normal lung cells and 8 microarray samples of lung cancer cells. Only 8 microarrays can be run in any given day, and there is a known day effect due to different technicians in charge of running the experiment. What is the best way to conduct the experiment that avoid confounding between biological condition and day effect?

Acknowledgements This work was supported by NHGRI R01 HG0033S4-01A2.

References

1. Alter, O., Brown, P.O., Botstein, D.: Singular value decomposition for genome-wide expression data processing and modeling. Proc Natl Acad Sci USA **97**, 10,101–10,106 (2000)
2. Armitage, P.: Fisher, Bradford Hill, and randomization. Int J Epidemiol **32**, 925–928 (2003)
3. Baldi, P., Long, A.D.: A Bayesian framework for the analysis of microarray expression data: Regularized t-test and statistical inferences of gene changes. Bioinformatics **17**, 509–519 (2001)
4. Benito, M., Parker, J., Du, Q., Wu, J., Xiang, D., Perou, C.M., Marron, J.S.: Adjustment of systematic microarray data biases. Bioinformatics **20**, 105–114 (2004)
5. Bolstad, B., Irizarry, R., Åstrand, M., Speed, T.: A comparison of normalization method for high density oligonucleotide array data based on variance and bias. Bioinformatics **19(2)**, 185–193 (2003)
6. Box, G.E.P., Hunter, W.G., Hunter, J.S.: Statistics for Experimenters: An Introduction to Design, Data Analysis, and Model Building. Wiley, New York, NY (1978)
7. Box, G.E.P., Wilson, K.B.: On the experimental attainment of optimum conditions (with discussion). J Roy Stat Soc B **13**, 1–45 (1951)

8. Breslow, N.E.: Statistics in epidemiology: The case-control study. J Am Stat Assoc **91**, 14–28 (1996)
9. Butte, A., Tamayo, P., Slonim, D., Golub, T., Kohane, I.: Discovering functional relationships between RNA expression and chemotherapeutic susceptibility using relevance networks. Proc Natl Acad Sci USA **97**, 12,182–12,186 (2000)
10. Chen, Y., Dougherty, E., Bittner, M.: Ratio-based decisions and the quantitative analysis of cDNA microarray images. J Biomed Opt **2**, 364–374 (1997)
11. Dobbin, K.K., Simon, R.M.: Sample size determination in microarray experiments for class comparison and prognostic classification. Biostatistics **6**(1), 27–38 (2005)
12. Dobbin, K.K., Simon, R.M.: Sample size planning for developing classifiers using high-dimensional DNA microarray data. Biostatistics **8**(1), 101–117 (2007)
13. Dobbin, K.K., Zhao, Y., Simon, R.M.: How large a training set is needed to develop a classifier for microarray data? Clin. Canc. Res. **14**(1), 108–114 (2008)
14. Draghici, S., Kuklin, A., Hoff, B., Shams, S.: Experimental design, analysis of variance and slide quality assessment in gene expression arrays. Cur. Opin. Drug Discov & Devel **4**, 332–337 (2001)
15. Dudoit, S., Yang, Y.H., Callow, M.J., Speed, T.P.: Statistical methods for identifying genes with differential expression in replicated cDNA microarrays experiments. Stat Sinica **12**, 111–139 (2001)
16. Efron, B., Storey, J.D., Tibshirani, R.: Empirical Bayes analysis of a microarray experiment. J Am Stat Assoc **96**, 1151–1160 (2001)
17. Friedman, N.: Inferring cellular networks using probabilistic graphical models. Science **303**, 799–805 (2004)
18. Golub, T.R., Slonim, D.K., Tamayo, P., Huard, C., Gaasenbeek, M., Mesirov, J.P., H. Coller, .M.L.L., Downing, J.R., Caligiuri, M.A., Bloomfield, C.D., Lander, E.S.: Molecular classification of cancer: Class discovery and class prediction by gene expression monitoring. Science **286**, 531–537 (1999)
19. Harris, L., Fritsche, H., R., M., Norton, L., Ravdin, P., Taube, S., Somerfield, M.R., Hayes, D.F., Bast Jr, R.C.: American society of clinical oncology 2007 update of recommendations for the use of tumor markers in breast cancer. J. Clin. Oncol. **25**(23), 5287–5312 (2007)
20. Ibrahim, J.G., Chen, M.H., Gray, R.J.: Bayesian models for gene expression with DNA microarray data. J Am Stat Assoc **97**, 88–99 (2002)
21. Jewell, N.P.: Statistics for Epidemiology. CRC/Chapman and Hall, Boca Raton (2003)
22. Johnson, W.E., Li, C., Rabinovic, A.: Adjusting batch effects in microarray expression data using empirical Bayes methods. Biostatistics **8**, 118–127 (2007)
23. Kerr, M.K.: Design considerations for efficient and effective microarray studies. Biostatistics **59**, 822–828 (2003)
24. Lander, E.S.: Array of hope. Nat Genet **21**, 3–4 (1999). Supplement
25. Lee, M.T., Kuo, F.C., Whitmorei, G.A., Sklar, J.: Importance of replication in microarray gene expression studies: Statistical methods and evidence from repetitive cDNA hybridizations. Proc Natl Acad Sci USA **18**, 9834–9839 (2000)
26. Mukherjee, S., Tamayo, P., Rogers, S., Rifkin, R., Engle, A., Campbell, C., Golub, T.R., Mesirov, J.P.: Estimating dataset size requirements for classifying DNA microarray data. J Comput Biol **10**, 119–142 (2003)
27. Naidoo, S., Denby, K.J., Berger, D.K.: Microarray experiments: Considerations for experimental design. S. Afr. J. Sci. **101**, 347–354 (2005)
28. Needham, C.J., Bradford, J.R., Bulpitt, A.J., Westhead, D.R.: A primer on learning in Bayesian networks for computational biology. PLoS Comput Biol **3**(8), e129 (2007)
29. Newton, M.A., Kendziorski, C.M., Richmond, C.S., Blattner, F.R., Tsui, K.W.: On differential variability of expression ratios: Improving statistical inference about gene expression changes from microarray data. J Comput Biol **8**(1), 37–52 (2001)
30. Page, D., Ong, I.M.: Experimental design of time series data for learning from dynamic Bayesian networks. In: Pacific Symposium on Biocomputing, pp. 267–278 (2006)
31. Pawitan, Y., Michiels, S., Koscielny, S., Gusnanto, A., Ploner, A.: False discovery rate, sensitivity and sample size for microarray studies. Bioinformatics **21(13)**, 3017–3024 (2005)

32. Quackenbush, J.: Microarray analysis and tumor classification. New Eng J Med **354**, 2463–72 (2006)
33. Ramoni, M., Sebastiani, P., Cohen, P.R.: Bayesian clustering by dynamics. Mach Learn **47**(1), 91–121 (2002)
34. Ravasz, E., Barabasi, A.L.: Hierarchical organization in complex networks. Phys Rev pp. E 67, 026,112 (2003)
35. R.Carandang, Wolf, P.A.: Trends in stroke over the past 50 years: the Framingham study. Cardiol. Rev. **24**, 23–25 (2007)
36. Schadt, E., Li, C., Su, C., Wong, W.: Analyzing high-density oligonucleotide gene expression array data. J. Cell. Biochem. **80**, 192–202 (2001)
37. Schildkraut, J.M.: Examining complex genetic interactions. In: Gene Mapping in Complex Human Diseases, pp. 379–410. John Wiley & Sons, New York (1998)
38. Sebastiani, P., Gussoni, E., Kohane, I.S., Ramoni, M.: Statistical challenges in functional genomics (with discussion). Stat Sci **18**, 33–70 (2003)
39. Sebastiani, P., Jeneralczuk, J., Ramoni, M.: Screening experiments with microarrays. In: A. Dean, S. Lewis (eds.) Screening, pp. 115–138. Springer Verlag, New York (2006)
40. Sebastiani, P., Xie, H., Ramoni, M.F.: Bayesian analysis of comparative microarray experiments by model averaging. Bayesian Anal. **1**, 707–732 (2006)
41. Shmulevich, I., Dougherty, E.R., Seungchan, K., Zhang, W.: Probabilistic boolean networks: A rule based uncertainty model for gene regulatory networks. Bioinformatics **18**, 261–274 (2002)
42. Simon, R., Radmacher, M.D., Dobbin, K.: Design of studies using DNA microarrays. Gen. Epidem. **23**, 21–36 (2002)
43. Simon, R., Radmacher, M.D., Dobbin, K., McShane, L.M.: Pitfalls in the use of DNA microarray data for diagnostic and prognostic classification. J. Natl. Cancer Inst. **95**(1), 14–18 (2003)
44. Snedecor, G.W., Cochran, W.G.: Statistical Methods. Ames: Iowa State Press (1988)
45. Tong, S.: Active learning for structure in Bayesian networks. In: International Joint Conference on Artificial Intelligence, pp. 863–869 (2001)
46. Trinh, Q.A., Hoang, T., Dorizzi, B., Asselain, B.: The effect of sample size and MLP architecture on Bayesian learning for cancer prognosis–a case study. Stud. Health Technol. Inform. **95**, 504–509 (2003)
47. Tusher, V.G., Tibshirani, R., Chu, G.: Significance analysis of microarrays applied to the ionizing radiation response. Proc Natl Acad Sci USA **98**, 5116–5121 (2000)
48. Van't Veer, L.J., Dai, H., Van de Vijver, M.J., He, Y.D., Hart, A.A.M., Mao, M., Peterse, H.L., Van der Kooy, K., Marton, M.J., Witteveen, A.T., Schreiber, G.J., Kerhoven, R.M., Roberts, C., Linsley, P.S., Bernards, R., Friend, S.H.: Gene expression profiling predicts clinical outcome of breast cancer. Nature **415**, 530–536 (2002)
49. Van de Vijver, M.J., He, Y.D., Van't Veer, L.J., Dai, H., Hart, A.A.M., Voskuil, D.W., Schreiber, G.J., Peterse, J.L., Roberts, C., Marton, M.J., Parrish, M., Atsma, D., Witteveen, A., Glas, A., Delahaye, L., Van der Velde, T., Bartelink, H., Rodenhuis, S., Rutgers, E.T., Friend, S.H., Bernards, R.: A gene-expression signature as a predictor of survival in breast cancer. New Eng J Med **347**, 1999–2009 (2002)
50. Wang, L., Montano, M., Rarick, M., Sebastiani, P.: Conditional clustering of temporal expression profiles. BMC Bioinformatics **9**(1), 147 (2008)
51. Wu, F.X., Zhang, W.J., Anthony, A.K.: Determination of the minimum number of microarray experiments for discovery of gene expression patterns. BMC Bioinformatics **7**(Suppl 4), S13 (2006)
52. Wu, F.X., Zhang, W.J., Kusalik, A.J.: Determination of the minimum sample size in microarray experiments to cluster genes using k-means clustering. In: Third IEEE Symposium on BioInformatics and BioEngineering (BIBE'03), p. 401 (2003)
53. Yang, Y., Sebastiani, P.: An evaluation of different experimental designs of temporal microarray experiments. In: Proceedings of the 2007 JSM (2007)

54. Yang, Y.H., Dudoit, S., Luu, P., Lin, D.M., Peng, V., Ngai, J., Speed, T.P.: Normalization for cDNA microarray data: A robust composite method addressing single and multiple slide systematic variation. Nucl Acid Res **30** (2002). E15
55. Yang, Y.H., Speed, T.P.: Design issues for cDNA microarray experiments. Nat Rev Genet **3**, 579–588 (2002)
56. Zien, A., Fluck, J., Zimmer, R., Lengauer, T.: Microarrays: How many do you need? J Comput Biol **10**(3-4), 653–67 (2003)

Matrix and Tensor Decompositions

Karthik Devarajan

Abstract Advances in high-throughput technologies such as gene and protein expression microarrays in the past decade have made it possible to simultaneously measure the expression levels of thousands of transcripts. This has resulted in large amounts of biological data requiring analysis and interpretation. Many methods for handling such large-scale data have been proposed in the literature. For example, consider a $p \times n$ gene expression matrix V consisting of observations on p genes from n samples representing different experimental conditions, phenotypes or time points. One could be interested in identifying clusters of genes with similar expression profiles across sub-groups of samples. Typically, this is accomplished via a decomposition of V into two or more matrices where each factored matrix has a distinct physical interpretation. Matrix decompositions have been successfully utilized in a variety of applications in computational biology such as molecular pattern discovery, class comparison, class prediction, functional characterization of genes, cross-platform and cross-species analysis, and biomedical informatics. In this chapter, we focus on available and commonly utilized methods for such matrix decompositions as well as survey other potentially useful methods for analyzing high-dimensional data.

1 Introduction

The past decade has witnessed a tremendous surge in high-throughput technologies that has resulted in large-scale biological data in the form of expression profiles of thousands of genes and proteins. Oftentimes, these studies involve only a handful of tissue samples and the number of observations (samples or data points) far exceeds the number of variables (genes or expression measurements). A primary goal of a high-throughput study such as gene expression microarrays is molecular pat-

Biostatistics & Bioinformatics, Fox Chase Cancer Center, Philadelphia, PA USA. e-mail: karthik.devarajan@fccc.edu

L.S. Heath and N. Ramakrishnan (eds.), *Problem Solving Handbook in Computational Biology and Bioinformatics*, DOI 10.1007/978-0-387-09760-2_14,

tern discovery. The emphasis is on the identification of clusters of genes that are associated with sub-groups of samples based on their expression profiles. Analyses of genome-wide expression patterns provide unique insights into the structure of genetic networks and into biological processes not yet understood at the molecular level. Dimensionality reduction and visualization are fundamental to effectively analyzing and interpreting the large-scale data in this setting. An unsupervised approach is appropriate when there is no *a priori* knowledge of the expected gene expression patterns for a given set of genes or for any phenotype or experimental condition.

In this chapter, we motivate some well known matrix decompositions within the framework of unsupervised clustering (i.e., clustering samples or genes) based on gene expression microarrays. Gene expression data from a set of microarray experiments is typically presented as a matrix in which the rows correspond to expression levels of genes, the columns to samples (which may represent distinct tissues, experiments or time points) and each entry to the expression level of a given gene in a given sample. For gene expression studies, the number of genes p is typically in the thousands, the number of samples n, is typically less than one hundred and the gene expression matrix V is of size $p \times n$, whose rows contain the expression levels of p genes in the n samples.

In order to facilitate ease of presentation, we focus on this specific application. However, the applicability of our methods remains broad since it is straightforward to extend this interpretation to other problems in computational biology involving large-scale data. First, we describe several well-known methods for decomposing high-dimensional data matrices arising in computational biology and bioinformatics. We then discuss and illustrate the properties of the various methods through a real-life example. We provide a unified representation of some well-known decompositions, as well as an interpretation of the factored matrices.

The remainder of the chapter is organized as follows. Using the unsupervised clustering framework, we describe non-negative matrix factorization and singular value decomposition in Sections 2 and 3, respectively. In addition, we discuss computational issues, implementation as well as variants of these methods in these sections. Section 4 describes the Bayesian decomposition and other matrix decomposition methods potentially applicable in large-scale biological data analysis. In Section 5, we compare and contrast these methods, and discuss some of their key properties. Section 6 outlines various methods for tensor decompositions, their applications as well as available computational tools. Finally in Section 7, we provide some concluding remarks. Section 8 contains some exercises covering many of the topics discussed.

2 Non-negative Matrix Factorization

Non-negative matrix factorization (NMF) was introduced in its current formulation by Lee and Seung [72, 73] as an unsupervised, parts-based learning paradigm, in which a nonnegative matrix V is decomposed into two nonnegative matrices $V \sim WH$ by a multiplicative updates algorithm. Their applications included text mining and facial pattern recognition. NMF has gained widespread recognition in recent years and has been used in a variety of areas including image processing and facial pattern recognition [14, 23, 35, 43, 44, 45, 71, 76, 99, 100, 104, 123], natural language processing such as in text mining and document clustering (see [22, 96, 106] and references therein), sparse coding [55, 56, 57, 71, 80], information retrieval [119, 125], speech recognition [11, 24, 89, 112], video summarization [25] and Internet research [81, 82] and computational biology. For a thorough review of its applications within the domain of computational biology, the interested reader is referred to Devarajan [32].

NMF has proved to be a very useful tool for dimensionality reduction while still providing a meaningful interpretation of the factored matrices. Given the $p \times n$ gene expression matrix V defined earlier, NMF finds a pre-specified number of metagenes, each defined as a nonnegative linear combination of the p genes. This is accomplished via a decomposition of V into two nonnegative matrices, $V \sim WH$, where W has size $p \times k$, with each of k columns defining a metagene and where H has size $k \times n$, with each of n columns representing the metagene expression pattern of the corresponding sample. The rank k of the factorization represents the number of latent factors in the decomposition (in our case, this corresponds to the number of clusters). It is generally chosen such that $(n+p)k < np$, i.e., a number less than $\min(n,p)$. The entry w_{ia} in the matrix W is the coefficient of gene i in metagene a and the entry h_{aj} in the matrix H is the expression level of metagene a in the sample j. There is also a dual view of the decomposition $V \sim WH$, which defines metasamples (rather than metagenes) and clusters the genes (rather than the samples) according to the entries of W.

The factorization in NMF is not unique, unlike methods like singular value decomposition which will be discussed later. The first step in obtaining an approximate factorization for V is to define cost functions that measure the distance (in some sense) between the observed matrix V and the product of the factored matrices WH. Various metrics have been proposed and utilized for NMF in the literature. Typically, these metrics are derived from Kullback-Leibler (KL) divergence or a generalization of it based on an assumed likelihood [12, 31, 32, 72, 73]. One commonly used metric is Euclidean distance $||V - WH||^2$ which can be derived as KL divergence between V and WH based on the Gaussian likelihood [27, 29, 31] (Exercise 1). Devarajan & Ebrahimi [28] introduced Renyi's divergence for NMF based on the Poisson likelihood of generating V from WH. For applications in image analysis involving count data, Renyi's divergence can be derived based on reconstruction of an image represented by V from WH by the addition of Poisson noise, i.e.,

$$V = WH + \varepsilon \tag{1}$$

where ε is a Poisson random variable. Renyi's divergence is indexed by a parameter $\alpha(\alpha \neq 1)$ and represents a continuum of distance measures that can be utilized for NMF, based on the choice of this parameter. It is given by

$$R_\alpha(V||WH) = \frac{1}{\alpha - 1}\sum_{i,j}\left[V_{ij}^{\alpha}(WH)_{ij}^{1-\alpha} - \alpha V_{ij} - (1-\alpha)(WH)_{ij}\right]. \quad (2)$$

Various well-known distance measures arise from Renyi's divergence as special cases [27, 28, 29, 31]. For example, in the limiting case $\alpha \to 1$, we obtain KL divergence given by

$$KL(V||WH) = \sum_{i,j}\left[V_{ij}\log\frac{V_{ij}}{(WH)_{ij}} - V_{ij} + (WH)_{ij}\right]. \quad (3)$$

Thus, we obtain the original formulation of Lee & Seung [73] based on (1) above (Exercise 2). Renyi's divergence is also applicable to other problems involving count data such as text mining and document clustering as well as in molecular pattern discovery where it provides a good approximation to gene expression profiles. Hence this generalization unifies various competing models into a unique framework for molecular pattern discovery using NMF [27, 28, 29, 30, 31, 32].

2.1 Convergence of the Algorithm

Once a metric is chosen, our goal is to minimize the distance measure such as in (2) or (3). Multiplicative update rules for W and H based on random initial values have been derived based on the EM algorithm. Specific details of the algorithm, its variants and their implementation can be found elsewhere [22, 31, 55, 57, 72, 76, 80]. These rules are applied simultaneously to W and H and guarantee convergence of the algorithm to a local minimum. However, the algorithm may not converge to the same solution on each run due to the stochastic nature of initial conditions. This requires that it be run multiple times, each based on a different random initialization for W and H. The gain in computational time due to the matrix representation of the NMF update rules is offset in part by the non-negativity constraints. These constraints require an algorithmically more complex implementation relative to methods such as singular value decomposition.

2.2 Model Selection

Model selection refers to the choice of the number of metagenes k. In the clustering context, we are interested in grouping the n samples into k clusters, where k is the pre-specified rank of the factorization. Despite its computational complexity, the stochastic nature of the NMF algorithm has proved to be rather useful in evaluating the consistency and robustness of its performance. It has been effectively utilized to assess whether a given rank k provides a biologically meaningful decomposition of the data. Studies have shown that 50-200 NMF runs are usually sufficient to provide stability to the clustering [12, 31]. A commonly used method for combining the information from multiple runs is consensus clustering [88]. In this approach, the class membership for each sample is determined based on the highest metagene expression profile [12, 31]. It quantifies the stability of the discovered clusters and can also be utilized to assess the sensitivity of a stochastic method like NMF to random initial conditions. Model selection procedures that quantify the robustness of the factorization via consensus clustering have been developed and applied to NMF [12, 29, 31]. For instance, Brunet et al. [12] propose a method for choosing the number of clusters k based on the cophenetic correlation coefficient. However, other approaches to handling the information across multiple runs are also possible [29, 66, 88].

2.3 Implementation

The implementation of the steps in any model selection procedure is computationally very intensive for any real large-scale biological data set. However, the stochastic nature of the algorithm enables each of these steps to be run independently and simultaneously. These steps can be repeated for multiple random initial conditions for W and H and the information from the independent runs combined via consensus clustering. Thus the NMF algorithm lends itself easily to a parallel implementation that would greatly increase speed and efficiency. Devarajan & Wang [30] outlined such a parallel implementation of this algorithm on a Message-Passing Interface/C++ platform (http://www-unix.mcs.anl.gov/mpi/mpich2/) using high-performance computing clusters. There have been other efforts to optimize the implementation of this algorithm. Lin [78] provides a tool based on nonnegative least squares using projected gradients. Okun & Priisalu [91] have reported faster convergence of the algorithm when feature scaling is applied to the original $p \times n$ data matrix V.

Brunet et al. [12] provide a MATLAB implementation for NMF based on KL divergence and consensus clustering. Their code is available for download at the Cancer Genomics page of MIT Broad Institute (http://www.broadinstitute.org/cgi-bin/cancer). Pascual-Montano et al. [93] provide an analytical tool called bio-NMF for simultaneous clustering of genes and samples. For more details, the interested reader is referred to http://www.dacya.ucm.es/apascual/bioNMF/. Lin's projected gradients approach [78] is available as MATLAB and Python implementations and can be downloaded at http://www.csie.ntu.edu.tw/∼cjlin/nmf/index.html.

3 Singular Value Decomposition

Singular Value Decomposition (SVD) is also known as the Karhunen-Loeve transform in machine learning, and as principal component analysis (PCA), the Hotelling transform and the empirical orthogonal function method in statistics. However, for clarity in our discussion here, we will make a slight distinction between SVD and PCA.

SVD is a linear decomposition of the $p \times n$ gene expression matrix V into matrices W, H and D such that

$$V' = WDH' \tag{4}$$

where W has size $n \times k$, H' has size $k \times p$ and D is a non-negative, diagonal matrix of size $k \times k$ where $k = \min(p,n)$. In our case, $n << p$ and hence we will take $k = n$. Each column of W represents a linear combination of the expression profiles of the genes showing the largest variance across the samples, and is termed an eigengene. Similarly, each row of H' represents a linear combination of the expression profiles of the samples showing the largest variance across the genes, and is termed an eigensample. The eigengenes and eigensamples represent the left and right singular vectors, respectively, in the decomposition [117]. Alter [4] is credited with coining the terms eigengene and eigenarray. However, we use the term eigensample instead of eigenarray to generalize applicability. We note that the formulation of SVD presented here is based on V' rather than V (as in [4]); however, it has an equivalent interpretation due to the symmetry underlying the SVD.

Unlike NMF, this decomposition is unique in that the eigengenes and eigensamples are unique, orthogonal superpositions of genes and samples based on the gene expression matrix V. The matrices W and H are both orthogonal such that

$$W'W = H'H = I \tag{5}$$

where I is the identity matrix. The entries of the matrix D are given by

$$d_{ij} = \sqrt{\lambda_i}\delta_{ij}, i,j = 1,...,n \tag{6}$$

where $\delta_{ij} = 1$ if $i = j$ and zero otherwise, i.e., the ith eigengene is expressed only in the corresponding i^{th} eigensample with magnitude $\sqrt{\lambda_i}$ and this expression level indicates its relative significance. This relative significance is given by

$$\psi_i = \frac{\lambda_i}{\sum_{i=1}^{n} \lambda_i}, i = 1,...,n \tag{7}$$

The λ_is represent the ordered eigenvalues such that $\lambda_1 \geq \lambda_2 \geq ... \geq \lambda_n$.

The expression of each eigengene does not depend on the expression levels of other eigengenes. In addition, the expression of each eigengene is also decorrelated from the other eigengenes [4]. These statements hold good in the case of eigenarrays as well. Alter et al. utilize these features of the decomposition and outline an approach for filtering the expression matrix V that circumvents the need to remove individual genes or samples within a data set. They suggest filtering out those eigengenes and the corresponding eigenarrays that may potentially represent noise simply by letting $\lambda_i = 0$ in D and reconstructing the matrix V using (4) above.

In terms of reducing the dimensionality of the data, PCA seeks to find a small number of orthogonal linear combinations or principal components (PC) that explain most of the variation in the data. This is achieved via the SVD of the gene expression matrix V' given by (4). In the least squares sense, PCA is the most optimal linear dimension reduction method. Let us assume that V has been standardized so that the covariance matrix does not depend on the scale of the expression measurements. The first PC $\mathbf{p_1}$ is the linear combination with the largest variance. It is obtained by solving the following equation

$$\mathbf{h_1} = argmax_{||\mathbf{h}=1||} Var(V'\mathbf{h}) \tag{8}$$

such that $\mathbf{p_1} = V'\mathbf{h_1}$ and $\mathbf{h}$ represents a column of H. The second PC $\mathbf{p_2}$ is the linear combination with the second largest variance and orthogonal to the first PC and so forth. Typically, the first few PCs explain most of the variation in the data and are usually sufficient. Using (4) and (5), it is easy to see that $V'H = WD$. The columns of $V'H$ are the PCs and they are ordered based on decreasing variance. An interesting property of SVD is that the total variation is equal to the sum of the eigenvalues of the $p \times p$ covariance matrix given by

$$\Sigma = \frac{1}{n}VV' = HD^2H'$$

where D^2 is a diagonal matrix of the ordered eigenvalues $\lambda_1 \geq \lambda_2 \geq ... \geq \lambda_n$ (Exercise 3). The columns of H are the eigenvectors of Σ and are known as the principal component directions of V' [37, 49, 108].

PCA is a second-order method due to its dependence on the covariance matrix of the data. The PCs are uncorrelated and have variances equal to the eigenvalues of Σ. Therefore, $\sum_{i=1}^{n} \lambda_i = trace(\Sigma) = \sum_{i=1}^{n} Var(\mathbf{p_i})$ and the relative significance ψ_i in (7) above can be re-written as $\psi_i = \frac{\lambda_i}{trace(\Sigma)}$. ψ_i gives the cumulative proportion of the variance explained by the first i PCs. A plot of ψ_i vs. i facilitates the choice of the appropriate number of PCs that are required to be kept in order to explain a given proportion of the total variation. Such a plot is known as a scree plot and a search for an elbow in the plot reveals the number of PCs that explain most of the variation.

Another useful exploratory tool is to graphically represent the correlations of the PCs with the genes. For example, consider the first two PCs, $\mathbf{p_1}$ and $\mathbf{p_2}$ that typically explain most of the variance. One could plot the correlations of the first PC with each gene against the correlation of the second PC with each gene. Alter et al. [4] recommend such an approach. The distance of each gene from the origin could be interpreted as its amplitude of expression in the subspace spanned by the two PCs. In addition, the angular distance of each gene from the x-axis could be interpreted as its phase in the transition between the expression patterns defined by $\mathbf{p_1}$ and $\mathbf{p_2}$.

3.1 Implementation

SVD is readily implemented in practically every available software package for statistical data analysis. Typically, tools for visualizing the principal components in two and three dimensions are also available. Some examples include, but are not limited to, the following: the open-source statistical language and environment R (www.r-project.org) and commercially available packages such as SAS (www.sas.com), Splus (www.insightful.com), STATA (www.stata.com), SPSS (www.spss.com), and MATLAB (www.mathworks.com).

3.2 Non-negative Sparse PCA

Zass & Shashua [126] described a non-negative, sparse variant of PCA (NSPCA) which creates PCs that maximize the variance as in regular PCA except that the PCs themselves are restricted to be non-negative. The imposition of non-negativity constraints facilitates a sparse representation just as in NMF, thus resulting in factors with a physical interpretation. An interesting outcome of imposing non-negativity constraints in PCA is that the resulting PCs are disjoint and each co-ordinate is non-zero in at most one PC. Each PC can then be viewed as a part. In addition to the non-negativity constraints, explicit sparseness constraints are also imposed where the desired amount of sparseness can be controlled with the choice of a tuning parameter. This approach thus compensates for two important shortcomings of PCA, namely lack of sparseness and non-negativity while still maintaining its maximal variance property.

In many applications, some amount of overlap among the parts is desirable. The authors [126] relax the disjointness of the PCs, thus allowing for some overlap among them. Using the same formulation as in section 3 above, we add appropriate constraints to incorporate non-negativity and sparseness to equation (8) as well as relax the disjointness to obtain the following objective function for NSPCA.

$$\max_{H} \frac{1}{2}||V'H||^2 - \frac{\alpha}{4}||I - H'H||^2 - \beta||H||_{L_0}, H \geq 0$$

The first term in the above expression is equivalent to the objective function in regular PCA (see equation (8) above), the second term is an orthonormality distance measure that represents the degree of overlap among PCs (i.e., it relaxes the disjointness of the PCs), and the last term represents explicit sparseness constraints. The orthonormality distance $||I - H'H||^2$ is non-negative and it is zero if and only if H is orthonormal. The parameter $\alpha > 0$ provides a tradeoff between reconstruction of the data matrix V and orthonormality and the parameter $\beta > 0$ controls the amount of additional sparseness required. Typically, this sparseness is imposed via the the L_0 norm by minimizing the number of non-zero elements in H. However, other constraints such as one based on the L_1 norm are also possible. Zass & Shashua [126] also provide an algorithm for maximizing this objective function and computing the corresponding non-negative, sparse PCs. Besides this method, there are other sparse variants of PCA that have been proposed. A brief discussion on topic is provided in [126].

4 Other Matrix Decompositions

There are several other matrix decomposition methods that abound in the literature. A thorough survey of various methods, including variants not discussed in this chapter, is presented in Fodor [37]. These include principal factor analysis, maximum likelihood factor analysis, projection pursuit, non-linear independent component analysis, random projections, principal curves and non-linear PCA. In this section, we outline some less commonly used methods that are variants of NMF or PCA. In particular, we cover Bayesian decomposition, factor analysis and independent component analysis briefly.

4.1 Bayesian Decomposition

Bayesian Decomposition (BD) was originally proposed by Ochs et al. [90] for spectroscopic analysis and later adapted for gene expression microarrays [86]. In BD, the $p \times n$ expression matrix V is decomposed into two matrices $V \sim WH$ where W has size $p \times k$ and H has size $k \times n$. The matrices W and H are referred to as the distribution and pattern matrices, respectively. Each row of the distribution matrix W represents the amplitude of each pattern within the corresponding gene (row of V), with each column of W being associated with a single pattern. Similarly, each row of the pattern matrix H represents the patterns that show the average behavior of the co-expressed genes across the samples. The rank k of the decomposition denotes the number of latent factors and is a number less than $\min(n, p)$. In BD, this corresponds to the number of patterns and represents the appropriate number of basis vectors (rows of H) required to reconstruct the matrix V from the product WH.

Mathematically, the decomposition in BD can be expressed as

$$V = WH + \varepsilon \tag{9}$$

where the gene expression matrix V is reconstructed based on the product WH by the addition of Gaussian noise. This is somewhat similar to the NMF formulation based on Euclidean distance but without the non-negativity constraints. Other distributions could also be used in the decomposition (9) above; however, the standard formulation of BD incorporates normally distributed errors (BD). The decomposition in (9) is not unique and a reasonable approximation to the distribution and pattern matrices, W and H respectively, are obtained via a Bayesian Markov Chain Monte Carlo (MCMC) procedure [70].

One of the attractive features of BD is its ability to capture the co-expression of a single gene in multiple groups that results in the identification of overlapping co-expression groups. The basis vectors in W and H are non-orthogonal and potentially represent the co-expression response due to different pathways or at different time points. Since many biological processes have overlapping transcriptional response profiles across experimental conditions (such as time points, phenotype etc.), BD is suitable for identifying such non-orthogonal patterns. For time-course gene expression data, for instance, these basis vectors represent time curves associated with a physical process. These processes could be progression through the cell cycle or activation or de-activation of a pathway due to a specific treatment.

Using Bayes' theorem on the decomposition $V \sim WH$, we obtain

$$P(W,H|V) \propto P(V|W,H)P(W,H) \tag{10}$$

Here, $P(V|W,H)$ is the likelihood of the data, $P(W,H)$ is the prior probability of the model and $P(W,H|V)$ is the posterior probability of the model given the data. The prior incorporates biological knowledge and specifies the probability of the model independent of the data. Starting with a suitable prior distribution in (9) and a pre-specified number of patterns k, the algorithm utilizes simulated annealing and iterates until convergence while updating W and H. Encoding the prior distribution and sampling from the posterior are done using a Bayesian MCMC approach implemented as a Gibbs sampler [70]. As with any Bayesian approach, encoding the prior distribution is an important first step in the process outlined above. In BD, prior encoding is used to incorporate prior biological knowledge in a number of different ways. Examples include the incorporation of class associations [69, 87] or co-regulation information from transcription factor databases [70] and encoding for non-negativity of expression measurements [70]. An important consideration in BD, as in other matrix decomposition methods, is the choice of the appropriate number of patterns k that adequately represents the data V. Moloshok et al. [86] and Kossenkov et al. [70] outline data driven approaches for selecting the number of patterns for a given data set.

4.2 Factor Analysis

Factor Analysis (FA) is a linear decomposition method based on second-order statistics similar to PCA. The fundamental assumption in FA is that the measured variables depend on a set of unknown and possibly unmeasurable common factors that contribute to the observed data. Assuming that the gene expression matrix V has been standardized so that the covariance matrix is scale-independent, the decomposition in FA is given by

$$\mathbf{v} = \Lambda \mathbf{f} + \mathbf{u} \tag{11}$$

where $\mathbf{v}$ is p-dimensional vector representing each column of the matrix V, Λ is a $p \times k$ matrix of constants, $\mathbf{f}$ is a $k \times 1$ matrix of random common factors and $\mathbf{u}$ is a $p \times 1$ matrix of specific factors [37]. Here k denotes the number of factors. The factors are assumed to be uncorrelated and the common factors are standardized to have zero mean and unit variance. We can re-write (11) as

$$v_i = \sum_{j=1}^{k} \lambda_{ij} f_j + u_i, i = 1, 2, ..., p \tag{12}$$

where λ_{ij} is the ij^{th} element of Λ, v_i is the ith element of $\mathbf{v}$, f_j is the jth element of $\mathbf{f}$ and u_i is the ith element of $\mathbf{u}$. Then,

$$Var(v_i) = \sigma_{ii} = \sum_{j=1}^{k} \lambda_{ij}^2 + \psi_{ii} \tag{13}$$

where the first and second terms denote the contributions of the common and specific factors to the total variance. Each term within the sum on the right hand side measures the magnitude of dependence of v_i on the common factor f_j. An important interpretation of the decomposition in (12) is that if several components of the vector $\mathbf{v}$ have high loadings λ_{ij} on a given factor f_j, these variables are deemed to measure the same unobservable quantity and considered redundant. The k-factor model is scale independent unlike SVD and also holds for orthogonal rotation of factors (Exercise 4). Factor analysis is implemented in standard statistical software packages such as SAS, SPSS and STATA, among others.

4.3 Independent Component Analysis

Independent Component Analysis (ICA) is a higher-order method that seeks linear projections that are not orthogonal to each other necessarily but are statistically independent. Statistical independence between components is a much stronger assumption than uncorrelatedness and requires higher-order statistics. This contrasts with PCA which requires only second-order statistics and orthogonal components. Independence implies uncorrelatedness except in the case of the normal distribution when they are equivalent, implying independent PCs. The independence requirement suggests that ICA may be appropriate for data that are not normally distributed such as those from gene expression microarrays. However, this method may not be suitable for analyzing large-scale genomic data due to strong correlations that are typically known to exist between clusters of genes and sub-groups of samples. In other words, identification of independent groups of genes may not be practically possible.

A survey of ICA can be found in [61], and advances in theory and applications are detailed in [42, 59, 74, 101]. Details on implementation of ICA and available software can be found in [15, 60, 105].

5 Comparison of the Methods

In this section, we compare and contrast the various matrix decomposition methods outlined above. We give particular attention to the most commonly used methods, NMF and SVD, and some of its variants. We provide an interpretation of the decomposition itself as well as the nonnegativity and orthogonality constraints.

5.1 Interpretation of the Decomposition

In the NMF representation, a single metagene expression pattern influences multiple samples. The metagenes provide a summary of the behavior of genes across the samples while the metagene expression patterns provide a summary of the behavior of samples across the genes (see Exercise 5) [32]. The metagene coefficient w_{ia} quantifies the influence of the ath metagene expression pattern h_{aj} on the gene expression of the ith sample, represented by the corresponding column of the gene expression matrix V. For a pre-specified rank k factorization, the relative magnitude of the non-zero entries in each of the k metagenes reflects the importance of the corresponding genes, and the expression pattern of each metagene across the n samples (represented by each row of H) reflects the importance of the corresponding latent factor. When a matrix whose columns represent data points in multi-dimensional space is decomposed, parts manifest as subsets of the data dimensions that take on values in a coordinated fashion [102]. This is generally relevant to any matrix decomposition, however, in NMF, there is strong evidence suggesting that the metagenes and the metagene expression patterns have a sparse, parts-based representation of the gene expression data [12, 16, 27, 31, 32, 40, 65, 66, 73, 93, 94, 95], potentially identifying local hidden variables or clusters. This is evidently due to the imposition of non-negativity constraints [32, 73, 126], which will be discussed further in the next section.

For clustering samples whose expression profiles are specified by columns of V, the parts identify homogeneous clusters and are represented by the expression patterns of metagenes across samples (or the rows of H). Moreover, genes with corresponding non-zero metagene coefficients represent groups that are co-expressed in samples. These parts provide a reduced representation of the original data, and their co-activation can be viewed as that corresponding to co-regulation or co-expression of groups of genes [32]. This interpretation of the metagene coefficients and their expression profiles across samples in NMF is similar to that in BD where the corresponding entities, the distribution and patterns matrices, are also non-orthogonal and represent the co-expression response due to different pathways or time points. In BD, each row of H can be viewed as measuring change across a set of pathways, indicating the strength of a transduced signal measured by the transcriptional response; and each column of W denotes the relative level of expression of genes to the associated pathways.

In text mining and document clustering, the matrix V is a summary of a corpus of documents where columns represent documents and contain word frequencies from these documents. Chagoyen et al. [22] discuss a novel application of NMF where existing information about the genes (or proteins) in a study can be used to establish putative relationships among subsets of these genes (or proteins) that characterize a subset of the data. In principle, we would like to identify subsets of semantic categories and cluster the documents based on their association with these categories. This is achieved via NMF by extracting literature profiles from a document corpus linked to large sets of genes using common semantic features extracted from the corpus. Genes are then represented as additive linear combinations of the semantic features which can be further used for studying their functional associations. The different semantic categories represent the parts in this application and the Poisson framework using the various metrics outlined in section 2 provides a solid statistical approach for modeling in this context.

Next, we interpret the decompositions in NMF and SVD using a real-life example. We make use of the leukemia microarray data available at the Cancer Genomics website of MIT Broad Institute (web link provided in section 2.3). This is one of the widely used publicly available data sets and has become a benchmark in the development, testing and illustration of methods for large-scale biological data. It consists of 5000 gene expression measurements each from 38 bone marrow samples from acute myelogenous leukemia (AML) and acute lymphoblastic leukemia (ALL). There are 27 ALL samples consisting of 19 B type and 8 T type, and 11 AML samples. In previous work, we used this data to demonstrate the factorization in NMF [32]. Here, we use it for further illustration and comparison of NMF and SVD. We applied a rank $k = 3$ factorization of the 5000×38 gene expression matrix using NMF based on KL divergence (equation (3)). Let w_1, w_2 and w_3 represent the three metagenes (columns of W) and let h_1, h_2 and h_2 represent the corresponding metagene expression profiles (rows of H). For the purpose of illustration, we use the factored matrices based on a single run of the algorithm. We also applied standard PCA to this gene expression matrix and computed the PCs.

The sparseness of the metagenes is clearly demonstrated by their box plots shown in Fig. 1(a). Each circle in this figure represents a gene. Of the 5000 genes in this data, the number of genes whose metagene coefficients exceed 10 are 38, 67 and 77, and correspond, respectively, to the metagenes w_1, w_2 and w_3. These genes may potentially behave in a strongly correlated fashion in a subset of the samples and this is determined by their metagene expression profiles across the 38 samples. We graphically illustrate this relationship between the metagenes and the gene expression profile for each sample by plotting their Pearson correlations. Panels (a)-(c) of Fig. 2 display these correlations for the three metagenes, plotted against the sample labels. Alternatively, one could plot these correlations for a given metagene with that of another metagene (shown in Fig. 2(d) for the first two metagenes). In this figure, "T", "B" and "M" denote an ALL-T, ALL-B and AML sample, respectively. The correlation between the gene expression profiles of samples and the first metagene shows a separation between ALL-T sub-type and the remainder of the samples. The second metagene, on the other hand, is able to distinguish between the two major

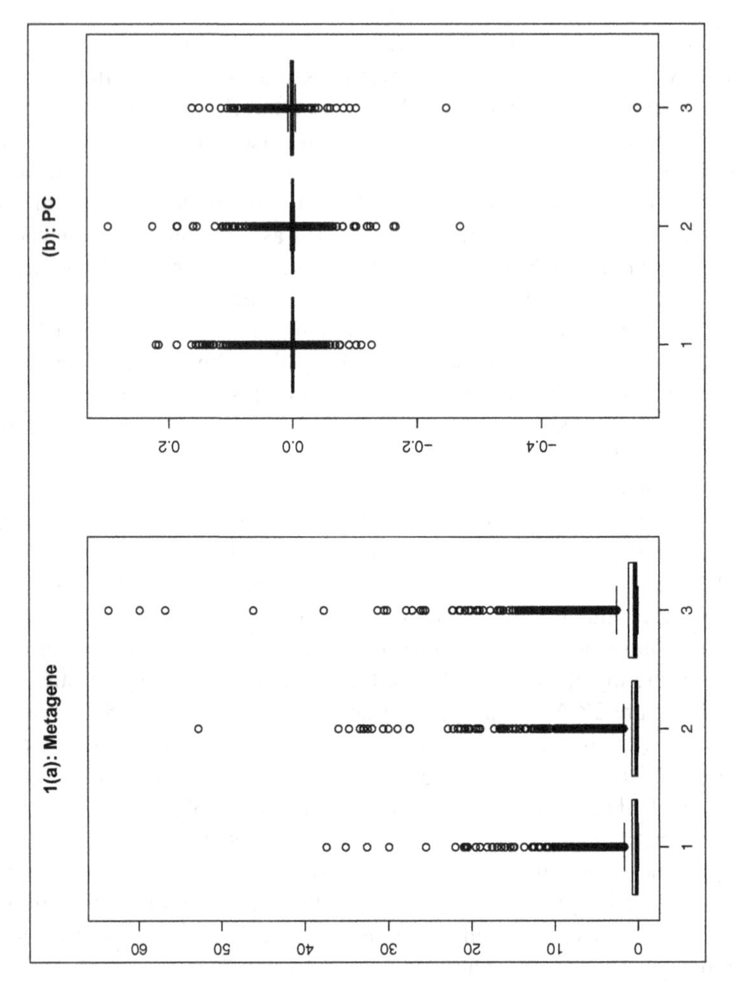

Fig. 1 Box plot of metagene projections.

classes. The combined effect of the first and second metagenes is seen in Fig. 2(d) where all three classes are separated. In these plots, the correlations themselves are not as important as identifying groups of samples showing similar correlations with the metagene of interest.

On the other hand, the distributed representation provided by PCA is seen in the box plots of the first three PCs shown in Fig. 1(b). It is evident from this figure that identifying potentially significant genes based on the PC profiles alone is not straightforward. In the case of PCA, plots analogous to those shown in Fig. 2 can be obtained by using the PCs (data not shown). We noted earlier in this section that the expression pattern of each metagene across the 38 samples reflected the importance of the corresponding class represented by it. This is demonstrated by a plot of the expression profiles for the first two metagenes shown in Fig. 3(a)-

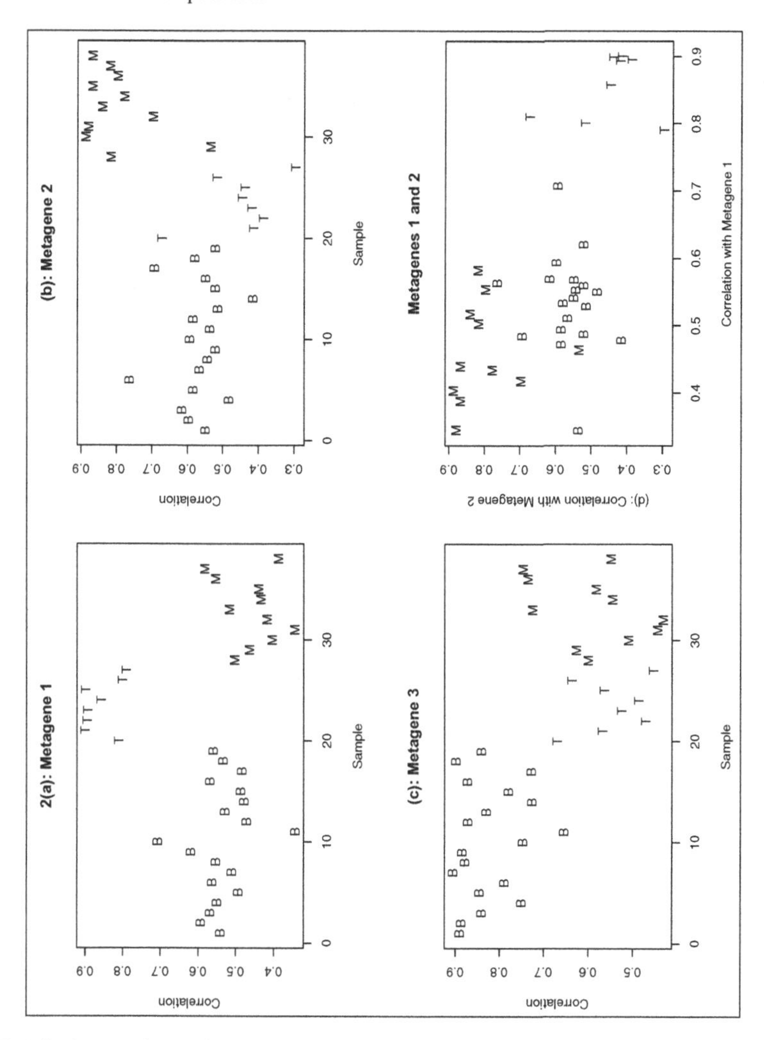

Fig. 2 Pairwise correlations between gene expression profiles of metagenes.

(b). The expression profile of the first metagene clearly distinguishes the ALL-T samples from the others while the expression profile of the second metagene shows the separation between the ALL and AML samples. These results are consistent with those observed in Fig. 2(a)-(b). The corresponding plots in the case of PCA are those based on the PCs and displayed in panels (c)-(d) of Fig. 3. The first PC (panel (c)) is neither able to provide a good separation between the ALL and AML classes nor between the two ALL sub-types, while the second PC is able to separate the two major classes. However, the third PC was able to delineate the three classes (data not shown). It is important to note that phenotypes are known in this data set and samples have been plotted by phenotype for illustration. Nevertheless, for any

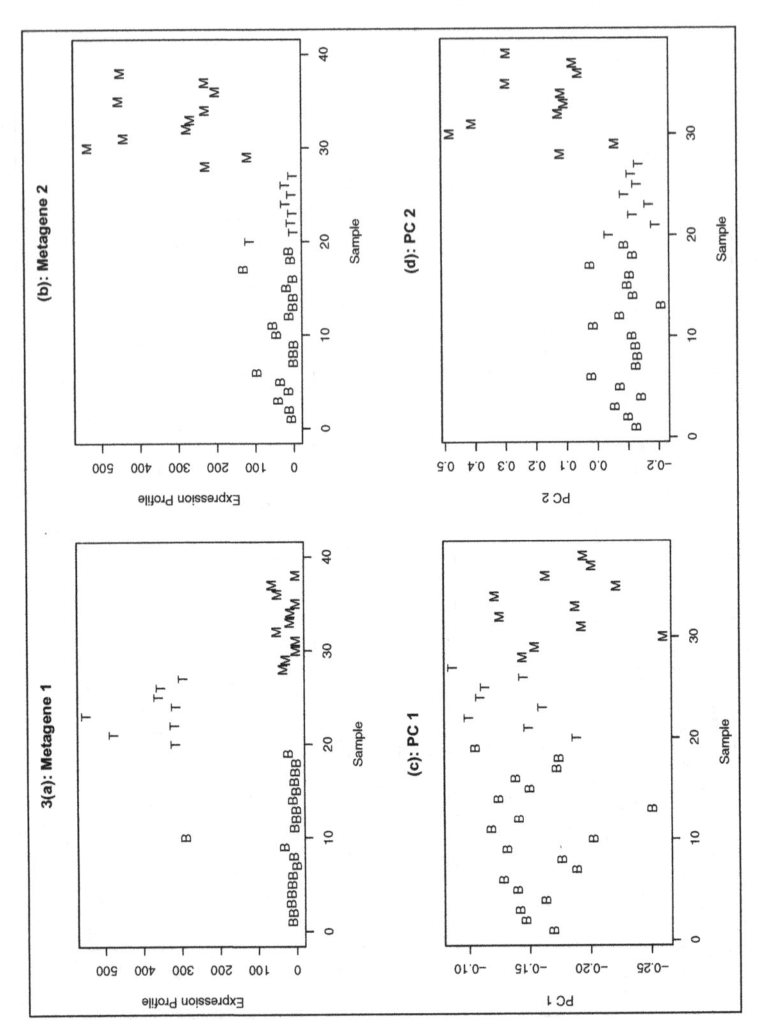

Fig. 3 Expression profiles of metagenes.

real data set with unknown phenotype, these plots are a useful tool for illustrating the underlying relationships between samples.

5.2 Interpretation of Constraints

In their seminal work, Lee & Seung [73] argued that the nonnegativity constraints in NMF are compatible with the intuitive notion of combining parts to form a whole, i.e., they provide a parts-based local representation of the data. They demonstrate this property using an example from facial pattern recognition. This unique property of NMF contrasts with a holistic representation of the data provided by a method such as vector quantization (VQ) and the distributed representation provided by PCA [73]. A parts-based model not only provides an efficient representation of the data but can potentially aid in the discovery of causal structure within it and in learning relationships between the parts [102]. The nonnegativity constraints in NMF result in a reconstruction of the original data by the addition of parts (metagenes and metagene expression profiles) while in SVD, it is a superposition of the orthogonal components with arbitrary signs (eigengenes and eigensamples) that lack intuitive meaning and physical interpretation. In many real-life applications, negative coefficients may contradict physical reality. In image analysis, for example, the pixels in a grey scale image with negative intensities cannot be meaningfully interpreted. Arguably, NMF provides a more intuitive interpretation of the data compared to SVD. However, it is inapplicable when the original data contains negative values. The basic principles underlying variations such as NSPCA and BD (outlined in sections 3.2 and 4.1, respectively) relax some of the restrictions imposed by NMF and SVD and offer flexible alternatives to these methods. NSPCA, for instance, can handle data that are not necessarily non-negative unlike NMF but still retains some of the useful features of NMF such as sparseness. As in section 4.1, we note that prior encoding in BD can be used to incorporate prior biological knowledge in a variety of ways and this includes encoding for non-negativity of expression measurements [70].

For gene expression data, one could interpret the nonnegative coefficients in each metagene as the relative contribution of genes, unlike PCA and VQ. This is clearly demonstrated in the example described in the previous section where a rank $k = 3$ factorization showed that only a small proportion of the genes corresponding to the three metagenes significantly contributed towards separation of the three classes (Figures 1(a) and 2). The identification of these small subsets of potentially active genes is possible only due to the nonnegativity constraints which is a requirement for such a parts-based representation. Such an interpretation would be virtually impossible with SVD where the eigengenes and eigensamples are orthogonal and can contain positive or negative coefficients (see Figure 1(b)). The perception of the whole is simply an additive linear combination of its parts represented in the metagenes and metagene expression profiles.

In their development of NSPCA, Zass & Shashua [126] observe that the addition of non-negativity alone ensures some sparseness of the PCs by inducing disjointness where each co-ordinate is non-zero in at most one PC. As noted in [32], the non-negativity constraints may be a necessary condition for a parts-based representation but they may not be sufficient to achieve sparseness. There is also some evidence that points to a parts-based but holistic (rather than local) representation produced by NMF [55, 56, 57, 76]. Thus, it may be desirable to explicitly enforce sparseness on the metagenes and the metagene expression patterns. This is similar to NSPCA in principle, and several variants of NMF imposing explicit sparseness constraints on the entries of H or W or both have appeared in the literature over the last several years [40, 55, 56, 57, 65, 76, 77, 80]. As in NSPCA, explicit sparseness is imposed by including an appropriate penalty term to the objective function of choice. A brief discussion of this topic including relevant literature and applications can be found in section 4.3 of [32].

Another important consideration is the orthogonality constraint that is so fundamental to PCA. In NMF, orthogonality of metagenes and metagene expression profiles may not be achievable in practice due to the nonnegativity constraints. This lack of orthogonality in NMF has been shown to be superior to SVD and other dimension reduction methods (see [106] and references therein). However, enforcing sparseness constraints decreases the overlap between metagenes as well as between metagene expression profiles. It facilitates the detection of sharp boundaries between different classes, and results in localized, disjoint groups of samples or genes [32]. Li et al. [75] have explicitly imposed orthogonality constraints and developed a variant of NMF with applicability in spectroscopy. However, non-orthogonality of the basis vectors can be extremely useful in gene expression studies since dependence among the gene expression profiles typically present in such studies can be captured by overlapping vectors. This property makes methods like NMF, BD and NSPCA particularly well-suited for the analysis of large-scale biological data where it is essential to capture relationships underlying inter-connected biological pathways or processes.

6 Tensor Decompositions

A tensor is a multi-dimensional or N-way array where N is the order of the tensor. The order N of a tensor is the number of dimensions and is also referred to as ways or modes. For example, a first-order tensor is a vector, a second-order tensor is a matrix and tensors with $N \geq 3$ are considered to be higher-order tensors. Some tensor decompositions are higher-order extensions of the well-known matrix version of SVD [68] while others are higher-order extensions of the matrix version of NMF [68]. Decompositions based on higher-order tensors possess an interesting uniqueness property that contrasts many matrix decompositions. Interpretation of the factor matrices arising from tensor decompositions is an important consideration but this may not always be possible. However, by imposing appropriate constraints on the model similar to those in the matrix versions of SVD and NMF, distinct interpretations of the factor matrices is possible.

In this section, we provide a summary of various methods in this area, their applications and related computational tools. For a thorough review, the interested reader is referred to [68].

6.1 Methods

Tensor decomposition was first introduced in 1927 by Hitchcock [53, 54] in which a tensor was expressed as a sum of a finite number of rank-one tensors. Later in 1944, Cattell [20, 21] proposed the concept of multiple axes for the analyses of high-dimensional data and parallel proportional analysis. However, the concept of tensor decomposition did not become popular until the 1960s when methods such as Canonical Decomposition (CANDECOMP) [19] and Parallel Factors (PARAFAC) [47] were introduced for applications in psychometrics. Decompositions based on these two methods factorize a tensor into a sum of rank-one tensors, and are also unique under weaker conditions compared to well-known matrix decompositions such as SVD. Another widely applied tensor decomposition method is the Tucker Decomposition (TD) [120]. TD is a form of higher-order PCA in which a tensor is decomposed into the product of a core tensor and a matrix along each mode. Unlike CANDECOMP and PARAFAC, this decomposition is not unique.

In addition to these methods, there are various other approaches to tensor decompositions that have appeared in the literature. For example, Individual Differences in Scaling (INDSCAL) is a special case of CANDECOMP and PARAFAC for three-way tensors that are symmetric in two modes [19]. Similarly, Canonical Decomposition with Linear Constraints (CANDELINC) is an extension of CANDECOMP and PARAFAC that imposes linear constraints on one or more of the factor matrices [18]. PARAFAC2 is another variant of CANDECOMP and PARAFAC that can be applied to a collection of matrices with the same number of columns but with different numbers of rows. In that sense, PARAFAC2 is not strictly a tensor decomposition. A distinct advantage of this method is that it can approximate data in a regular three-way tensor with fewer constraints that the standard CANDECOMP and PARAFAC approach. Another family of tensor decompositions is Decomposition into Directional Components (DEDICOM) [48] that is applicable for skew-symmetric data and data measured over time. A method that combines CANDECOMP and PARAFAC and is also a generalization of DEDICOM is PARATUCK2 [46] that considers the interactions between two different sets of objects. For example, the transition from one object to another is accounted for in this analysis. Such a consideration is not possible in standard factor analysis involving matrices. For more details, see [68] and references cited therein. There are also several higher-order extensions of NMF known broadly as non-negative tensor factorizations (NTF). These are comprised of non-negative versions of CANDECOMP and PARAFAC as well the non-negative Tucker decomposition (NNT).

6.2 Applications

Decompositions of higher-order tensors have found applications in a variety of areas such as signal processing, chemometrics, data mining, neuroscience, computer vision and numerical analysis. Despite their usefulness for high-dimensional data, the potential of tensor decompositions is yet to be fully realized in the areas of bioinformatics and computational biology. CANDECOMP and PARAFAC are also known as the Topographic Components Model in neuroscience [85] where these methods have found a variety of applications in brain imaging, analyses of functional magnetic resonance imaging data, electro-encephelogram (EEG) spectra and of epileptic seizures [83, 84]. Other areas of applications include chemometrics [6], image analysis [5, 98, 39, 10], signal processing and telecommunications [26, 109, 110, 111] and in data mining such as Internet research [2, 3] and text mining [8]. TD has found applications in chemical analysis, psychometrics, signal processing, image analysis, computer vision, video analysis, Internet research, text mining, handwriting analysis and environmental modeling [68]. CANDELINC, PARATUCK2 and PARAFAC2 have been used to handle multicollinearity in chemometrics data. PARAFAC2 has also been used in clustering documents across multiple languages. DEDICOM has been successfully applied to model data on email communications over time. Higher-order extensions of NMF such as NTF and NNT have found applications in image analysis, audio analysis and the analysis of EEG data [36, 50, 107, 124].

6.3 Implementation

Computational tools for tensor decompositions include high-level programming environments such as MATLAB, Mathematica and Maple as well as FORTRAN and C++ libraries. In particular, MATLAB provides a variety of tools for tensor manipulation and decomposition. The base package supports element-wise manipulation of tensors, and toolboxes such as N-way, CuBatch, PLS and Tensor provide algorithms for computing various tensor decompositions, handling missing data as well as non-negativity and orthogonality constraints. The PLS toolbox provides various multi-dimensional models for data analysis and is well-suited for applications in chemometrics. The Multilinear Engine [92] is a FORTRAN library that supports CANDECOMP, PARAFAC and PARAFAC2, among others. The HUJI Tensor Library [126] and the Boost Multidimensional Array Library [41] are C++ libraries of classes for expressing tensors. These libraries support various operations involving tensors and sparse tensors except tensor multiplication.

7 Concluding Remarks

In this chapter, we have described several methods for matrix decompositions that are useful for a variety of applications in computational biology and bioinformatics. Even though we focused on dimensionality reduction and unsupervised clustering of gene expression data for the purpose of illustration, the extension of the methods to other applications is straightforward. For instance, methods such as principal components regression [49], supervised principal components [9, 117] and partial least squares [49] incorporate PCA within the supervised learning framework where one is interested in identifying genes that are correlated with an outcome variable such as patient response to a drug or survival time. There is a wealth of literature in the area of supervised learning (please see [49] and references cited therein). The application of NMF to other problems in computational biology has been surveyed in [32]. Methods such as NMF and NSPCA demand greater applicability in large-scale biological data analysis due to their sparse, parts-based representation and nicer interpretability. The concept of matrix decomposition methods extends to the multi-dimensional setting in which tensors are higher-dimensional analogues of matrices and represent data in higher dimensions. Tensor decompositions have gained popularity recently, both in terms of methodological development as well as applications [68]. However, tensor decompositions are still in their fledgling state in the area of bioinformatics and their potential is yet to be realized.

8 Exercises

1. Consider the matrices V, W and H described in Section 1. Using the Kullback-Leibler divergence between two probability density (or mass) functions f and g given by

$$KL(f:g) = \int_{\Re} \log\left(\frac{f(x)}{g(x)}\right) dF(x), \tag{14}$$

show that the divergence between V and WH based on the Gaussian likelihood (assuming equal variance) is equivalent to the Euclidean norm $||V - WH||^2$.
2. Using (14) above, derive the quantity in equation (3) based on the Poisson likelihood of generating V from the product WH.
3. Show that the total variation explained by the principal components in SVD is equal to the sum of the eigenvalues of the covariance matrix $\Sigma = \frac{1}{n}VV'$.
4. In factor analysis, show that the k-factor model is scale independent and holds for orthogonal rotation of factors.
5. Graphically illustrate that the representation in NMF ensures that a single metagene expression pattern influences multiple samples.

Acknowledgements

Research supported in part by NIH grant P30 CA 06927 and an appropriation from the Commonwealth of Pennsylvania.

References

1. Abdallah, E. E., Hamza, A. B. and Bhattacharya, P.: MPEG video watermarking using tensor singular value decomposition, in Image Analysis and Recognition, vol. 4633 of Lecture Notes in Computer Science Springer, pp. 772-783 (2007)
2. Acar, E., Camtepe, S. A. and Yener, B.: Collective sampling and analysis of high order tensors for chatroom communications, in ISI 2006: Proceedings of the IEEE International Conference on Intelligence and Security Informatics, vol. 3975 of Lecture Notes in Computer Science—, Springer, pp. 213-224 (2006)
3. Acar, E., Camtepe, S. A., Krishnamoorthy, M. S. and Yener, B.: Modeling and multiway analysis of chatroom tensors, in ISI 2005: Proceedings of the IEEE International Conference on Intelligence and Security Informatics, vol. 3495 of Lecture Notes in Computer Science Springer, pp. 256-268 (2005)
4. Alter, O., Brown, P.O.,Botstein, D.: Singular value decomposition for genoe-wide expression data processing and modeling, Proceedings of the National Academy of Sciences 97(18):10101-10106. (2000)
5. Andersen, C. M. and Bro, R.:Practical aspects of PARAFAC modeling of fluorescence excitation-emission data, Journal of Chemometrics 17, pp. 200-215 (2003)
6. Appellof, C. J., and Davidson, E. R.: Strategies for analyzing data from video fluorometric monitoring of liquid chromatographic effuents, Analytical Chemistry 53 , pp. 2053-2056(1981)
7. Ashburner, M., Ball, C.A., Blake, J.A., Botstein,D., Butler, H., Cherry, J.M., Davis, A.P., Dolinski, K., Dwight, S.S., Eppig, J.T., Harris, M.A., Hill, D.P., Issel-Tarver, L., Kasarskis, A., Lewis, S., Matese, J.C., Richardson, J.E., Ringwald, M., Rubin, G.M., Sherlock, G.: Gene Ontology: tool for the unification of biology. The Gene Ontology Consortium, Nature Genetics 25(1):25-29(2000)
8. Bader, B. W., Berry, M. W., and Browne, M.: Discussion tracking in enron email using PARAFAC, in Survey of Text Mining: Clustering, Classification, and Retrieval Second Edition, M. W. Berry and M. Castellanos, eds., Springer, pp.147-162 (2007)
9. Bair, E., Hastie, T., Paul, D., Tibshirani, R.: Prediction by supervised principal components. Journal of the American Statistical Association, 101:119 (2006)
10. Bauckhage, C.: Robust tensor classifiers for color object recognition, in Image Analysis and Recognition, vol. 4633 of Lecture Notes in Computer Science Springer, pp. 352-363 (2007)
11. Behnke, S.: Discovering hierarchical speech features using convolutional non-negative matrix factorization, Proceedings of the International Joint Conference on Neural Networks vol. 4, pp. 2758-2763, Portland, Oregon, USA (2003)
12. Brunet, J-P., Tamayo, P., Golub, T., Mesirov, J.: Metagenes and molecular pattern discovery using nonnegative matrix factorization, Proceedings of the National Academy of Sciences USA 101: 4164-4169 (2004)
13. Buchsbaum, G., Bloch, O.: Color Categories Revealed by Non-negative Matrix Factorization of Munsell Color Spectra, Vision Research 42, 559-563 (2002)
14. Buciu, I., Pitas, I.: Application of non-negative and local non negative matrix factorization to facial expression recognition, Proceedings of the 17th International Conference on Pattern Recognition vol. 1, pp. 288-291, Cambridge, UK (2004)
15. Cardoso, J-F., http://www.tsi.enst.fr/∼cardodo/icacentral
16. Carmona-Saez, P., Pascual-Marqui, R.D., Tirado, F., Carazo, J.M., Pascual-Montano, A.: Biclustering of gene expression data by non-smooth non-negative matrix factorization, BMC Bioinformatics 7:78 (2006)
17. Carrasco, D.R., Tonon, G., Huang, Y., Zhang, Y., Sinha, R. et al :High resolution genomic profiles define distinct clinico-pathogenic subgroups of multiple myeloma patients, Cancer Cell 9:313-325 (2006)
18. Carroll, J. D., Pruzansky,S. and Kruskal,J. B.: CANDELINC: A general approach to multidimensional analysis of many-way arrays with linear constraints on parameters, Psychometrika 45 , pp. 324 (1980)

19. Carroll, J. D.and Chang, J. J.: Analysis of individual differences in multidimensional scaling via an N-way generalization of 'Eckart-Young' decomposition, Psychometrika 35 , pp. 283-319 (1970)
20. Cattell, R. B.: Parallel proportional profiles and other principles for determining the choice of factors by rotation, Psychometrika 9 , pp. 267283 (1944)
21. Cattell, R. B.: The three basic factor-analytic research designs - their interrelations and derivatives, Psychological Bulletin 49 , pp. 499-452 (1952)
22. Chagoyen, M., Carmona-Saez, P., Shatkay, H., Carazo, J.M., Pascual-Montano, A.: Discovering semantic features in the literature: a foundation for building functional associations, BMC Bioinformatics 7:41 (2006)
23. Chen, X., Gu, L., Li, S-Z., Zhang, H-J.: Learning representative local features for face detection, Proceedings of the IEEE Computer Society Conference on Computer Vision and Pattern Recognition vol. 1, pp. I-1126-I-1131, Kauai, Hawaii, USA (2001)
24. Cho, Y-C., Choi, S., Bang, S-Y.: Non-negative component parts of sound for classification, Proceedings of the 3rd IEEE International Symposium on Signal Processing and Information Technology pp. 633-636, Darmstadt, Germany (2003)
25. Cooper, M., Foote, J.: Summarizing video using nonnegative similarity matrix factorization, Proceedings of the IEEE Workshop on Multimedia Signal Processing pp. 25-28, St.Thomas, Virgin Islands, USA (2002)
26. De Lathauwer, L. and Castaing, J.: Tensor-based techniques for the blind separation of DS-CDMA signal, Signal Processing, 87, pp. 322-336 (2007)
27. Devarajan, K., Ebrahimi, N.: Class discovery via nonnegative matrix factorization, American Journal of Management and Mathematical Sciences, 28(3&4):457-467 (2008)
28. Devarajan, K., Ebrahimi, N.: Molecular pattern discovery using non-negative matrix factorization based on Renyi's information measure, Proceedings of the XII SCMA InternationalConference, Auburn University, Auburn, Alabama (2005); http://atlas-conferences.com/c/a/q/t/98.htm
29. Devarajan, K., Wang, G., Ebrahimi, N.: A generalized approach to non-negative matrix factorization with applications, Technical Report, Division of Population Science, Fox Chase Cancer Center, 2009
30. Devarajan, K., Wang, G.: Parallel implementation of non-negative matrix factorization algorithms using high-performance computing cluster, Proceedings of the 39th Symposium on the Interface: Computing Science and Statistics, Theme: Systems Biology, Temple University, Philadelphia, Pennsylvania (2007). Available at http://sbm.temple.edu/interface07/
31. Devarajan, K.: Nonnegative matrix factorization - A new paradigm for large-scale biological data analysis, Proceedings of the Joint Statistical Meetings, Seattle, Washington (2006)
32. Devarajan, K.: Nonnegative matrix factorization: An analytical and interpretive tool in computational biology. PLoS Computational Biology, 4(7), July (2008)
33. Dietterich, T. G., Becker, S. and Ghahramani, Z. (Eds.).: Advances in neural information processing systems , 14:897-904, MIT Press, Cambridge, MA, USA (2002)
34. Donoho, D., Stodden, V.: When does nonnegative matrix factorization give a correct decomposition into parts?, Advances in neural Information Processing Systems 16, MIT Press (2003)
35. Feng, T., Li, S-Z., Shum, H-Y., and Zhang, H-Y.: Local nonnegative matrix factorization as a visual representation, Proceedings of the 2nd International Conference on Development and Learning, pp. 178-183, Cambridge, Massachusetts, USA (2001)
36. FitzGerald, D., Cranitch, M. and Coyle,E.: Non-negative tensor factorisation for sound source separation, in ISSC 2005: Proceedings of the Irish Signals and Systems Conference (2005)
37. Fodor, I.K.: A survey of dimension reduction methods. LLNL technical report. UCRL-ID 148494 (2002)
38. Fogel, P., Young, S.S., Hawkins, D.M., Ledirac, N.: Bioinformatics, Inferential, robust non-negative matrix factorization analysis of microarray data, 23(1):44-49 (2007)
39. Furukawa, R., Kawasaki, H., Ikeuchi, K., and Sakauchi, M.: Appearance based object modeling using texture database: acquisition, compression and rendering, in EGRW'02: Proceedings of the 13th Eurographics workshop on Rendering, Airela-Ville, Switzerland, Switzerland, Eurographics Association, pp. 257-266 (2002)

40. Gao, Y., Church, G.: Improving molecular cancer class discovery through sparse non-negative matrix factorization, Bioinformatics, 21(21):3970-3975 (2005)
41. Garcia, R. and Lumsdaine, A.: MultiArray: A C++ library for generic programming with arrays, Software: Practice and Experience, 35, pp. 159-188 (2004)
42. Girolami, M.: Advances in Independent Component Analysis. Perspectives in Neural Computing, Springer (2000)
43. Guillamet, D., Vitri'a, J., Schiele, B.: Introducing a weighted non-negative matrix factorization for image classification, Pattern Recognition Letters, vol. 24, no. 14, pp. 2447-2454 (2003)
44. Guillamet, D., Vitri'a, J.: Evaluation of distance metrics for recognition based on non-negative matrix factorization, Pattern Recognition Letters, vol. 24, no. 9-10, pp. 1599-1605 (2003)
45. Guillamet, D., Vitri'a, J.: Discriminant basis for object classification, Proceedings of the 11th International Conference on Image Analysis and Processing, pp. 256-261, Palermo, Italy (2001)
46. Harshman, R. A. and Lundy, M. E.: Uniqueness proof for a family of models sharing features of Tucker's three-mode factor analysis and PARAFAC and CANDECOMP, Psychometrika, 61, pp. 133-154 (1996)
47. Harshman, R. A.: Foundations of the PARAFAC procedure: Models and conditions for an "explanatory" multi-modal factor analysis, UCLA working papers in phonetics, 16 , pp. 184 (1970). Available at http://publish.uwo.ca/∼harshman/ wpppfac0.pdf
48. Harshman, R. A.: Models for analysis of asymmetrical relationships among N objects or stimuli, in First Joint Meeting of the Psychometric Society and the Society for Mathematical Psychology, McMaster University, Hamilton, Ontario, August (1978) Available at http://publish.uwo.ca/∼harshman/asym1978.pdf
49. Hastie, T., Tibshirani, R., Friedman, J.: The elements of statistical learning. Springer-Verlag, New York (2001)
50. Hazan, T. Polak, S. and Shashua, A.: Sparse image coding using a 3D nonnegative tensor factorization, in ICCV 2005: Proceedings of the 10th IEEE International Conference on Computer Vision, vol. 1, IEEE Computer Society, pp. 50-57 (2005)
51. Heger, A., Holm, L.: Sensitive pattern discovery with 'fuzzy' alignments of distantly related proteins, Bioinformatics, 19(1):i130-i137 (2003)
52. Hiisila, H., Bingham, E.: Dependencies between Transcription Factor Binding Sites: Comparison between ICA, NMF, PLSA and Frequent Sets, Proceedings of the Fourth IEEE International Conference on Data Mining, 114-121 (2004)
53. Hitchcock, F. L.: Multiple invariants and generalized rank of a p-way matrix or tensor, Journal of Mathematics and Physics, 7 , pp. 39-79 (1927)
54. Hitchcock, F. L.: The expression of a tensor or a polyadic as a sum of products, Journal of Mathematics and Physics, 6 , pp. 164-189 (1927)
55. Hoyer, P.O.: Nonnegative matrix factorization with sparseness constraints, Journal of Machine Learning Research, 5:1457-1469 (2004)
56. Hoyer, P.O.: Modeling receptive fields with nonnegative sparse coding, Neurocomputing, 52-54:547-552 (2003)
57. Hoyer, P.O.: Nonnegative sparse coding, Proceedings of the IEEE Workshop on Neural Networks for Signal Processing, Neural Networks for Signal Processing XII, 557-565, Martigny, Switzerland (2002)
58. HUJI tensor library. http://www.cs.huji.ac.il/∼zass/htl/(2006)
59. Hyvarinen, A., Karhunen, J. and Oja, E.: Independent Component Analysis. Series on Adaptive and Learning Systems for Signal Processing, Communications and Control, Wiley (2001)
60. Hyvarinen, A.: http://www.cis.hut.fi/∼aapo
61. Hyvarinen, A.: Survey on independent component analysis, Neural Computing Surveys, 2, 94-128 (1999)
62. Isakoff, M.S., Sansam, C.G., Tamayo, P., Subramanian, A., Evans, J.A., Fillmore, C.M., Wang, X., Biegel, J.A., Pomeroy, S.L., Mesirov, J.P., Roberts, C.S.: Inactivation of the Snf5 tumor suppressor stimulates cell cycle progression and cooperates with p53 loss in oncogenic transformation, Proceedings of the National Academy of Sciences USA, 102:17745:17750(2005)

63. Jung, I., Lee, J., Kim, H., Lee, S-Y. et al: Improving profile-profile alignment feature for fold-recognition using nonnegative matrix factorization, Proceedings of the Seventh International Conference of the Korean Society for Bioinformatics, 22-27 (2006)
64. Kelm, B.M., Menze, B.H., Zechmann, C.M., Baudendistel, K.T., Hamprecht, F.A.: Automated estimation of tumor probability in prostate magnetic resonance spectroscopic imaging: pattern recognition vs. quantification, Magnetic Resonance in Medicine, 57:150-159 (2007)
65. Kim, H., Park, H.: Sparse Non-negative matrix factorizations via alternating non-negativity-constrained least squares, Proceedings of the IASTED International Conference on Computational and Systems Biology, pp. 95-100, Dallas, Texas (2006)
66. Kim, P., Tidor, B.: Subsystem identification through dimensionality reduction of large-scale gene expression data, Genome Res, 13:1706-1718 (2003)
67. Kim, S.P., Rao, Y.N., Erdogmus, D., Sanchez, J.C., Nicolelis, M.A.L. et al: Determining patterns in neural activity for reaching movements using nonnegative matrix factorization, EURASIP Journal on Applied Signal Processing, 19:3113-3121 (2005)
68. Kolda, T.G., Bader, B.W.: Tensor decompositions and applications, SIAM Review, 51(3) (2009)
69. Kossenkov, A.V., Bidaut, G., Ochs, M.F. Genes associated with prognosis in adenocarcinoma across studies at multiple institutions, In K.F. Johnson and S.M. Lin editors, Methods of Microarray Data Analysis IV, p. 239, Kluwer Academic, Boston (2005).
70. Kossenkov, A.V., Bidaut, G., Ochs, M.F. Estimating cellular signaling from transcription data, In K-A Do, P. Muller, M. Vannicci editors, Bayesian Inference for Gene Expression and Proteomics, pp.366-384, Cambridge University Press, New York, (2006).
71. Lawrence J, Rusinkiewicz S, Ramamoorthi R (2004) Efficient BRDF importance sampling using a factored representation," ACM Transactions on Graphics, 23(3):496-505
72. Lee, D.D., Seung, S.H.: Algorithms for nonnegative matrix factorization, Advances in Neural Information Processing Systems, 13:556-562 (2001)
73. Lee, D.D., Seung, S.H.: Learning the parts of objects by nonnegative matrix factorization, Nature, 401:788-791 (1999)
74. Lee, T-W.: Independent Component Analysis: Theory and Applications, Kluwer Academic Publishers (2001)
75. Li, H., Adali, T., Wang, W., Emge, D., Cichocki, A. Non-negative matrix factorization with orthogonality constraints and its application to Raman spectroscopy, Journal of VLSI Signal Processing, 48:83-97 (2007).
76. Li, S.Z., Hou, X., Zhang, H., Cheng, Q.: Learning spatially localized, partsbased representations, Proceedings of the IEEE Conference on Computer Vision and Pattern Recognition, 1:207-212 (2001)
77. Li, Y., Cichocki, A.: Sparse representation of images using alternating linear programming, Proceedings of the 7th International Symposium on Signal Processing and Its Applications, vol. 1, pp. 57-60, Paris, France (2003)
78. Lin, C-J.: Projected gradient methods for non-negative matrix factorization, Neural Computation, 19, 2756-2779 (2007)
79. Liu, N., Zhang, B., Yan, J., Chen, Z., Liu,W., Bai, F. and Chien, L.: Text representation: From vector to tensor, in ICDM 2005: Proceedings of the 5th IEEE International Conference on Data Mining, IEEE Computer Society, pp. 725-728 (2005)
80. Liu, W., Zheng, N., Lu, X.: Non-negative matrix factorization for visual coding, Proceedings of the IEEE International Conference on Acoustics, Speech and Signal Processing 3:293-296 (2003)
81. Lu, J., Xu, B., Yang, H.: Matrix dimensionality reduction for mining Web logs, Proceedings of the IEEE/WIC International Conference on Web Intelligence, pp. 405-408, Halifax, Nova Scotia, Canada (2003)
82. Mao, Y., Saul, L.K.: Modeling distances in large-scale networks by matrix factorization, Proceedings of the ACM Internet Measurement Conference, pp. 278-287, Sicily, Italy (2004)
83. Martinez-Montes, E., Valdes-Sosa, P. A., Miwakeichi, F., Goldman, R. I. and Cohen, M. S.: Concurrent EEG/fMRI analysis by multiway partial least squares, NeuroImage, 22 , pp. 1023-1034 (2004)

84. Miwakeichi,F., Martinez-Montes, E. P., Valds-Sosa, A., Nishiyama, N., Mizuhara, H. and Yamaguchi, Y.:Decomposing EEG data into space-time-frequency components using parallel factor analysis, NeuroImage, 22 , pp. 1035-1045 (2004)
85. Mocks, J.: Topographic components model for event-related potentials and some biophysical considerations, IEEE Transactions on Biomedical Engineering, 35, pp. 482484 (1988)
86. Moloshok, T.D., Klevecz, R.R., Grant, J.D., Manion, F.J., Speier, W.F., Ochs, M.F. Application of Bayesian decomposition for analyzing microarray data, Bioinformatics, 18(4):566-575 (2002).
87. Moloshok, T.D., Datta, D., Kossenkov, A.V., Ochs, M.F. Bayesian decomposition classification of the project normal data set, In K.F. Johnson and S.M. Lin editors, Methods of Microarray Data Analysis III, pp. 211-232, Kluwer Academic, Boston, (2003).
88. Monti, S., Tamayo, P., Golub, T.R., Mesirov, J.P.: Consensus clustering: A resampling-based method for class discovery and visualization in gene expression microarray data, Machine Learning Journal, 52:91-118 (2003)
89. Novak, M., Mammone, R.: Use of non-negative matrix factorization for language model adaptation in a lecture transcription task, Proceedings of the IEEE International Conference on Acoustics, Speech, and Signal Processing, vol. 1, pp. 541-544, Salt Lake City, Utah, USA (2001)
90. Ochs, M.F., Stoyanova, R., Arias-Mendoza, A., Brown, T.R. A new method for spectral decomposition using a bilinear Bayesian approach, Journal of Magnetic Resonance Imaging, 137:161-176 (1999).
91. Okun, O., Priisalu, H.: Fast nonnegative matrix factorization and its application for protein fold recognition, EURASIP Journal on Applied Signal Processing, Article ID 71817 (2006)
92. Paatero,P.: The multilinear engine: A table-driven, least squares program for solving multilinear problems, including the n-way parallel factor analysis model, Journal of Computational and Graphical Statistics, 8 (1999), pp. 854-888
93. Pascual-Montano, A., Carmona-Saez, P., Chagoyen, M., Tirado, F., Carazo, J.M., Pascual-Marqui, R.D.: bioNMF: a versatile tool for non-negative matrix factorization in biology, BMC Bioinformatics, 28(7):366 (2006)
94. Pascual-Montano, P., Carazo, J.M., Kochi, K., Lehmann, D., Pascual-Marqui, R.: Nonsmooth nonnegative matrix factorization, IEEE Transactions on Pattern Analysis and Machine Intelligence, 28(3):403-415 (2006)
95. Pascual-Montano, P., Carazo, J.M., Kochi, K., Lehmann, D., Pascual-Marqui, R.: Two-way clustering of gene expression profiles by sparse matrix factorization, Proceedings of the Computational Systems Bioinformatics Conference,Workshops and Poster Abstracts, 103-104 (2005)
96. Pauca, P., Shahnaz, F., Berry, M., and Plemmons, R.: Text mining using nonnegative matrix factorizations, Proceedings of the Fourth SIAM International Conference on Data Mining, Lake Buena Vista, Florida (2004)
97. Pehkonen, P.,Wong, G., Toronen, P.: Links Theme discovery from gene lists for identification and viewing of multiple functional groups, BMC Bioinformatics, 6:162 (2005)
98. Qi, L., Sun, W. and Wang, Y.: Numerical multilinear algebra and its applications, Frontiers of Mathematics in China, 2 , pp. 501-526 (2007)
99. Rajapakse, M., Wyse, L.: NMF vs ICA for face recognition, Proceedings of the 3rd International Symposium on Image and Signal Processing and Analysis, vol. 2, pp. 605-610, Rome, Italy (2003)
100. Ramanath, R., Snyder, W.E., Qi, H.: Eigenviews for object recognition in multispectral imaging systems, Proceedings of the 32nd Applied Imagery Pattern Recognition Workshop, pp. 33-38, Washington, DC, USA (2003)
101. Roberts, T., Everson, R.: Independent Components Analysis: Principles and Practice, Cambridge University Press, Cambridge, United Kingdom (2000)
102. Ross, D.A., Zemel, R.S.: Learning parts-based representations of data, Journal of Machine Learning Research, 7:2369-2397 (2006)

103. Sajda, P., Du, S,. Brown, T.R., Stoyanova, R., Shungu, D.C. et al: Nonnegative matrix factorization for rapid recovery of constituent spectra in magnetic resonance chemical shift imaging of the brain, IEEE Transactions on Medical Imaging, 23:1453-65 (2004)
104. Saul, L.K., Lee, D.D.: Multiplicative updates for classification by mixture models, Advances in Neural and Information Processing Systems (2002)
105. Sejnowski, T.: http://www.cnl.salk.edu/∼tewon/ica.cnl.html
106. Shahnaz, F., Berry, M.: Document clustering using nonnegative matrix factorization, Information Processing and Management: An International Journal, 42(2): March 2006)373-386 (2006)
107. Shashua, A. and Hazan, T.: Non-negative tensor factorization with applications to statistics and computer vision, in ICML: Proceedings of the 22nd International Conference on Machine Learning, 2005, pp. 792-799 (2005)
108. Shlens, J.: A tutorial on principal component analysis Available at http://www.snl.salk.edu/∼schlens/pub/notes/pca.pdf (2009)
109. Sidiropoulos, N. and R. Budampati, Khatri-Rao space-time codes, IEEE Transactions on Signal Processing, 50 (2002), pp. 2396-2407
110. Sidiropoulos, N., Bro, R. and Giannakis, G.: Parallel factor analysis in sensor array processing, IEEE Transactions on Signal Processing, 48 , pp. 2377-2388 (2000)
111. Sidiropoulos, N., Giannakis, G., and Bro, R.: Blind PARAFAC receivers for DSCDMA systems, IEEE Transactions on Signal Processing, 48, pp. 810-823 (2000)
112. Smaragdis, P., Brown, J.C.: Non-negative matrix factorization for polyphonic music transcription, Proceedings of the IEEE Workshop on Applications of Signal Processing to Audio and Acoustics, pp. 177-180, New Paltz, NY, USA (2003)
113. Sun, J., Papadimitriou, S. and Yu, P. S.: Window-based tensor analysis on highdimensional and multi-aspect streams, in ICDM 2006: Proceedings of the 6th IEEE Conference on Data Mining, IEEE Computer Society, pp. 1076-1080 (2006)
114. Sun, J., Tao D., and Faloutsos, C.: Beyond streams and graphs: Dynamic tensor analysis, in KDD '06: Proceedings of the 12th ACM SIGKDD International Conference on Knowledge Discovery and Data Mining, ACM Press, pp. 374-383 (2006)
115. Sun, J.-T., ZengH.-J., Liu,H., Lu, Y. and Chen,Z.: Cube SVD: A novel approach to personalized Web search, in WWW2005: Proceedings of the 14th International Conference on World Wide Web, ACM Press, pp. 382-390 (2005)
116. Tamayo, P., Scanfield, D., Ebert, B.L., Gillette, M.A., Roberts, C.W.M., Mesirov, J.P.: Metagene projection for cross-platform, cross-species characterization of global transcriptional states, Proceedings of the National Academy of Sciences, 104(14): 5959-5964 (2007)
117. Tibshirani, R. and Bair, E.: Improved detection of differential gene expression through the singular value decomposition. Available at http://www-stat.stanford.edu/∼ tibs/ftp/eric.pdf (2003)
118. Tresch, M.C., Cheung, V.C., d'Avella, A.: Matrix factorization algorithms for the identification of muscle synergies: evaluation on simulated and experimental data sets, Journal of Neurophysiology Apr;95(4):2199-2211 Epub 2006 Jan 4 (2006)
119. Tsuge, S., Shishibori, M., Kuroiwa, S., Kita, K.: Dimensionality reduction using nonnegative matrix factorization for information retrieval, Proceedings of the IEEE International Conference on Systems, Man, and Cybernetics, vol. 2, pp. 960- 965, Tucson, Arizona, USA (2001)
120. Tucker, L. R.: Implications of factor analysis of three-way matrices for measurement of change, in Problems in Measuring Change, C.W. Harris, ed., University of Wisconsin Press,, pp. 122-137 (1963)
121. Vasilescu, M. A. O. and Terzopoulos, D.: Tensortextures: multilinear image based rendering, ACM Transactions on Graphics, 23 , pp. 336-342 (2004)
122. Wang, G., Kossenkov, A.V., Ochs, M.F.: LS-NMF: A modified non-negative matrix factorization algorithm utilizing uncertainty estimates, BMC Bioinformatics, 7:175 (2005)

123. Wang, Y., Jia, Y., Hu, C., Turk, M.: Fisher non-negative matrix factorization for learning local features, Proceedings of the 6th Asian Conference on Computer Vision, pp. 806-811, Jeju Island, Korea (2004)
124. Welling, M. and Weber,M.: Positive tensor factorization, Pattern Recognition Letters, 22, pp. 1255-1261 (2001)
125. Xu, B., Lu, J., Huang, G.: A constrained non-negative matrix factorization in information retrieval, Proceedings of the IEEE International Conference on Information Reuse and Integration, pp. 273-277, Las Vegas, NV, USA (2003)
126. Zass, R. and Shashua,A.: Nonnegative sparse PCA. Advances in Neural Information Processing Systems (2006)

Practical Applications of the Gene Ontology Resource

Rachael P. Huntley, Emily C. Dimmer, and Rolf Apweiler

Abstract The Gene Ontology (GO) is a controlled vocabulary that represents knowledge about the functional attributes of gene products in a structured manner and can be used in both computational and human analyses. This vocabulary has been used by diverse curation groups to associate functional information to individual gene products in the form of annotations. GO has proven an invaluable resource for evaluating and interpreting the biological significance of large data sets, enabling researchers to create hypotheses to direct their future research. This chapter provides an overview of the Gene Ontology, how it can be used, and tips on getting the most out of GO analyses.

1 Introduction

With the advent of the sequencing age and advances in high-throughput experimental methodologies, there has been an explosion in the amount of data obtained from biological research. Researchers must now be able to manage, manipulate, and interpret large data sets, and the Gene Ontology resource has proven an invaluable aid for helping researchers achieve this.

Traditionally, the naming of biological concepts has been inconsistent, with biologists from different communities producing both multiple names for identical concepts and identical names for multiple concepts. Depending on what field of biology you work in, you may interpret the meaning of 'bud' in many different ways; an anatomist may think of a tooth bud, a botanist would contemplate leaf or flower

Rachael P. Huntley
European Bioinformatics Institute, e-mail: huntley@ebi.ac.uk

Emily C. Dimmer
European Bioinformatics Institute, e-mail: edimmer@ebi.ac.uk

Rolf Apweiler
European Bioinformatics Institute, e-mail: apweiler@ebi.ac.uk

L.S. Heath and N. Ramakrishnan (eds.), *Problem Solving Handbook in Computational Biology and Bioinformatics*, DOI 10.1007/978-0-387-09760-2_15,

buds, and a yeast geneticist would see a yeast cell budding to form a new cell. Although a human can make use of the context of such descriptions, this is not possible in computational analyses. The GO Consortium (GOC, No. 1 in Table 1) was founded in 1998 to tackle the problem of describing functional information. The founding members of FlyBase, the Berkeley Drosophila Genome Project, the Saccharomyces Genome Database and the Mouse Genome Database stated their goal as being "to produce a structured, precisely defined, common, controlled vocabulary for describing the roles of genes and gene products in any organism" [2]. Since then, the Consortium has grown to include 16 databases covering a diverse range of species and as of September 2008, the GO consisted of over 25,000 terms.

The GO has three hierarchies of terms: **molecular function** terms describe the biochemical activity of a gene product; **biological process** terms describe a series of functions ending in a biological objective; and **cellular component** terms describe where in the cell a gene product is located. Figure 1 illustrates part of the Biological Process ontology and the relationships between the terms.

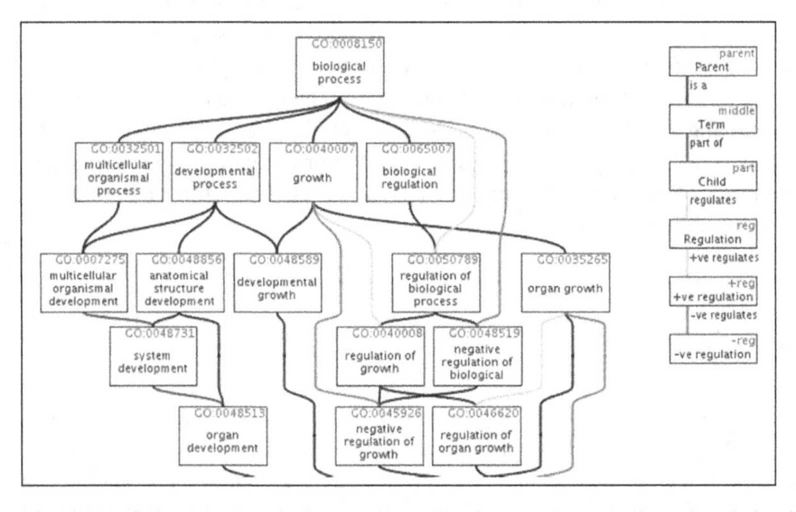

Fig. 1 A section of the Biological Process ontology. Lines between boxes indicate the relationships between the terms as described in the key.

2 GO Annotations

The GOC annotation groups also create associations (known as 'annotations') between gene/gene product identifiers and GO terms. An annotation is a specific association between a GO term identifier and sequence identifier and has a distinct evidence source which supports the association. A single well-characterised gene

Table 1 Useful web sites for GO documentation, data and tools.

		Databases
1	GO Consortium	http://www.geneontology.org/
2	GOA project	http://www.ebi.ac.uk/goa/
		Resources
3	Evidence code guide	http://www.geneontology.org/GO.evidence.shtml
4	Qualifier documentation	http://www.geneontology.org/GO.annotation.conventions.shtml#qual
5	AmiGO browser	http://amigo.geneontology.org/cgi-bin/amigo/go.cgi
6	QuickGO browser	http://www.ebi.ac.uk/QuickGO
7	GO Consortium tools listing	http://www.geneontology.org/GO.tools.shtml
8	GO current annotations	http://www.geneontology.org/GO.current.annotations.shtml
9	GOA file downloads	http://www.ebi.ac.uk/GOA/downloads.html
10	GO Online SQL Environment (GOOSE)	http://www.berkeleybop.org/goose
11	InterProScan	http://www.ebi.ac.uk/Tools/InterProScan/
12	GO slims	http://www.geneontology.org/GO.slims.shtml
13	AmiGO slimmer	http://amigo.geneontology.org/cgi-bin/amigo/slimmer
14	SGD GO slim mapper	http://db.yeastgenome.org/cgi-bin/GO/goSlimMapper.pl
15	Map2slim Perl script	http://search.cpan.org/ cmungall/go-perl/scripts/map2slim
16	UniProtKnowledgeBase (UniProtKB) ID mapper	http://www.uniprot.org/
17	Protein Identifier Cross-Reference service	http://www.ebi.ac.uk/Tools/picr/
18	GO newsletter August 2008	http://www.geneontology.org/newsletter/archive/200808.pdf
19	AmiGO term enrichment tool	http://amigo.geneontology.org/cgi-bin/amigo/term_enrichment
20	Cardiovascular GO annotation initiative	http://www.ucl.ac.uk/silva/cardiovasculargeneontology
21	GO Friends mailing list subscription	http://www.geneontology.org/GO.list.gofriends.shtml
22	GOA feedback form	http://www.ebi.ac.uk/GOA/contactus.html

product can be annotated to multiple GO terms at different levels of the three GO hierarchies.

Annotations can be created either computationally or manually — both methods are strictly controlled to produce high-quality GO annotation and both require highly trained biologists and software engineers [7]. Currently, there are 20 annotation groups providing manual and electronic annotation to over 160,000 different taxonomic groups.

At present, two main methods for producing electronic annotation are applied by GOC groups. The first is the mapping of appropriate terms from external vocabularies to equivalent GO terms producing 'translation tables', which are then applied to database entries annotated with the external vocabulary generating electronic annotations for GO. Examples of this include: Enzyme Commission numbers to GO, Swiss-Prot Keywords to GO and InterPro to GO. The latter example alone provides 64% of UniProtKnowledgeBase (UniProtKB) proteins with at least one electronic GO annotation (as of September 2008). The second electronic method involves the projection of manually assigned GO terms from proteins in one species to the orthologous proteins in other species, for example the projection of annotations between proteins from species in the Ensembl Compara ortholog set.

Electronic annotation is essential to providing annotation to the rapidly increasing numbers of proteins being identified but is particularly useful for organisms which have not been extensively studied on an experimental level, as annotations may be transferred to them from model organism species based on sequence similarity, thus providing functional information about the novel proteome. Electronic annotations are distinguished from manual annotations by their 'IEA' (Inferred from Electronic Annotation) evidence code.

The Gene Ontology Annotation (GOA) group is the principal provider of electronic GO annotations to the GOC. For a more detailed description of each of the electronic annotation methods performed by the GOA group, please see the GOA web site (No. 2 in Table 1).

Although electronic annotation provides fast, large-scale assignment of less detailed GO terms — as of September 2008, 98.5% of the GO annotations in UniProtKB were produced using electronic methods — there is also a need for highly detailed functional information about gene products, and this is why manual curation is necessary.

Manual curation is a very expensive and time-consuming process involving highly trained curators reading published experimental literature to find evidence for protein function and location. The resulting annotation is both more detailed and more accurate than any computational method could achieve. An example of this is the annotation associated with the human APOA4 protein (Figure 2). The InterPro2GO electronic mapping has predicted the term 'lipid transport' whereas a curator who has studied the scientific literature for APOA4 has assigned the more specific GO term'reverse cholesterol transport' which is a child term of 'lipid transport'.

There are, at present, 20 annotation groups contributing manual annotation to the GO Consortium covering almost 950 taxonomic groups. Depending on the re-

DB	ID	Alt	Symbol	Taxon	Qualifier	GO ID	GO Term name	Reference	Ev	With	A	Date	From
						Process							
UniProtKB/SwissProt	P06727		APOA4	9606		GO:0006869	lipid transport	spkw	IEA	KW-0445	P	20080616	UniProt
UniProtKB/SwissProt	P06727		APOA4	9606		GO:0006810	transport	spkw	IEA	KW-0813	P	20080616	UniProt
UniProtKB/SwissProt	P06727		APOA4	9606		GO:0042157	lipoprotein metabolic process	interpro	IEA	IPR000074	P	20080613	UniProt
UniProtKB/SwissProt	P06727		APOA4	9606	A	GO:0006869	lipid transport	interpro	IEA	IPR000074	P	20080613	UniProt
UniProtKB/SwissProt	P06727		APOA4	9606		GO:0002227	innate immune response in mucosa	15254593	IDA		P	20080107	BHF-UCL
UniProtKB/SwissProt	P06727		APOA4	9606		GO:0007159	leukocyte adhesion	15254593	IDA		P	20080107	BHF-UCL
UniProtKB/SwissProt	P06727		APOA4	9606	B	GO:0043691	reverse cholesterol transport	3095477	IDA		P	20080107	BHF-UCL
UniProtKB/SwissProt	P06727		APOA4	9606		GO:0044240	multicellular organismal lipid catabolic process	3095477	IDA		P	20080107	BHF-UCL
UniProtKB/SwissProt	P06727		APOA4	9606		GO:0033700	phospholipid efflux	11162594	IDA		P	20080402	BHF-UCL

Fig. 2 GO annotations associated with human APOA4 protein. Box A indicates the less specific term predicted by the InterPro2GO electronic mapping and box B indicates the more specific term chosen by the curator.

sources of each group, some species may have more comprehensive annotation than others. For example, the Mouse Genome Informatics (MGI) database has provided over 55,000 manual annotations to mouse genes, whereas there are only just over 1,400 manual annotations for pig, a species that does not have a dedicated annotation group. Manual curation also allows the addition of extra information into an annotation such as experimental evidence codes and qualifiers, which are explained in the next sections.

2.1 Evidence codes

An important component of a GO annotation is the evidence code selected to describe the type of support that exists for an association between a GO term and a gene/protein identifier. There are currently 17 evidence codes applied by curators, which fall into three main groups: annotations based on published experimental data (such as an enzyme assay, e.g. Inferred by Direct Assay, IDA); non-experimental statements provided by an author or inferred by a curator (for instance inferring a nuclear localisation for an in vitro-characterised transcription factor, e.g. Non-traceable Author Statement, NAS); and finally evidence from computational predictions (Inferred by Electronic Annotation, IEA) (No. 3 in Table 1).

Evidence codes are useful for evaluating the sources of information that were available for a particular annotation set. When analysing data sets, users should be aware that, for less well-annotated genomes such as chicken, bovine, or even human, they may need to use both electronic and manual annotation sets to ensure that their sequences of interest are sufficiently populated with annotation data, whereas for well-annotated genomes such as yeast, a user could choose to include only manually created annotations.

2.2 Qualifiers

An optional addition to a manual annotation is a qualifier. These provide extra information on the relationship between a protein and its associated GO term (No. 4 in Table 1). Three qualifiers are currently available: 'co-localizes with' (to indicate a transient or peripheral association of the protein with an organelle or complex); 'contributes to' (where a function of a protein complex is facilitated, but not directly carried out by one of its subunits); and 'NOT' (to indicate conflicting published data, or where in contrast to previous assumptions, a protein is not found to have a particular activity, location, or process involvement).

It is vital that users are aware of such qualifiers, for although they are rarely used they can change the meaning of the associated annotation considerably. Most importantly the 'NOT' qualifier produces the most drastic change in the interpretation of an annotation, and users of large data sets are advised to ensure 'NOT' annotations are appropriately considered in their analysis (see Section 6 'The effective use of GO for large-scale analyses').

3 Viewing the GO and its annotations

3.1 Browsing the GO

Several web-based tools are freely available to search, browse, and view the GO hierarchy and annotations. The official GOC browser, AmiGO (No. 5 in Table 1), enables users to search and browse both GO terms and the manual GO annotation supplied by all the GO Consortium members. There are also browsers developed by individual annotation groups such as MGI's GO browser, which displays annotation to mouse proteins and the QuickGO browser, produced by the GOA group at the European Bioinformatics Institute (EBI) (No. 6 in Table 1).

QuickGO allows users to both a) search individual proteins or groups of proteins and view all the associated manual and electronic annotation and b) search individual or multiple GO terms and view all the associated proteins — the annotation sets can be further filtered in a variety of ways, including by taxonomic group, evidence code, and GO term identifier, among others, to create a customised subset of annotations, which can then be downloaded.

Each GO browser has unique features, so it is worth comparing a few to find one which suits your requirements. A list of GO browsers is supplied on the GO Consortium web site (No. 7 in Table 1).

3.2 Bulk retrieval of GO annotations

In addition to viewing single or small groups of annotations, users can also download entire sets of annotations produced by the individual GOC annotation groups. The GO Consortium supplies GO annotations in 15 column tab-delimited 'gene association files'. All the information required for a complete GO term-gene product association is contained in one row of the file and includes information such as the identifier of the sequence annotated, the GO term identifier, evidence code, and the reference used (Table 2).

Table 2 Format of the Gene Association File.

Column	Name	Description
1	Database	The database from which the Object ID is drawn.
2	Object ID	A unique identifier in the database for the sequence being annotated.
3	Object Symbol	A gene symbol or ORF name where possible.
4	Qualifier	Flags that modify the interpretation of an annotation, e.g. 'NOT', 'contributes to', 'co-localizes with'.
5	GO ID	The GO identifier for the term attributed to the Object ID.
6	Reference	The source cited as an authority for the attribution of the GO ID to the Object ID, e.g. PubMed ID or a database record.
7	Evidence	A code indicating the type of evidence that supports the GO annotation.
8	With or From	Required only for some evidence codes, values can include database gene ID, sequence ID or GO ID.
9	Aspect	One of P (Biological Process), F (Molecular Function) or C (Cellular Component) to match the specific GO ID used in column 5.
10	Object Name	Name of gene or gene product.
11	Object Synonym	Any synonym of a gene or gene product.
12	Object Type	The entity that is being annotated, e.g. protein, gene.
13	Taxon ID	The ID of the species or strain encoding the gene product. In certain cases, such as in annotations describing interactions between organisms, two taxon IDs can be piped together.
14	Date	The date on which the annotation was made.
15	Source DB	The database group which has created the annotation.

The files are stored by the GO Consortium in a publicly available central repository (No. 8 in Table 1). It is important to note that the member databases annotate to a variety of gene or protein identifiers, so depending on what identifiers are required by the user, a mapping between identifiers may need to be performed (see Exercise 2 in Section 7 for an example of how to map identifiers).

Gene association files can also be downloaded directly from member database web sites, which can sometimes contain different groupings of annotations. Examples include: the 914 proteomes gene association files from the GOA web site (No. 9 in Table 1), species-specific files produced for all species whose genomes have been fully sequenced, is publicly available and has more than 25% GO annotation coverage (these are much smaller in size than the multi-species UniProtKB gene association file and so easier to work with); the gene association file produced by the

GOA group in collaboration with the British Heart Foundation-funded Cardiovascular Gene Ontology Annotation Initiative (BHF-UCL), which contains GO annotations to human proteins implicated in cardiovascular development or disease; and the TAIR gene association file, which contains annotations to *Arabidopsis thaliana* gene identifiers.

Alternatively, the GO database can be directly queried using the GO Online SQL Environment (GOOSE) (No. 10 in Table 1).

4 Use of GO in the literature

From the most simplistic study of an individual protein's function to the more involved analysis of microarray data or interaction network predictions, GO has been used in many and varied ways by the scientific community. Some examples of studies that have used GO data in their analyses are summarised below.

4.1 Microarray/Proteomics studies

One of the most common uses of GO is to analyse results of microarray and proteomic studies. Here users can start to form biological hypotheses by determining whether genes/proteins with similar expression patterns are also annotated to the same types of GO terms, which could indicate that a particular cellular process has been activated. Numerous examples of this type of study exist in the literature, including one by Wertheim *et al.*, [22] in which a microarray experiment was performed to follow changes in gene expression after a parasitoid attack in *Drosophila*. Among the genes that were differentially expressed there were a significant number that had GO annotation to defense or immune responses, response to bacteria and proteolytic processes, however there were also a large number of differentially expressed genes that had not previously been associated with defense responses, suggesting a different set of genes are involved in the parasitic response compared to the well-studied antimicrobial responses.

4.2 Functional characterisation of gene products

With a goal of determining the functions of siRNA targets from *C. elegans*, Asikainen *et al.* [3] looked for GO terms that were enriched in the target genes. The genes were grouped according to the length of their siRNA, i.e. 18- to 22-mers, 23-mers and 24- to 26-mers. The authors found that the different lengths of siRNA were enriched with different types of GO terms: the targets of the 18- to 22-mers were involved in embryonic development; the 23-mer targets were associated with post-embryonic development; and the targets of the 24- to 26-mers were involved in phosphorus metabolism or protein modification.

4.3 Biomarker discovery

The Gene Ontology has also proved useful in finding markers for disease that could potentially become targets for therapy. Gorter *et al.* [15] studied a rat model for temporal lobe epilepsy and used GO data to determine which terms were enriched in the genes that were differentially expressed during the three phases of epileptogenesis. Genes involved in many different processes were found to display significant expression changes, such as stress response, apoptosis and synaptic transmission, although terms associated with the immune response provided the most significant change. This study was used to help the authors identify possible targets for antiepileptic drugs.

4.4 Functional overview of proteomes/genomes using GO slims

Lin *et al.* [18] compared functional classification of singleton genes and paralogous families in rice and *Arabidopsis* using plant GO slim categories (see Section 5.2.1 for more information on GO slims). They found enrichment of 12 GO slim categories including DNA binding, transcription factor activity and hydrolase activity in rice paralogous family proteins but a substantial reduction in five categories including receptor activity, kinase activity and carbohydrate binding. A similar result was found between *Arabidopsis* singleton and paralogous genes. They concluded that while some paralogous protein families have conserved functions others have been subject to evolutionary pressure causing functional divergence.

4.5 Interactome studies

Brown and Jurisica [5] studied the conservation of interaction networks between human and yeast using 'Interologs' which describe the process of mapping the interactome of one organism to another, such that when orthologous proteins exist in two species the interactions between proteins in one species can be inferred in the other species. The authors created a database of orthologs among six organisms and used these to map the interactomes between species. They noted that protein complexes are preferentially conserved compared to more transient interactions and went on to use GO to determine the functional category of interacting proteins. Interactions were transferred from humans to yeast, and the coatomer complex was studied as an example. Using protein binding data mapped from the human interaction network, the yeast protein GCS1 was suggested to be involved in retrograde transport because of its, so far unreported, physical interaction with the transport and sensing protein COPA and also the COPB-interacting protein, ERD2.

4.6 Validation of purification techniques

Scientists have also found the GO resource useful for evaluating the results from certain experimental techniques, e.g., for purification of proteins from a particular cell organelle. By seeing what gene products are in their purified fractions and what GO cellular component annotations they are associated with, they can determine whether or not the technique was effective. An example of this is Cao *et al.* [8] who developed a two-phase partition method for purification of rat liver plasma membrane proteins. They found that 67% of the purified proteins had been annotated as integral membrane or membrane-associated proteins, and that a low proportion of proteins had been annotated to mitochondrion and endoplasmic reticulum, the main contaminating compartments, suggesting they had obtained a highly purified fraction of membrane proteins.

5 Popular methods for supplementing and grouping GO annotations

As we have seen, GO data has proven to be useful in many different studies from the literature. This section will explore in detail two common uses of GO; how to obtain functional information for gene products that have not been extensively studied and, secondly, how to get a broad overview of the functional information associated with a set of gene products using GO slims.

5.1 Functional prediction of uncharacterised sequences

One of the strengths of GO is that terms are defined so that they are species-neutral, therefore the GO annotation for an experimentally well-studied species can be transferred to species for which little data is available. Several tools are available for automatically transferring GO annotations between species that exploit either phylogenetic data (SIFTER [14], GOAnno [9]), signal or transit peptide patterns (SignalP [4]), TargetP [13]), or simple sequence similarity (Blast2GO [10]). In addition, the EBI-based InterProScan tool [19] (No. 11 in Table 1) provides predictions of GO terms by allowing users to query protein sequences (genomic sequences can also be queried in local installations of InterProScan) against the protein family database

InterPro. Results provide both a list of protein domains found in the query sequence and the GO terms associated with particular InterPro domains.

An example of an InterProScan query is illustrated in Figure 3 using a GAF modulated sigma54 specific transcriptional regulator from the proteobacterium *Burkholderia phymatum* (UniProtKB accession number: B2JN45). The protein sequence of interest is pasted into the query box on the InterProScan home page and the job submitted (Figure 3a). The results are displayed in 'Picture View' by default, which illustrates the InterPro domains predicted to be present in the query sequence. The 'Table View' (Figure 3b) lists those domains together with the GO terms to which they have been associated. The results suggest the *B. phymatum* sequence may act to modulate transcription.

5.2 Using GO slims for summarising functional data

5.2.1 GO slims

It is the relationships between terms that make the GO so useful for summarising results from large data sets (see Figure 1). Terms linked in the ontology by a relationship must obey the 'true path rule': if a term used in an annotation describes an attribute of the gene product, then all of the parent terms through to the top-level of the hierarchy must also be able to describe that gene product.

Although there are currently over 25,000 terms, users can select a number of high-level terms and 'map up' the annotations that have been made to more descriptive, descendant terms. This is a useful way of providing a broad overview of the ontology content without the detail of the specific, more refined terms. Many tools that analyse GO terms use this feature of the ontology to interactively determine which ancestor term provides the most significant relationship with the selected subset of terms. Such term subsets, or GO slims, are especially useful for summarising the results of GO annotation of a genome or list of genes from a microarray experiment when broad classification of gene product function is required.

Although several pre-defined GO slims are available, such as the GO Consortium generic slim or the species-specific slims from the TAIR (plant) and SGD (yeast) groups (all available on the GOC web site (No. 12 in table 1)), many users need to create their own GO slim to answer specific biological questions. Slims can be made especially for an individual species or to particular areas of the ontologies. When the slim is applied you will find that multiple annotations for one gene product can map up to a single GO slim term if they are in the same path up to that term but also one annotation can map up to multiple slim terms. This is because of the way the GO is constructed; there can be multiple relationships between different terms so often a GO term can have more than one path up to the root nodes ('Molecular_Function', 'Biological_Process', 'Cellular_Component'), if this path up the GO passes through several terms selected for a slim then an annotation to the GO term could map to several slim terms. However, if a parent term to the first slim term encountered in

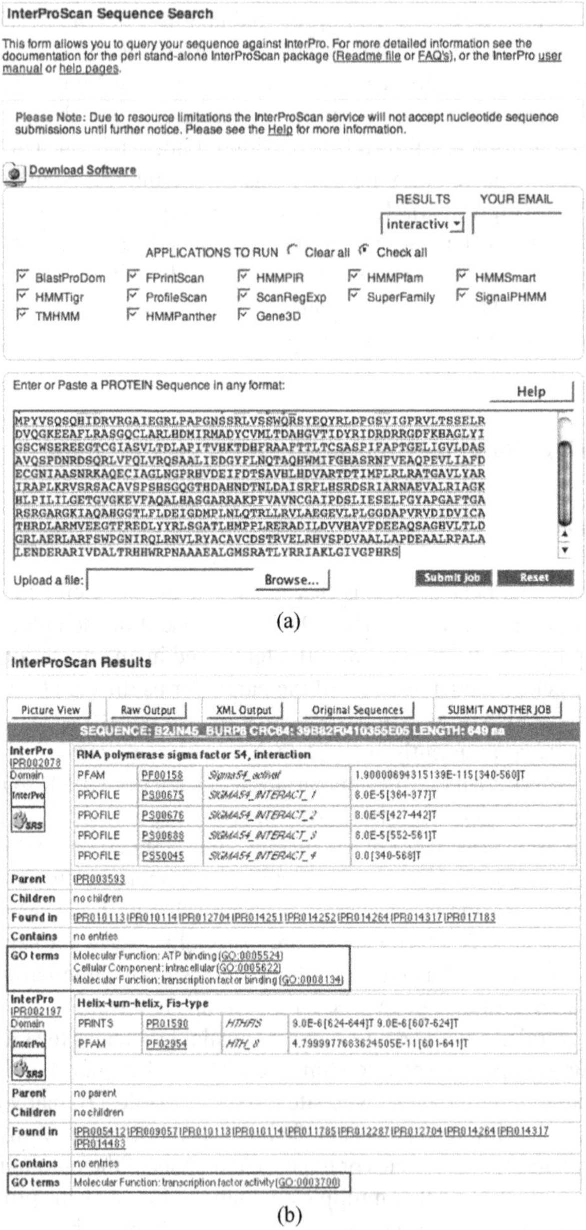

InterProScan Sequence Search

This form allows you to query your sequence against InterPro. For more detailed information see the documentation for the perl stand-alone InterProScan package (Readme file or FAQ's), or the InterPro user manual or help pages.

Please Note: Due to resource limitations the InterProScan service will not accept nucleotide sequence submissions until further notice. Please see the Help for more information.

Download Software

RESULTS: interactive YOUR EMAIL:

APPLICATIONS TO RUN: Clear all / Check all

BlastProDom, FPrintScan, HMMPIR, HMMPfam, HMMSmart, HMMTigr, ProfileScan, ScanRegExp, SuperFamily, SignalPHMM, TMHMM, HMMPanther, Gene3D

Enter or Paste a PROTEIN Sequence in any format: Help

```
MPYVSQSQHIDRVRGAIEGRLPAPGNSSRLVSSWQRSYEQYRLDPGSVIGPRVLTSSELR
DVQGKEEAFLRASGQCLARLHDMIRMADYCVMLTDAHGVTIDYRIDRDRRGDFKHAGLYI
GSCWSEREEGTCGIASVLTDLAPITVHKTDHFRAAFTTLTCSASPIFAPTGELIGVLDAS
AVQSPDNRDSQRLVFQLVRQSAALIEDGYFLNQTAQHWMIFGHASRNFVEAQPEVLIAFD
ECGNIAASNRKAQECIAGLNGPRHVDEIFDTSAVHLHDVARTDTIMPLRLRATGAVLYAR
IRAPLKRVSRSACAVSPSHSGQGTHDAHNDTNLDAISRFLHSRDSRIARNAEVALRIAGK
HLPILILGETGVGKEVFAQALHASGARRAKPFVAVNCGAIPDSLIESELFGYAPGAFTGA
RSRGARGKIAQAHGGTLFLDEIGDMPLNLQTRLLRVLAEGEVLPLGGDAPVRVDIDVICA
THRDLARMVEEGTFREDLYYRLSGATLHMPPLRERADILDVVHAVFDEEAQSAGHVLTLD
GRLAERLARFSWPGNIRQLRNVLRYACAVCDSTRVELRHVSPDVAALLAPDEAALRPALA
LENDERARIVDALTRHHWRPNAAAEALGMSRATLYRRIAKLGIVGPHRS
```

Upload a file: Browse... Submit Job Reset

(a)

InterProScan Results

Picture View | Raw Output | XML Output | Original Sequences | SUBMIT ANOTHER JOB

SEQUENCE: B2JN45_BURP8 CRC64: 39B82F0410355E05 LENGTH: 649 aa

InterPro IPR002078 Domain — **RNA polymerase sigma factor 54, interaction**

PFAM	PF00158	*Sigma54_activat*	1.90000694315139E-115[340-560]T
PROFILE	PS00675	*SIGMA54_INTERACT_1*	8.0E-5[364-377]T
PROFILE	PS00676	*SIGMA54_INTERACT_2*	8.0E-5[427-442]T
PROFILE	PS00688	*SIGMA54_INTERACT_3*	8.0E-5[552-561]T
PROFILE	PS50045	*SIGMA54_INTERACT_4*	0.0[340-568]T

Parent	IPR003593
Children	no children
Found in	IPR010113 IPR010114 IPR012704 IPR014251 IPR014252 IPR014264 IPR014317 IPR017183
Contains	no entries
GO terms	Molecular Function: ATP binding (GO:0005524) Cellular Component: intracellular (GO:0005622) Molecular Function: transcription factor binding (GO:0008134)

InterPro IPR002197 Domain — **Helix-turn-helix, Fis-type**

PRINTS	PR01590	*HTHFIS*	9.0E-6[624-644]T 9.0E-6[607-624]T
PFAM	PF02954	*HTH_8*	4.7999977683624505E-11[601-641]T

Parent	no parent
Children	no children
Found in	IPR005412 IPR009057 IPR010113 IPR010114 IPR011785 IPR012287 IPR012704 IPR014264 IPR014317 IPR014483
Contains	no entries
GO terms	Molecular Function: transcription factor activity (GO:0003700)

(b)

Fig. 3 (a) The *B. phymatum* sequence entered into the InterProScan query box. (b) Results from an InterPro scan query, showing a list of InterPro domains predicted to be present in the query sequence and the GO terms which have been associated to any of the domains (only two of the six resulting domains are shown for clarity). From these results we can deduce that the *B. phymatum* sequence may act to modulate transcription.

the path also exists in the slim ontology then the parent will be discarded and the more specific term kept. It is worth noting that as GO is often modified by editors, old GO slim subsets may need to be checked to ensure they are not out of date with the current ontology.

5.2.2 Use of QuickGO for making GO slims

A number of tools are now available which enable users to create their own GO slim sets and map annotations up to these terms. Tools include the AmiGO GO slimmer, the SGD GO slim mapper, the 'map2slim' Perl script, and QuickGO. (No's 13, 14, 15 and 6 in Table 1). The QuickGO browser allows users to both define their own sets of GO terms for use as a GO slim or use/modify one of the pre-defined slims. Once users have developed their slim, they can map up annotations for a list of gene products that may have been obtained from, for example, a microarray or proteomics study.

An example of the latter is illustrated in Figures4 and 5 using a list of cardiovascular-associated proteins to which we wanted slimmed-up annotation. The example uses the GO slim provided by the GOA group, consisting of 64 terms, which is one of the slims that can be selected directly from the QuickGO home page (Figure4a). There are multiple ways to use a subset of GO terms (Figure4b), but this example will demonstrate their use as a GO slim to produce an overview of annotation to a list of proteins. Selecting 'Use these terms as a GO slim' will present the user with a view of the QuickGO annotation table, where the selected slim terms have been used to categorise all proteins in the GOA database. The proteins that are mapped to the slim can be further restricted by adding a list of accession numbers in the Filter ID field (box 'A' in Figure 5a). There are also options for filtering annotations by Evidence Code or by Taxonomic Identifier. The percentage and count of proteins associated with each of the slim GO terms can be viewed in the GO ID statistics field (box 'B' in Figure 5a). A useful way of showing the distribution of GO terms among the selected set of proteins is by pie chart as illustrated in Figure 5b. This figure shows the distribution of Biological Process terms for the cardiovascular-associated proteins where we can see that a significant proportion of these proteins are involved in regulation of biological processes.

5.2.3 Using GO slims for proteome comparison

GO slims are also useful for comparing proteomes, as illustrated in Figure 6 for the *Drosophila* and *Anopheles* proteomes. We can see that the two proteomes are fairly similar regarding the biological processes that they are involved in, except the *Drosophila* proteome has more proteins involved in developmental processes. When comparing such proteomes, users need to take into consideration the types of annotation that the proteomes have, in this example 99% of *Anopheles* GO annotation has been predicted by electronic methods whereas *Drosophila* has a greater

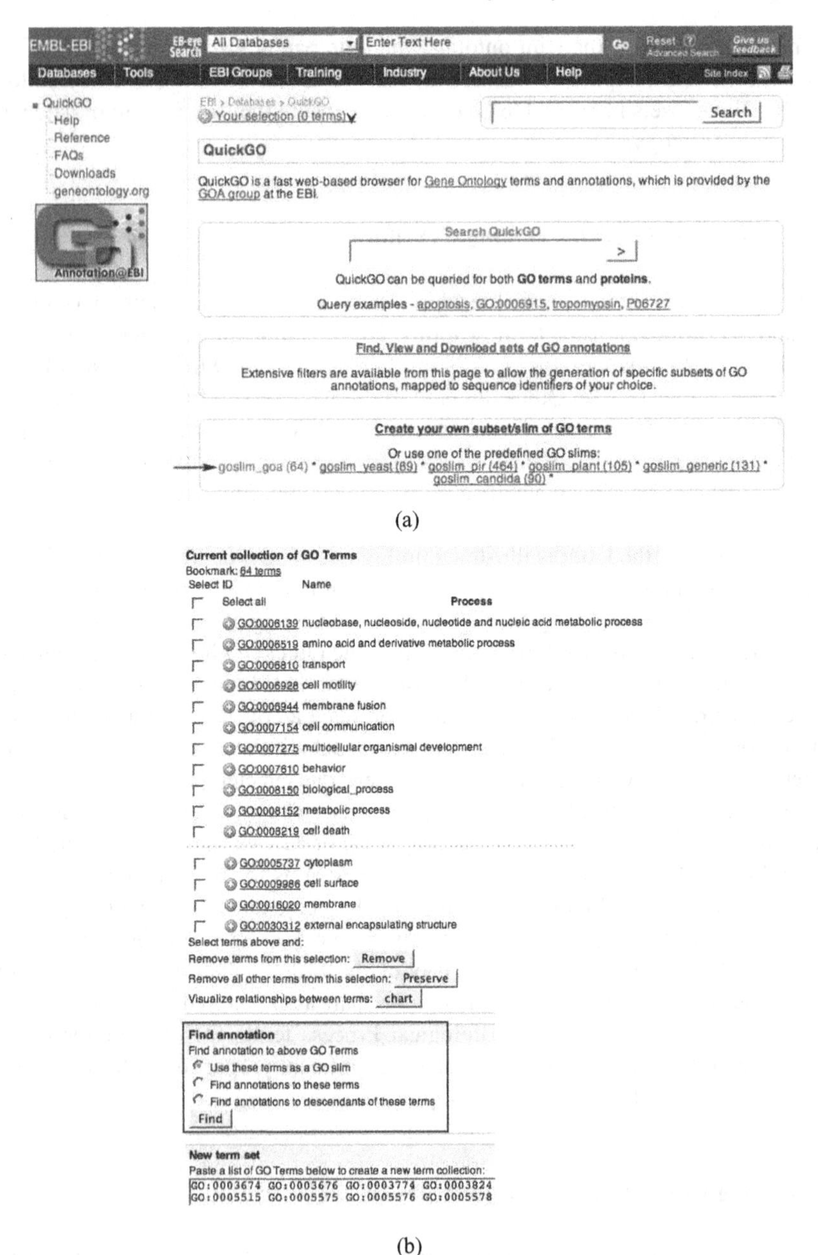

Fig. 4 Using GO slims in QuickGO (a) QuickGO home page showing the starting point for using GO slims. Clicking on the GOA slim in the QuickGO home page (indicated by a black arrow) will bring the user to the 'GO slim' page, where users can modify and view the structure of their slim. (b) GO slim page in QuickGO showing the terms in the chosen slim and the options for associating annotations to them (black box).

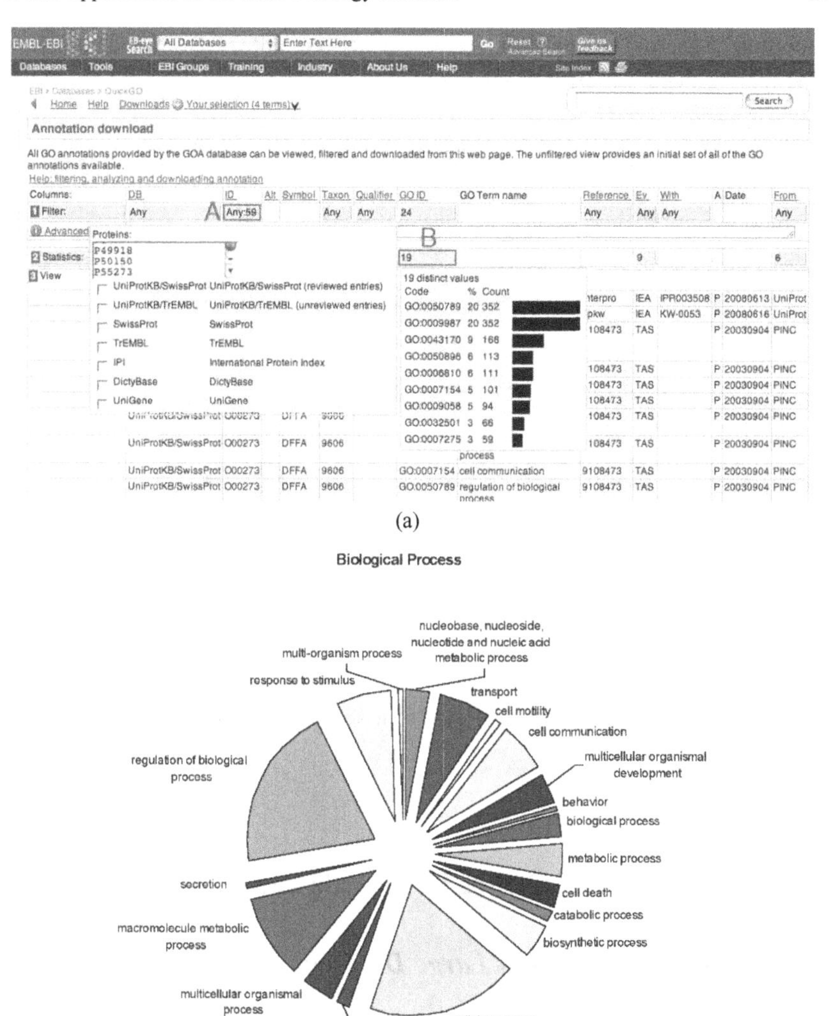

Fig. 5 Results of slimming annotations in QuickGO (a) Annotation page in QuickGO showing slimmed up annotations. From this page the user can edit the list of proteins used in the GO slim (box 'A') as well as view statistics for the number and percentage of proteins associated with each GO ID (box 'B'). (b) Pie charts can be constructed using the GO ID statistics from QuickGO, this example shows the distribution of Biological Process terms for cardiovascular-associated proteins. A significant proportion of these proteins are involved in regulation of biological processes.

proportion of manual annotations; this may well account for the different annotation subsets being displayed by the GO slim.

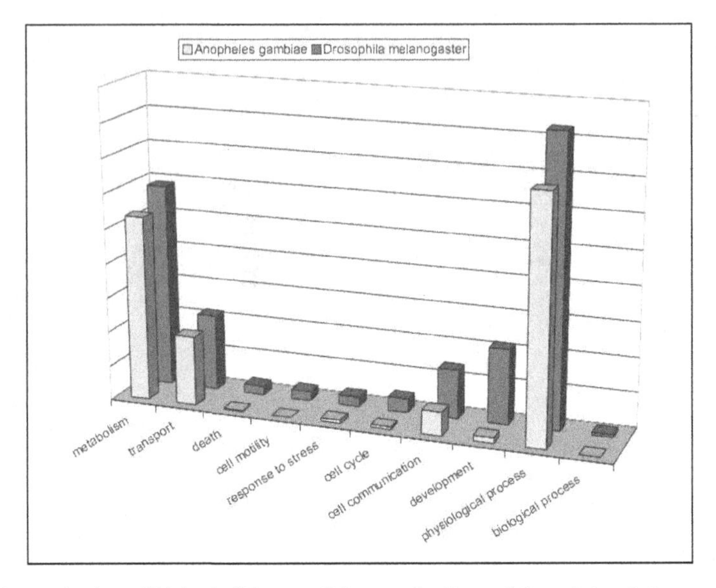

Fig. 6 Categorisation of Biological Process GO terms for *Drosophila* and *Anopheles* proteomes.

6 The effective use of GO for large-scale analyses

6.1 Functional Analysis of Large Datasets

One of the most effective ways of analysing large data sets obtained from high-throughput experiments is by using analysis tools that have been developed specifically for this purpose. These allow users to query GO annotation data with lists of gene, protein or probe identifiers. There exists a plethora of tools to assist biologists with these analyses, all offering different functionalities and features. A list of some of the tools available can be found on the GO Consortium tools web site (No. 7 of Table 1). A quick overview of four of these tools, Blast2GO [10], FatiGO [1], The Ontologizer [21], and Onto-Express [16] have been compiled with the assistance of the tool developers in the online supplementary material of Dimmer *et al.*, [11], which is a useful starting point for scientists wishing to familiarise themselves with such tools.

For reviews on available GO annotation tools and how to use GO annotations, see references [17] and [20].

6.2 Key considerations for large-scale analysis

There are a number of aspects of GO that should be considered before embarking on an analysis of your data set to ensure that you get the most reliable and meaningful results possible. Table 3 describes these considerations when analysing your set of genes or proteins with GO data.

7 Exercises

7.1 Exercise 1

This exercise will familiarise you with searching and browsing the Gene Ontology by using one of the available GO browsers. There is a list of GO browsers on the GO Consortium web site (No. 7 in Table 1). AmiGO (No. 5 in Table 1) is the official GO Consortium browser and QuickGO (No. 6 in Table 1) is the browser developed by the EBI.

The following exercise can be performed using any GO browser, but they all differ slightly in how they display the GO data (for example as of September 2008, AmiGO does not display electronic annotation) and data filtering options may vary between browsers.

1. In your chosen browser, search for the GO term 'Immune response'.

Q. How many child terms does the term 'Immune response' have?

2. View the annotations associated with this term.

3. Filter the annotations to view only *Drosophila melanogaster* (Taxon ID:7227) annotations.

Q. How many annotations exist to Drosophila melanogaster *proteins?*

Q. Which evidence codes have been most frequently applied in annotations to these terms?

4. Filter the evidence codes to view only the manual annotations made using the experimental evidence code 'IDA' (Inferred from Direct Assay).

Q. How many annotations have been made to D. melanogaster *gene products using the IDA code?*

Q. Which databases have made these annotations?

5. Now look at the annotations made directly to 'Immune response', i.e. not to any of it's child terms.

Table 3 Considerations for high-throughput data analyses. Key points to consider when using GO based analysis tools. Reproduced with permission from [11].

Key Questions	Reasons
Does the tool enable the hierarchical structure of GO to be exploited?	GO analysis tools should be designed to improve the identification of functional groups within a data set by allowing the user to manually consolidate genes associated with highly specific (child) GO terms to those with the higher (parent) GO terms in order to formulate and test specific biological hypotheses.
What is the release date of the data used by the tool?	Each month an average of 240 GO terms is added to the Gene Ontology and 1,500 GO annotations are added to the human GO data set. Tools that infrequently download GO data will restrict analyses. Remember to include the release date(s) of the GO annotation data set and ontology file used by the tool in any resulting publications. The tool should provide this information.
Does the tool correctly treat the GO annotations with the qualifier 'NOT' ?	This qualifier reverses the meaning of an annotation (see 'Qualifiers' section), so should be either removed from the analysis or used to calculate the amount of evidence against certain hypotheses involving the terms annotated with NOT.
Does the analysis tool enable concomitant functional profiling for all three GO categories?	In addition, an increasing number of tools also display other annotation data such as TRANSFAC regulatory motifs, BioCarta, KEGG and Reactome pathways.
Is the type of identifier used in the assay directly accepted by the tool (e.g. probe IDs, RefSeq protein IDs, etc.) or will it be necessary to map one identifier type to another?	You may need to convert the identifiers of your gene list into those accepted by the tool. This can be an important source of errors since up to 20% of the identifiers can be routinely lost or incorrectly mapped during identifier translations [12]. Even if the type of identifier is directly accepted, is this the native identifier used in the analysis or is an internal identifier mapping being performed? If internal identifier mappings are being carried out, what are the sources of data and their release dates?
Does the tool test for both enrichment AND depletion of the GO terms?	Some tools only test for over-representation of the differentially expressed genes within the given GO term. However, both significantly enriched as well as depleted GO terms can be biologically meaningful.
Does the tool enable the user to submit their own GO annotation data set or select specific evidence code-supported annotations for the analysis?	In many cases, it will be appropriate for the user to define the background set of proteins used in the analysis. However, there are only a few species where filtering out certain evidence code supported annotations is appropriate (see Section 2.1).
What is the statistical model used and are there several alternative models that the user may choose from?	One serious and widely neglected problem in GO profiling is that the same data submitted to different tools can provide widely different results for the same GO terms. Having the ability to specify the model allows the user to eliminate one variable to verify their analysis.
What choice of correction factors is available?	To compensate for the propagation of gene associations from each GO term to all their parent GO terms many tools give a choice of correction factors, such as Bonferroni, Holmes, false discovery rate (FDR) and Ŝidák [20]. Bonferroni or Ŝidák are suitable when less than 50 unrelated GO categories are involved, Holmes is more appropriate for larger numbers of unrelated GO categories and FDR is a good choice if several GO categories are related, e.g. contain several GO terms with a common parent.

Q. How many annotations are now available?

7.2 Exercise 2

You have a list of gene product identifiers with which you want to carry out a term enrichment analysis, however, the analysis tool you want to use does not accept the identifiers you have, so you need to map them to another identifier type.

This exercise will explain how to convert Entrez Gene IDs to UniProtKB accession numbers and Exercise 3 will cover how to perform term enrichment on the resulting list of accession numbers. There are a number of ID mapping services available, for example UniProtKB (No. 16 in Table 1) provides one as does PICR (No. 17 in Table 1). For a list of further services, please see the 'ID mapping' article in the August 2008 edition of the GO newsletter (No.18 in Table 1). This exercise will use the UniProtKB ID mapper.

1. Go to the UniProtKB web site, click on the tab entitled 'ID mapping' and paste the following list of Entrez Gene IDs into the query box:

27040, 7454, 7535, 7126, 54900, 8542, 665, 8741

2. In the 'From' drop-down menu select 'Entrez Gene (Gene ID)' and in the 'To' drop-down menu select 'UniProtKB AC' and click on 'Map'.

Note, for several Gene IDs there may be more than one UniProtKB AC returned. The duplicates are generally unreviewed entries from the UniProtKB/TrEMBL database so will only be found in term enrichment tools if they include electronic (IEA) annotations in their analyses.

3. Use the link to download the target list as a text file.

7.3 Exercise 3

This exercise will demonstrate how to do term enrichment analysis on the list of UniProtKB accession numbers obtained from Exercise 2. There are several analysis tools which are capable of term enrichment and a list of these is on the GO Consortium web site (No. 7 in Table 1). The AmiGO term enrichment tool will be used for this exercise (No. 19 in Table 1).

1. Open the AmiGO term enrichment tool and paste in the target list you saved in Exercise 2.

2. Select the database filter 'UniProtKB', by default the tool will use the same database as a background set so there is no need to select one. Click on 'Submit Query'.

Q. What Biological Process terms are most commonly associated with your list of proteins?

3. In AmiGO you can choose to see your results in graphical form by clicking on the 'visualize tool' link near the top of the page. The more significant the GO term is, the darker the shading in the box.

8 Further Reading

More on annotation methods can be found in [6].

Available GO analysis tools and use of GO annotations are reviewed in [17, 20].

More information about the Cardiovascular Gene Ontology Annotation Initiative can be found on the web site (No. 20 in Table 1).

If you are interested in the GO project and would like to receive public announcements related to the project you can join the GO Friends mailing list (No. 21 in Table 1).

If you have any questions or comments about GOA or our resources, please use our feedback form (No. 22 in Table 1).

Acknowledgements The GOA project is funded by a P41 grant from the National Human Genome Research Institute (HG002273), core EMBL funding and a grant from the British Heart Foundation (SP/07/007/23671).

References

1. Al-Shahrour, F., Díaz-Uriarte, R., Dopazo, J.: FatiGO: a web tool for finding significant associations of Gene Ontology terms with groups of genes. Bioinformatics **20**(4), 578–580 (2004)
2. Ashburner, M., Ball, C.A., Blake, J.A., Botstein, D., Butler, H., Cherry, J.M., Davis, A.P., Dolinski, K., Dwight, S.S., Eppig, J.T., Harris, M.A., Hill, D.P., Issel-Tarver, L., Kasarskis, A., Lewis, S., Matese, J.C., Richardson, J.E., Ringwald, M., Rubin, G.M., Sherlock, G.: Gene Ontology: tool for the unification of biology. Nature Genetics **25**(1), 25–29 (2000)
3. Asikainen, S., Heikkinen, L., Wong, G., Storvik, M.: Functional characterization of endogenous siRNA targets in *Caenorhabditis elegans*. BMC Genomics **9**(270), 1–10 (2008)
4. Bendtsen, J.D., Neilsen, H., von Heijne, G., Brunak, S.: Improved prediction of signal peptides: SignalP 3.0. Journal of Molecular Biology **340**(4), 783–795 (2004)
5. Brown, K.R., Jurisica, I.: Unequal evolutionary conservation of human protein interactions in interologous networks. Genome Biology **8**, R95 (2007)

6. Camon, E., Magrane, M., Barrell, D., Binns, D., Fleischmann, W., Kersey, P., Mulder, N., Oinn, T., Maslen, J., Cox, A., Apweiler, R.: The Gene Ontology Annotation (GOA) project: Implementation of GO in SWISS-PROT, TrEMBL, and InterPro. Genome Research **13**(4), 662–672 (2003)
7. Camon, E.B., Barrell, D.G., Dimmer, E.C., Lee, V., Magrane, M., Maslen, J., Binns, D., Apweiler, R.: An evaluation of GO annotation retrieval for BioCreAtIvE and GOA. BMC Bioinformatics **6**, S17 (2005)
8. Cao, R., Li, X., Liu, Z., Peng, X., Hu, W., Wang, X., Chen, P., Xie, J., Liang, S.: Integration of a two-phase partition method into proteomics research on rat liver plasma membrane proteins. Journal of Proteome Research **5**(3), 634–642 (2006)
9. Chalmel, F., Lardenois, A., Thompson, J.D., Muller, J., Sahel, J.A., Léveillard, J.A., Poch, O.: GOAnno: GO annotation based on multiple alignment. Bioinformatics **21**(9), 2095–2096 (2005)
10. Conesa, A., Götz, S., García-Gómez, J.M., Terol, J., Talón, M., Robles, M.: Blast2GO: A universal tool for annotation, visualization and analysis in functional genomics research. Bioinformatics **21**(18), 3674–3676 (2005)
11. Dimmer, E.C., Huntley, R.P., Barrell, D.G., Binns, D., Draghici, S., Camon, E.B., Hubank, M., Talmud, P.J., Apweiler, R., Lovering, R.: The Gene Ontology — Providing a functional role in proteomic studies. Proteomics (2008)
12. Draghici, S., Sellamuthu, S., Khatri, P.: Babel's tower revisited: A universal resource for cross-referencing across annotation databases. Bioinformatics **22**(23), 2934–2939 (2006)
13. Emanuelsson, O., Neilsen, H., Brunak, S., von Heijne, G.: Predicting subcellular localization of proteins based on their N-terminal amino acid sequence. Journal of Molecular Biology **300**(4), 1005–1016 (2000)
14. Engelhardt, B.E., Jordan, M.I., Muratore, K.E., Brenner, S.E.: Protein molecular function prediction by Bayesian phylogenomics. PLoS Computational Biology **1**(5), 432–445 (2005)
15. Gorter, J.A., van Vliet, E.A., Aronica, E., Breit, T., Rauwerda, H., da Silva, F.A.L., Wadman, W.J.: Potential new antiepileptogenic targets indicated by microarray analysis in a rat model for temporal lobe epilepsy. The Journal of Neuroscience **26**(43), 11,083–11,110 (2006)
16. Khatri, P., Bhavsar, P., Bawa, G., Draghici, S.: Onto-Tools: An ensemble of web-accessible, ontology-based tools for the functional design and interpretation of high-throughput gene expression experiments. Nucleic Acids Research **32**, W449–456 (2004)
17. Khatri, P., Draghici, S.: Ontological analysis of gene expression data: Current tools, limitations, and open problems. Bioinformatics **21**(18), 3587–3595 (2005)
18. Lin, H., Ouyang, S., Egan, A., Nobuta, K., Haas, B.J., Zhu, W., Gu, X., Silva, J.C., Meyers, B.C., Buell, C.R.: Characterization of paralogous protein families in rice. BMC Plant Biology **8**(18) (2008)
19. Quevillon, E., Silventoinen, V., Pillai, S., Harte, N., Mulder, N., Apweiler, R., Lopez, R.: InterProScan: Protein domains identifier. Nucleic Acids Research **33**, W116–120 (2005)
20. Rhee, S.Y., Wood, V., Dolinski, K., Draghici, S.: Use and misuse of the Gene Ontology annotations. Nature Review Genetics **9**(7), 509–515 (2008)
21. Robinson, P.N., Wollstein, A., Böhme, U., Beattie, B.: Ontologizing gene-expression microarray data: Characterizing clusters with Gene Ontology. Bioinformatics **20**(6), 979–981 (2004)
22. Wertheim, B., Kraaijeveld, A.R., Schuster, E., Blanc, E., Hopkins, M., Pletcher, S.D., Strand, M.R., Partridge, L., Godfray, H.C.J.: Genome-wide gene expression in response to parasitoid attack in *Drosophila*. Genome Biology **6**(11), R94 (2005)

Index

E

F

G

H

I

J

K

L

M

T

U

W

Y